Nanostructure Science and Technology

Series Editor

David J. Lockwood, FRSC
National Research Council of Canada
Ottawa, ON, Canada

Nanostructure science and technology now forms a common thread that runs through all physical and materials sciences and is emerging in industrial applications as nanotechnology. The breadth of the subject material is demonstrated by the fact that it covers and intertwines many of the traditional areas of physics, chemistry, biology, and medicine. Within each main topic in this field there can be many subfields. For example, the electrical properties of nanostructured materials is a topic that can cover electron transport in semiconductor quantum dots, self-assembled molecular nanostructures, carbon nanotubes, chemically tailored hybrid magnetic-semiconductor nanostructures, colloidal quantum dots, nanostructured superconductors, nanocrystalline electronic junctions, etc. Obviously, no one book can cope with such a diversity of subject matter. The nanostructured material system is, however, of increasing significance in our technology-dominated economy and this suggests the need for a series of books to cover recent developments.

The scope of the series is designed to cover as much of the subject matter as possible – from physics and chemistry to biology and medicine, and from basic science to applications. At present, the most significant subject areas are concentrated in basic science and mainly within physics and chemistry, but as time goes by more importance will inevitably be given to subjects in applied science and will also include biology and medicine. The series will naturally accommodate this flow of developments in the sciences and technology of nanostructures and maintain its topicality by virtue of its broad emphasis. It is important that emerging areas in the biological and medical sciences, for example, not be ignored as, despite their diversity, developments in this field are often interlinked. The series will maintain the required cohesiveness from a judicious mix of edited volumes and monographs that while covering subfields in depth will also contain more general and interdisciplinary texts.

Thus the series is planned to cover in a coherent fashion the developments in basic research from the distinct viewpoints of physics, chemistry, biology, and materials science and also the engineering technologies emerging from this research. Each volume will also reflect this flow from science to technology. As time goes by, the earlier series volumes will then serve as reference texts to subsequent volumes.

More information about this series at https://link.springer.com/bookseries/6331

Muhammad Ikram · Ali Raza · Salamat Ali

2D-Materials for Energy Harvesting and Storage Applications

Springer

Muhammad Ikram
Solar Cell Applications Research Lab
Department of Physics
Government College University Lahore
Lahore, Pakistan

Ali Raza
Department of Physics
University of Sialkot (USKT)
Sialkot, Pakistan

Salamat Ali
Department of Physics
Riphah Institute of Computing and Applied Science (RICAS)
Riphah University Lahore Campus
Lahore, Pakistan

ISSN 1571-5744 ISSN 2197-7976 (electronic)
Nanostructure Science and Technology
ISBN 978-3-030-96020-9 ISBN 978-3-030-96021-6 (eBook)
https://doi.org/10.1007/978-3-030-96021-6

This Springer imprint is published by the registered company Springer Nature Switzerland AG
The registered company address is: Gewerbestrasse 11, 6330 Cham, Switzerland

Preface

Securing sustainable green energy is becoming increasingly vital due to the depletion of current fossil resources and pollution concerns. To minimize this issue, various complementing methods, including solar cells, fuel cells, and thermal power generation, are being investigated. Also, everything becomes user-centered, and society becomes more data-dependent. A modest, semipermanent power source is necessary to analyze huge amounts of data and deliver personalized services to consumers. We need new ultracompact and lightweight energy-producing devices since present technologies have limits in terms of life span, size, and mechanical characteristics. Two-dimensional (2D) layered materials have shown potential as unusual energy materials. They are a subset of 2D materials having almost planar shape. Since the discovery of graphene in 2004, the development of 2D materials has received unprecedented attention from both industry and academia. A lot of studies have gone into developing new materials and exploring their characteristics in many technological applications, from electronics to healthcare. These researches revealed that altering 2D materials' dimensions, crystallographic structure, flaws, or doping with heteroatoms may efficiently adjust their physical and chemical characteristics. This versatility enables the creation of 2D materials for energy harvesting and storage applications.

The purpose of this book is to discuss graphene-like materials at their most advanced level of development. These debates center on the relationship between synthesis procedures and increased performance in energy-related applications. A depth of knowledge ranging from the principles of 2D materials to application is addressing challenges and opportunities. We review the state-of-the-art improvements in ultrathin 2D materials, focusing on their most significant achievements, as well as recent emergence and numerous potential applications. The following table summarizes the contents of this book: Chaps. 1 and 2 briefly introduced the speciality of 2D materials, ranging from the major types of 2D family currently available to related opportunities and research highlights from the last four years; Chap. 3 provided compositions and structural chemistries, as well as explain how they facilitate the 2D structure over others. Additionally, Chap. 4 discusses numerous ways for understanding 2D nanomechanics and determining which method is most appropriate for a certain energy-related application. Moreover, Chap. 5 will explore 2D heterostructures in detail, including fabrication methods and uses with current understanding of how heterostructure can be stable and increase device stability and dependability. One step ahead, the most effective applications for energy harvesting will be described, including mechanical (piezoelectricity), solar (solar cell, hydrogen evolution, and CO_2 resection), electrochemical (oxygen evolution), and storage applications like batteries and supercapacitors. The final portion will discuss the most advanced sorts of work, such as wearable and flexible energy gadgets. Our objective was to establish a link between current strategies and structural modification with regard to energy harvesting and material storage capability.

Lahore, Pakistan

Muhammad Ikram
Ali Raza
Prof. Salamat Ali

Contents

About the Authors

Dr. Muhammad Ikram obtained his Master degree (M.Phil. Physics) from BZU Multan, Pakistan, in 2010. He obtained his Ph.D. degree in Physics from the Department of Physics, Government College University (GCU), Lahore, through the Pak-US joint project between the Department of Physics, GCU Lahore, Pakistan and the University of Delaware, USA, in 2015. In 2017, Ikram joined the Department of Physics, GC University Lahore as Assistant Professor of Physics. Ikram received the Seal of Excellence Marie Skłodowska-Curie Individual Fellowship in 2017 and 2020. In 2021, Ikram was included in the 2% scientists announced by Stanford University. His research interest involves the synthesis and characterization of inorganic semiconductor nanomaterials, 2D materials for water treatment and optoelectronic applications.

Mr. Ali Raza obtained his B.S. (Physics) degree from the University of Gujrat (Punjab, Pakistan) in 2016 and his Master degree in Material Physics from Riphah International University, Islamabad Pakistan in 2019. Raza completed his M.Phil. research work in the Solar Cell Applications Lab at GCU Lahore (Punjab, Pakistan) and later worked as Research Associate in the same lab on the field of catalytic and energy harvesting applications of pure and engineered 2D-Materials (TMDCs, Graphene, h-BN, and MXenes). Currently, he is serving as Lecturer (Research) at the Department of Physics, University of Sialkot, Punjab, and has enrolled as a Ph.D. scholar at CAS Key Laboratory of Design and Assembly of Functional Nanostructures, Fujian Institute of Research on the Structure of Matter, CAS. His current research directions are the fabrication of functional 2D-Materials for catalytic and other energy-harvesting applications.

Prof. Dr. Salamat Ali is working as Professor of Physics, RICAS, Riphah International University Lahore Campus. He has completed his Ph.D. in 1996 from the University of Durham, UK, in Superconductivity and Magnetism and his Post-Doc. in 2006 from the K. F. University of Graz, Austria, Specialization in Nanotechnology. He also completed a training in Solar Cell Technology from the University of Delaware, USA, in 2012. His current research interests include water purifications, sensor technology, development of medicines for cancer treatment. He has 35 years experience in material science.

Abbreviations

0D	Zero-dimensional
1D	One-dimensional
2D	Two-dimensional
2DM	Two-dimensional material
2DM-WSs	2D material-based wearable supercapacitors
3D	Three-dimensional
AFM	Atomic force microscopy
AgNW	Silver nanowire
ALE	Assisted liquid exfoliation
AN	Acetonitrile
AQY	Apparent quantum yield
AR-STM	Atomic-resolution STM
BH_2NH_2	Aminoborane
BL	Bilayer
BNNS	BN nanosheet
B_o	Magnetic field
BP	Black phosphorus
BPQD	Black phosphorus quantum dot
BTO	Barium titanate
BV	Benzyl viologen
b-WSe_2	Bilayered WSe_2
C	Carbon
C_2H_7NO	Aminoethanol
C_{60}	Fullerene
CB	Conduction band
CBM	Conduction band minimum
CC-STM	Constant-current STM
CDW	Charge density wave
CE	Counter electrode
CMOS	Complementary metal-oxide-semiconductor
CNT	Carbon nanotube

COF	Covalent organic framework
CP	Coordination polymer
CS	Centrosymmetry
C-TAB	Cetyltrimethylammonium bromide
DA	Diels-Alder
DCE	1,2 dichloroethane
DFT	Density functional theory
DFTB	Density functional tight-binding
DHLA	Dihydrolipoic acid
DMF	Dimethylformamide
DMSO	Dimethyl sulfoxide
DOS	Density of state
DSSC	Dye-sensitized solar cell
DTBP	Di tert-butyl peroxide
EDLC	Electric double-layer capacitor
EELS	Electron energy loss spectroscopy
E_{ex}	External electric field
E_g	Bandgap energy
EG	Extended graphite
EIS	Electrochemical impedance spectroscopy
EM-CCG	Electrolyte-mediated chemically converted graphene
EMIMBF4/AN	1-ethyl-3-methylimidazolium tetrafluoroborate/AN
ETL	Electron transport layer
ETM	Electron transporting material
F4-TCNQ	2,3,5,6-tetrafluoro-7,7,8,8-tetracyanoquinodimethane
Fcc	Face-centered cubic
FET	Field-effect transistor
FETT	Field-effect tunneling transistors
FF	Fill factor
FL	Few layers
FM	Ferromagnetic or ferromagnet
f-rGO	Functionalized reduced graphene oxide
FTO	Fluorine-doped tin oxide
g_{11}	Piezoelectric voltage coefficient
GB	Grain boundary
GNR	Graphene nanoribbon
GO	Graphene oxide
GR	Graphene
$H_8N_2MoS_4$	Ammonium thiomolybdate
h-BN	Hexagonal boron nitride
Hcp	Hexagonal close-packed
HER	Hydrogen evolution reaction
HGF	Holey graphene framework
HI	Hot-injection
HJ	Heterojunction

HMDS	Hexamethyldisilazane
HS	Heterostructure
HTL	Hole transporting layer
HTM	Hole transporting material
IM	Interface-mediated
IP	In-plane
IR	Infrared
ITO	Indium tin oxide
J_{sc}	Photocurrent density
KIB	Potassium-ion batteries
L-	Lateral
LB	Langmuir–Blodgett
LCD	Liquid crystal display
LDH	Layered double hydroxide
LED	Light-emitting diode
LFP	$LiFePO_4$
LIB	Lithium-ion battery
LiPS	Lithium polysulfide
LiSB	Lithium-sulfur battery
LiTFSI	Lithium bis-trifluoromethyl sulfonamide
LPE	Liquid-phase exfoliation
LTO	$Li_4Ti_5O_{12}$
ML	Multilayers
MOF	Metal–organic framework
MPTMS	(3-mercaptopropyl)trimethoxysilane
MPX_3	Metal phosphorus trichalcogenide
MR	Magnetoresistance
MWCNT	Multiwalled Carbon Nanotubes
m-WSe_2	Monolayer WSe_2
n-BuLi	n-butyl-lithium
NCS	Non-centrosymmetry
NF	Nanoflakes
NG	Nanogenerator
NIR	Near infrared
NMP	N-methyl-pyrrolidone
NPs	Nanoparticles
NS	Nanosheet
o/p	Output
ODTS	Octadecyltrichlorosilane
OER	Oxygen evolution reaction
OLA	Oleylamine
OP	Out-plane
ORR	Oxygen reduction reaction
OSC	Organic solar cell
P(VDF-TrFE)	Poly(Vinylidenefluoride-Trifluoroethylene)

PA6	Polyamide-6
PAS	Phenyl acetylene silver
PBE-SOC	Perdew–Burke–Ernzerhof with spin-orbit coupling
PC	Propylene carbonate
PCA	Photocatalytic activity
PCE	Power conversion efficiency
P_D	Power density
PDMS	Polydimethylsiloxane
PE	Piezoelectric
PECVD	Plasma-enhanced CVD
PEDOT	Poly(3,4-ethylenedioxythiophene)
PEI	Plyethylenimine
PEN	Polyethylene naphthalate
PENG	Piezoelectric nanogenerator
PEO	Poly(ethylene oxide)
PET	Polyethylene terephthalate
PF	Polyfurfural
PFM	Piezoelectric force microscopy
PL	Photoluminescence
PMMA	Poly(methyl methacrylate)
p-MSB	1,4-bis(4-methylstyryl)benzene
PPC	Polypropylene carbonate layer
PR	Potential reversal
PS	Potentiostatic
PSCs	Perovskite solar cells
PTAA	Poly[bis(4-phenyl)(2,4,6-trimethylphenyl)amine
PTEG	Piezo/triboelectric nanogenerator
PTENG	Photo thermoelectric nanogenerator
PTFE	Polytetrafluoroethylene
PTP	Photopolymerization
p-TSA	p-toluene sulfonic acid
PU	Polyurethane
PVA	Poly(vinyl alcohol)
PVDF	Poly(vinylidene fluoride)
PVP	Polyvinylpyrrolidone
PZT	Lead zirconate titanate
QD	Quantum dot
QE	Quantum efficiency
QHE	Quantum Hall effect
QSs	Quantum sheets
QY	Quantum yield
RE	Rare earth
R_{ex}	External resistance
rGO	Reduced graphene oxide
RH	Relative humidity

R_L	Load resistance
R_{sh}	Shunt resistant
S	Sulfur
SAM	Self-assembled monolayer
SBH	Schottky barrier height
SBU	Secondary-building unit
SHG	Second-harmonic generation
SL	Single-layer
SSA	Specific surface area
SSE	Solid-state electrolyte
STS	Scanning tunneling spectroscopy
SWCNT	Single-walled carbon nanotubes
TBAP	Tetrabutylammonium hexafluorophosphate
T_c	Curie temperature
TCE	Transparent conductive electrode
TCNQ	7,7,8,8-tetracyanoquinodimethane
TEA BF4/AN	tetraethylammonium tetrafluoroborate/acetonitrile
TENG	Triboelectric nanogenerator
TETA	Triethylenetetramine
TFET	Tunnel field-effect transistors
THAB	Tetraheptylammonium bromide
THF	Tetrahydrofuran
TM	Transition metal
TMDC	Transition metal dichalcogenide
TMO	Transition metal oxide
T_N	Neel temperature
TPA	tetrapropylammonium
Tpy	Hexafunctionalterpyridine
TTRAM	Terminal tunneling random access memory
UPS	Ultraviolet photoelectron spectroscopy
UV	Ultraviolet
UVO	Ultraviolet ozone
V-	Vertical
VB	Valence band
VBM	Valence band maximum
vdW	van der Waals
vF	Fermi velocity
V_g	Gate voltage
V_{oc}	Open-circuit voltage
WC	Wet-chemical
ZIF	Zeolite imidazole framework
ΔU_P	Piezoelectric potential
ΔU_t	Electric potential difference

Chapter 1
Introduction

The discovery of graphene (GR) and its amazing characteristics spawned a new material category called "2D materials". Inspired by success of GR synthesis, researchers are now set their focus on other layered and non-layered two-dimensional materials (2DMs) with distinct physical and chemical characteristics. The dimensionality effect and modulability of their band structure are responsible for these expected characteristics. Numerous applications in electronics/optoelectronics, electrocatalysis, batteries, supercapacitors, and sensing platforms are current hot researched topics. This chapter introduces the 2D framework and discusses the distinctiveness of 2D family.

1.1 Background

The collection of scientific knowledge and technology appears to have expanded exponentially throughout time. If present trends continue, the frequency of technological revolutions will grow progressively in the twenty-first century and beyond. Nanotechnology, a new scientific and technology age with applications ranging from nanoparticle (NPs)-based cancer therapy [1] to high-performance composites [2–4], provides new paths for innovation in health, electronics, materials, and other fields [5, 6]. Aside from new technologies, the growth of nanoscale sciences is marked by a novel class of low-dimensional systems. For example, quantum dots (QDs) in the 0D domain, nanoribbons, nanotubes, and nanowires in the 1D domain, single-atom-thick materials in the 2D domain, and nanoballs and nanocones in the 3D domain [7]. Dimensionality is a key material characteristic that distinguishes various nanostructure types, influences atomic structure and qualities of a substance [8]. A chemical element or compound might have distinct characteristics in different dimensions.

Carbon (C) materials have made a significant contribution to the progress of modern science and technology. As a matter of fact, C-nanostructures are recognized

M. Ikram et al., *2D-Materials for Energy Harvesting and Storage Applications*, Nanostructure Science and Technology,
https://doi.org/10.1007/978-3-030-96021-6_1

as game-changing materials in the field of nanotechnology. With the discovery of fullerene (C_{60}) and single-walled carbon nanotubes (SWCNT), 2D C-nanomaterials-based research has been attempted to separate 2D graphitic material. Recently, the technique of synthesizing 1D nanoribbon from 2D crystals has been undergoing investigation. Since 2004, when Geim and Novoselov announced the first isolation and electrical characterization of GR transistors [9], efforts have begun to synthesize and characterize more nanostructures. The majority of these nanostructures are composed entirely of sp^2 hybridized C atoms, forming a variety of structures ranging from C_{60} to the single-atom-thick plane of C called GR to the "rolled up," tubular sheets of GR called carbon nanotube (CNT), as illustrated in Fig. 1.1 [10, 11]. Although these GR-like structures are made of sp^2-C, their physicochemical characteristics are unlike their atomic structure. Starting with theoretical studies of GR and graphite's electrical characteristics, it has been evident that graphite possesses certain unique features [12].

GR-based materials such as graphene oxide (GO) and reduced graphene oxide (rGO) have previously been described in terms of their physicochemical characteristics and applications [13, 14]. Many academics were surprised to learn that less than a decade after the discovery of GR, and over a dozen 2DMs were separated and investigated. The advancement of GR research has sparked interest in other 2DMs, with metal nitrides and carbides forming the first materials to investigate immediately after GR. Additionally, 2D nanostructures are a prominent research

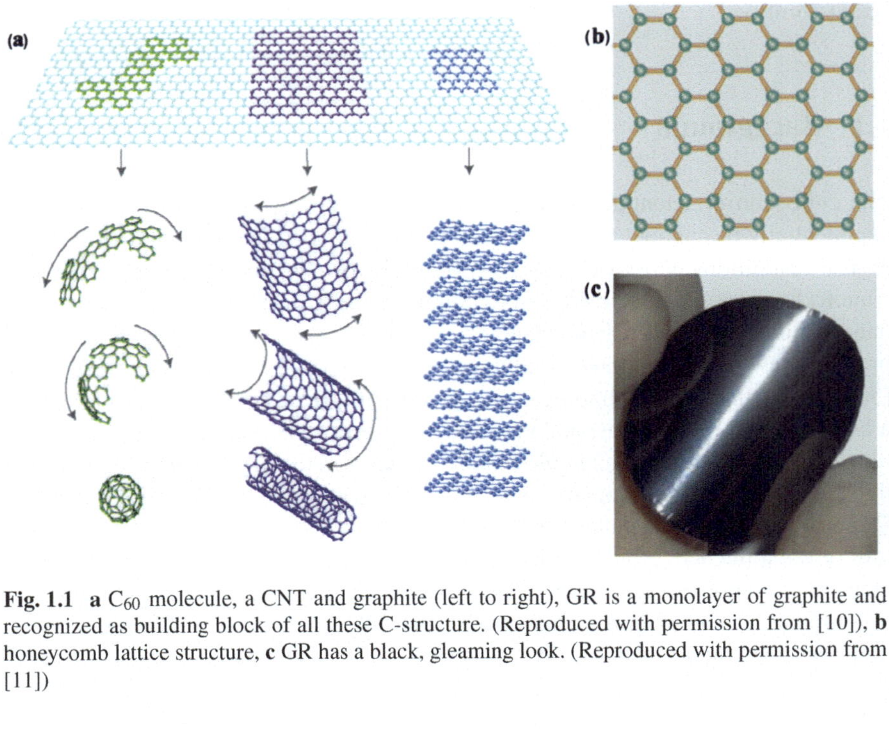

Fig. 1.1 **a** C_{60} molecule, a CNT and graphite (left to right), GR is a monolayer of graphite and recognized as building block of all these C-structure. (Reproduced with permission from [10]), **b** honeycomb lattice structure, **c** GR has a black, gleaming look. (Reproduced with permission from [11])

material owing to unique physical singularities that emerge when heat and charge are linked together to a plane. Further, 2D nanostructures are predicted to have a substantial influence on many applications have excellent properties, ranging from high-performance sensors to electronics, gas separation or storage, catalysis, inert coatings, and support membranes, among others. The 2DMs family comprises Transition metal dichalcogenide (TMDCs) and layered metal oxides along with C materials. The 2DMs utilization in electrical devices is one of the most exciting applications [15, 16]. Electrical characteristics permitting, a new range of superconductors, metallic materials, semimetals, solar cells, and insulators may be formed from 2DMs.

Although GR is the last member of C-family to be separated, it has provided a strong building block for other members of the family with varying dimensions. Due to their varied characteristics and natural abundance, single layer (SL) of TMDCs has garnered considerable interest. Despite the similarities in the chemical formula, MX_2, where M is generally from transition metals (TMs) groups IV-X, (Mo, Nb, W, Ni, V, or Re) and X is a chalcogen (commonly Se, Te, or S) (See Chap. 3 for details) [17, 18]. While the trilayers are covalently linked, adjacent sheets are covalently bonded to form 3D crystals via van der Waals forces (vdW). At the moment, 40 distinct kinds of TMDCs with various chalcogen atoms are available. TMDCs can be classified as insulators (HfS_2), semiconductors (MoS_2), semimetals ($TiSe_2$), or real metals ($NbSe_2$), based on their coordination and oxidation state of the metal atoms [19, 20]. Additionally, Group IV 2D oxides have recently gained popularity. The electronic structure of these various forms of nanostructures, whether oxides or TMDCs has been investigated experimentally and computationally and is comparable to GR [21].

Figure 1.2 summarizes the present membership of the family of 2D layered materials [22]. Stability, on the other hand, is a vital aspect of any functional material. For multilayers (MLs), the blue-colored materials remain constant at room temperature in air. Perhaps compounds with a greenish color are stable in air, whereas those with a pink color are only stable in an inert environment. Gray shading indicates that the ML was exfoliated, but no more information other than an Atomic Force Microscope (AFM) has been supplied. A comprehensive periodic table of crystalline materials exists, each with unique mechanical, transport, and electronic characteristics, and the possibility of fabricating single or multiple-atom polyhedral thick 2D sheets from any material remains. Furthermore, these layered 2DMs exhibit an excellent combination of electronic [23], mechanical [24], optical [25], and thermal properties [26], suggesting that can be used in place of Si in electronics, photonics, and nanoelectromechanical systems in the coming century or beyond. Most academics envision 2DMs being employed in real-world applications like this.

2DMs in nano and microscale have a limited 1D structure that comes into close contact with one another to establish boundaries. Numerous techniques are now being employed to synthesize 2D nanostructures and quantum features with uniform and comparable thickness and dimensions (See Chap. 4). It was not until 2004 that scientists isolated individual GR planes using sticky tape, resulting in flakes with novel electrical characteristics [27]. The finding piqued the scientific community's

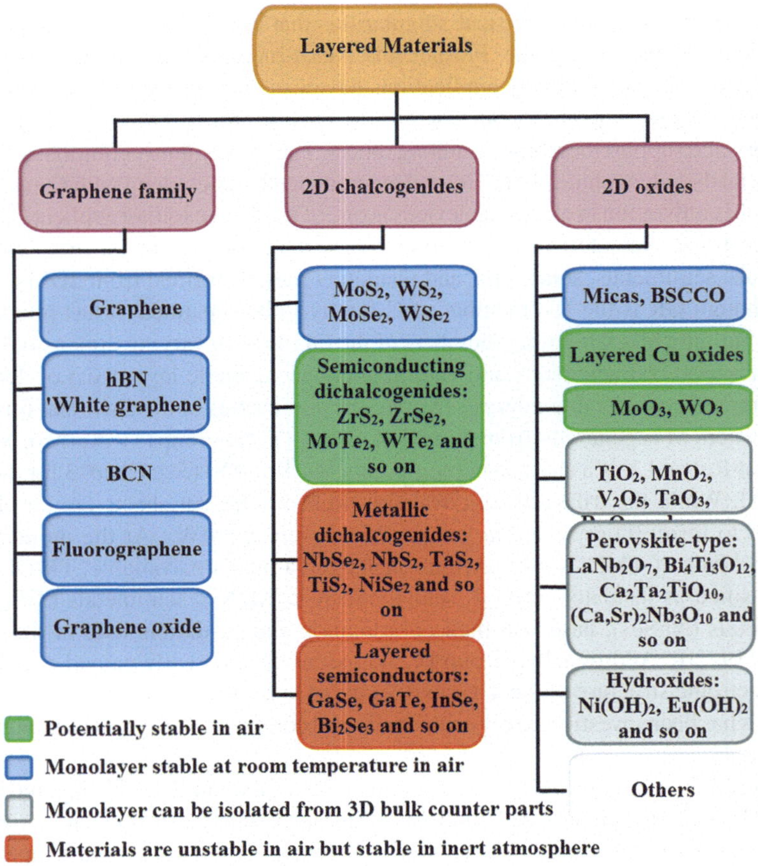

Fig. 1.2 2DMs library (Modified from [22])

interest in large-scale manufacturing of 2DMs. Chemical and mechanical exfoliation of vdW materials into an SL or few layers (FLs) has been shown throughout the previous decade [22]. Numerous scholars have been encouraged by this exploratory effort to obtain and differentiate these layers in order to uncover novel characteristics. Numerous techniques for fabricating, detecting, transferring, characterizing, and manipulating the characteristics of 2D layered materials have been described in the field of GR research [28, 29]. Furthermore, new synthesis techniques as solvothermal, solution-based, and epitaxial methods, have opened up new avenues for the fabrication of novel SL or FL and vdW materials. These well-established approaches have aided in quick establishment of 2DMs field other than GR. Soon after these findings, several GR "cousins" hinted that 2D elements in groups IV and II–VI, as silicene [30–32] and germanene [33, 34], may exhibit GR-like characteristics. Moreover, these 2D semiconductor materials are extremely valuable structural agents that may be altered and combined with other materials to produce composites for various purposes.

Numerous layered materials exist with strong in-plane (IP) chemical connections and poor coupling between layers. Individual freestanding atomic layers can be formed in this layered architecture. Once the layered material thickness is lowered to an SL or FLs, remarkable changes in its electrical characteristics occur. GR proposed building a paradigm for studying 2DMs not only physically in terms of synthesis, transfer, and characterization, but theoretically in terms of electrical structure, quantum Hall effect (QHE), and spintronics. Nanofilms covered with tiny layers of inorganic GR-like materials offer a variety of possible uses, ranging from lightweight anticorrosive materials to electron field emitters [35, 36]. Moreover, the semiconducting characteristics of inorganic nanofilms provide them suitable materials for further optoelectronic material downsizing. Lastly, nonlinear optics and photovoltaics technologies benefit from GR-like materials' hexagonal structure, since these conducting nanofilms can rapidly and efficiently detect self-inductance and resistance. Examples are the sharp tips on scanning spectroscopy probes and other magnetic, electronic detecting devices [37, 38]. Further research and possible practical uses expanded with the discovery of other unique properties of these materials.

1.2 Why 2D Materials?

Numerous questions may be asked regarding 2DMs. What distinguishes 2DMs? Why is it they are more appealing? Why do the characteristics of 2DMs differ from those of their bulk form? Is there a potential that 2DM will become the next revolutionary technology? In recent years, 2D inorganic materials have gotten a lot of interest because of their unusual physical features, which are related to their nanoscale thickness due to the quantum size effect. Some noteworthy changes in the electrical and optical characteristics of layered materials happen as a consequence of electron confinement and as a result of the layers being in close proximity to one another to create interfaces. This lack of interlayer contact plays a critical role in determining their band shape. The chemical and mechanical reaction also contributes significantly to the modification of their characteristics, owing to quantum effects and a high surface area to volume ratio.

Are the special features of 2DMs worth the inevitable difficulties of switching technologies? To develop 2DMs appropriate for industrial uses, research and development activities on 2DMs have progressed tremendously already. The fast evolution of 2DMs is one notable feature, with advances from laboratory random creation of micro to large-scale flakes, and now to roll-to-roll processing of nanostructures reaching meter-scale. There are various reports of how 2DMs may be used to develop new materials. The surface nature of 2DMs allow for surface treatments to alter their properties (by chemical functionalization [39, 40]). Also, 2DMs may be easily produced as a dispersion of nanoflakes [41]. The resulting composite materials have many of the same remarkable characteristics as the originals.

The abundance of unique characteristics found is one reason for the rapid advancement of 2DM research. 2DMs are ideal for future electrical devices due to their one-atom thickness. These materials have 2D nature, chiral charge carrier characteristics, strong carrier mobility at room and low temperature, and high thermal conductivity. Furthermore, SL-2D crystals have features similar to relativistic Dirac particles [42], whereas FLs of electrons have qualities similar to Dirac particles but with parabolic dispersion. Both allotropes can be gapless semiconductors. FLs, unlike SL, are more flexible: a transverse "E" can open a gap [43, 44], and a small strain can qualitatively modify the low-energy band structure [45]. Due to electrons' speculative nature, exploring the relativistic effects of 2DMs in condensed matter research is extremely interesting. Understanding 2D transport characteristics requires understanding how they behave under a strong magnetic field. This behavior displays the fundamental QHE [46, 47] is the exact quantization of the device's Hall resistance. Once more layers (more than three) are layered; the unusual band structure becomes frequent.

It is also known as zero-gap semiconductor/semimetal because the valence and conduction bands (VB and CB) touch in 2DMs [48]. Due to their brittleness in single point contact, the band structure of 2DMs is extremely sensitive to changes including mechanical deformations, external electric field (E_{ex}), doping, and adsorbents which are required for sensor applications. Likewise, an am-bipolar portion onto 2DMs is suggested, in which the charge carrier can include either es^- or hs^+ depending on the dopant material [27]. However, doping is not always required to affect the charge carrier's sign; nearby electric fields can also do so. This unique characteristic of 2DM is critical for the development of high-mobility p–n junction transistors from the basis of current CMOS (complementary metal–oxide–semiconductor) technology.

However, structural study of 2DMs demonstrates that they may form crystals of high quality at micron sizes. Electrons travel easily without scattering in defect-free materials, even on rough surfaces. The absence of scattering results in high charge carrier mobilities that shows the major QHE [49]. Moreover, field-effect transistors (FET) employ a direct bandgap SL-2DM, achieving on/off ratios of 10^8. It possesses outstanding electrostatic integrity enables sub-nanometer-thick, flexible electronics, and consumes up to 100,000 times less power in sleep mode relative to conventional silicon (Si) transistors [50].

In view of the great mobility of the charge carriers in 2DMs, these materials provide strong candidates for high-gain photodetection. Thus, this material's conductivity is extremely sensitive to an electrostatic disturbance caused by photogenerated charge carriers close to top layer (surface). This provides 2DMs especially favorable for detecting exceptional gain photons via the photogating phenomenon. An enhanced photodetection using 2DMs will pave the way for applications in biomedical imaging, integrated optoelectronic devices, quantum information technology, remote sensing, and optical communications. Similarly, the 2DMs thickness allows for the measurement of field effects in the sample conductance is critical for sensor and optoelectronics applications. Furthermore, because 2DMs are one atom thick, their electron waves are amenable to comprehensive surface characterization methods.

Electronic band structures in 2DMs are extremely fascinating; zero bandgap of these materials enables to absorb light throughout a broad range of wavelengths, from infrared (IR) to ultraviolet (UV) regions [51]. This offers up a vast array of electronic transition possibilities, culminating in high-performance optoelectronic devices, other fundamental optics and photonics research [52]. A FLs of 2DM exhibit absorption of less than 10%, whereas an SL has a value of 2.3% attributed to their thickness and transparency [53]. This remarkable transparency, along with superior conductivity, enables them ideal for use as transparent electrodes in a wide variety of electrical devices, including solar cells and liquid crystal displays [54–56]. While this is an intriguing subject, the optical characteristics of these materials may be unknown, and the connection between optical and electrical properties will open up new paths of research in 2DMs. Practical application of 2DMs needs an understanding of their performance in real devices, along with their endurance in ambient and severe environments. A dedicated effort is be required to investigate the dependability of 2DMs-based devices, electric or thermal stress tests, as well as device lifespan. To maintain device function, some protection of the 2DM and metals will very certainly be required to reduce environmental impacts.

The strong IP covalent bonding and atomic thickness frequently provide them with great mechanical strength and flexibility. As a key example, GR has a measured breaking strength and Young's modulus of 42 Nm^{-1} and 1.0 TPa, respectively [57]. Nonetheless, some 2D nanomaterials have demonstrated superior mechanical characteristics. A typical example is an SL-MoS_2 nanosheet (NS) has Young's modulus of 270 GPa higher than bulk MoS_2 (240 GPa) or steel (205 GPa) [58, 59]. Due to the fact that all exhibit comparable 2D structural features to GR, it is plausible to expect that all 2DMs exhibit superior mechanical capabilities. Apart from mechanical strength and flexibility, the atomic thickness can confer exceptional optical transparency improves with decreasing thickness. Thus, atomic thickness ensures maximal optical transparency. GR absorbs only 2.3% of white light and reflects only 0.1%, indicating it can transmit over 97% of white light [53], this result established GR's superior optical transparency. Numerous more 2DMs with equivalent or better optical transparency than GR are predicted. The combination of superior mechanical strength, high flexibility, extraordinary light transmittance, and high-performance electronic/optoelectronic devices require materials with extraordinary electronic characteristics that are wearable, flexible, and transparent [60, 61]. Without a rupture, it is simple to locate creases and pleats. However, because these materials are thin, they are brittle and break with higher force than glass.

Further, arising from their large lateral (L-) dimension, 2DMs have a high specific surface area (SSA) and ultimate exposure of their surface atoms provides them extremely attractive for a variety of practical surface-related applications [62, 63]. According to the estimated conclusion, GR theoretically has an SSA of 2630 m^2g^{-1} [64] and proved to be extremely appealing for surface-related applications due to high SSA. GR thin films, when used as the electrode in supercapacitors, have a significantly higher specific capacitance than graphite due to their SSA. [53]. Owing to this property 2D nanomaterials demonstrated outstanding catalytic activity in a variety of catalytic processes, including organic catalysis, electrocatalysis, and photocatalysis [62]. In summary, the 2DM's architect combined with its chemical bond strength enables an intriguing mechanical material; chemical alteration of 2DMs is also significant. In this regard, 2D functionalization can convert metal to insulator [65], and chemical doping with metal ions [66]. Mechanical characteristics can be adjusted by flaws or functional groups [67]. These chemical characteristics allow the creation of all 2DM devices with distinct areas with different properties.

References

1. Davis ME, Zuckerman JE, Choi CHJ, Seligson D, Tolcher A, Alabi CA, Yen Y, Heidel JD, Ribas A (2010) Nature 464:1067
2. Singh V, Diaz R, Balani K, Agarwal A, Seal S (2009) Acta Mater 57:335
3. Keshri AK, Huang J, Singh V, Choi W, Seal S, Agarwal A (2010) Carbon 48:431
4. Zou J, Liu J, Karakoti AS, Kumar A, Joung D, Li Q, Khondaker SI, Seal S, Zhai L (2010) ACS Nano 4:7293
5. Karakoti AS, Tsigkou O, Yue S, Lee PD, Stevens MM, Jones JR, Seal S (2010) J Mater Chem 20:8912
6. Joung D, Singh V, Park S, Schulte A, Seal S, Khondaker SI (2011) J Phys Chem C 115:24494
7. Tiwari JN, Tiwari RN, Kim KS (2012) Prog Mater Sci 57:724
8. Mas-Ballesté R, Gómez-Navarro C, Gómez-Herrero J, Zamora F (2011) Nanoscale 3:20
9. Novoselov KS, Geim AK, Morozov SV, Jiang D, Katsnelson MI, Grigorieva IV, Dubonos SV, Firsov AA (2005) Nature 438:197
10. Geim AK, Novoselov KS (2007) Nat Mater 6:183
11. Li D, Kaner RB (2008) Science 320:1170
12. Wallace PR (1947) Phys Rev 71:622
13. Singh V, Joung D, Zhai L, Das S, Khondaker SI, Seal S (2011) Prog Mater Sci 56:1178
14. Yang N, Zhai J, Wang D, Chen Y, Jiang L (2010) ACS Nano 4:887
15. Wang QH, Kalantar-Zadeh K, Kis A, Coleman JN, Strano MS (2012) Nat Nanotechnol 7:699
16. Wilson JA, Di Salvo FJ, Mahajan S (1974) Phys Rev Lett 32:882
17. Mak KF, Lee C, Hone J, Shan J, Heinz TF (2010) Phys Rev Lett 105: 136805
18. Canadell E, LeBeuze A, El Khalifa MA, Chevrel R, Whangbo MH (1989) J Am Chem Soc 111:3778
19. Huang YH, Peng CC, Chen RS, Huang YS, Ho CH (2014) Appl Phys Lett 105: 093106
20. Jeong S, Yoo D, Jang J-T, Kim M, Cheon J (2012) J Am Chem Soc 134:18233
21. Xu M, Liang T, Shi M, Chen H (2013) Chem Rev 113:3766
22. Geim AK, Grigorieva IV (2013) Nature 499:419
23. Eda G, Fujita T, Yamaguchi H, Voiry D, Chen M, Chhowalla M (2012) ACS Nano 6:7311
24. Li H et al (2013) Small 9:1974
25. Li H, Lu G, Yin Z, He Q, Li H, Zhang Q, Zhang H (2012) Small 8:682
26. Tongay S, Zhou J, Ataca C, Lo K, Matthews TS, Li J, Grossman JC, Wu J (2012) Nano Lett 12:5576
27. Novoselov KS, Geim AK, Morozov SV, Jiang D, Zhang Y, Dubonos SV, Grigorieva IV, Firsov AA (2004) Science 306:666
28. Vishnoi P, Pramoda K, Rao CNR (2019) ChemNanoMat 5:1062
29. Roldán R, Chirolli L, Prada E, Silva-Guillén JA, San-Jose P, Guinea F (2017) Chem Soc Rev 46:4387
30. Li W, Qian X, Li J (2021) Nat Rev Mater

31. Mannix AJ, Kiraly B, Hersam MC, Guisinger NP (2017) Nat Rev Chem 1:0014
32. Zhao J et al (2016) Prog Mater Sci 83:24
33. Li C et al (2020) J Mater Chem C 8:16318
34. Acun A et al (2015) J Phys: Condensed Matter 27: 443002
35. Husain E, Narayanan TN, Taha-Tijerina JJ, Vinod S, Vajtai R, Ajayan PM (2013) ACS Appl Mater Interfaces 5:4129
36. He W, Zhu L, Chen H, Nan H, Li W, Liu H, Wang Y (2013) Appl Surf Sci 279:416
37. Cao L et al (2013) Small 9:2905
38. Zhan Y, Liu Z, Najmaei S, Ajayan PM, Lou J (2012) Small 8:966
39. Boukhvalov DW, Katsnelson MI (2009) J Phys: Condensed Matter 21: 344205
40. Kuila T, Bose S, Mishra AK, Khanra P, Kim NH, Lee JH (2012) Prog Mater Sci 57:1061
41. Hernandez Y et al (2008) Nat Nanotechnol 3:563
42. Geim AK (2009) Science 324:1530
43. McCann E, Fal'ko VI (2006) Phys Rev Lett 96: 086805
44. Zhang Y, Tang T-T, Girit C, Hao Z, Martin MC, Zettl A, Crommie MF, Shen YR, Wang F (2009) Nature 459:820
45. Mohiuddin TMG et al (2009) Phys Rev B 79: 205433
46. Heersche HB, Jarillo-Herrero P, Oostinga JB, Vandersypen LMK, Morpurgo AF (2007) Nature 446:56
47. Li X, Wang X, Zhang L, Lee S, Dai H (2008) Science 319:1229
48. Williams JR, DiCarlo L, Marcus CM (2007) Science 317:638
49. Novoselov KS et al (2007) Science 315:1379
50. Radisavljevic B, Radenovic A, Brivio J, Giacometti V, Kis A (2011) Nat Nanotechnol 6:147
51. Reina A, Jia X, Ho J, Nezich D, Son H, Bulovic V, Dresselhaus MS, Kong J (2009) Nano Lett 9:30
52. Mao N, Chen Y, Liu D, Zhang J, Xie L (2013) Small 9:1312
53. Nair RR, Blake P, Grigorenko AN, Novoselov KS, Booth TJ, Stauber T, Peres NMR, Geim AK (2008) Science 320:1308
54. Stanford MG, Zhang C, Fowlkes JD, Hoffman A, Ivanov IN, Rack PD, Tour JM (2020) ACS Appl Mater Interfaces 12:10902
55. You R, Liu Y-Q, Hao Y-L, Han D-D, Zhang Y-L, You Z (2020) Adv Mater 32:1901981
56. Han T-H, Kim H, Kwon S-J, Lee T-W (2017) Mater Sci Eng R Rep 118:1
57. Lee C, Wei X, Kysar JW, Hone J (2008) Science 321:385
58. Bertolazzi S, Brivio J, Kis A (2011) ACS Nano 5:9703
59. Tan C et al (2017) Chem Rev 117:6225
60. Akinwande D, Petrone N, Hone J (2014) Nat Commun 5:5678
61. Nathan A et al (2012) Proc IEEE 100:1486
62. Deng D, Novoselov KS, Fu Q, Zheng N, Tian Z, Bao X (2016) Nat Nanotechnol 11:218
63. Mendoza-Sánchez B, Gogotsi Y (2016) Adv Mater 28:6104
64. Stoller MD, Park S, Zhu Y, An J, Ruoff RS (2008) Nano Lett 8:3498
65. Gierz I, Riedl C, Starke U, Ast CR, Kern K (2008) Nano Lett 8:4603
66. Gómez-Navarro C, Burghard M, Kern K (2008) Nano Lett 8:2045
67. Yazyev OV, Louie SG (2010) Phys Rev B 81: 195420

Chapter 2
Advances in Ultrathin 2D Materials

In recent years, a significant number of articles concentrating on GR, TMDCs, MXenes, and other 2DMs have been documented because of their novel characteristics and applications that arise from 2D confinement. In this chapter, we summarize recent progress and new significant breakthroughs in 2DMs from ancient to modern era (GR and beyond). We also address the newest family of 2DMs as well as recent advancements in fabrication to application. Recent four years of research highlights in the context of energy-related applications solar cells, batteries, supercapacitors, H_2 evolution, piezoelectricity, hybrid devices, flexible electronics, and so on are presented at the end of section.

2.1 Revolution of 2D Materials

Recently, 2DMs have been rise as the material world's superstars. Due to their extraordinary physical and electrical characteristics which stem from their physical size, 2DMs have the potential to transform everything from digital computing to energy harvesting and storage. These crystalline solids are virtually all surface, consisting of layers only a few atoms thick. This quality allows them not only to provide a large surface area for interaction with molecules and reacting to physical forces but also ensures they are subject to quantum effects, imparting characteristics that can be quite different from those of their 3D counterparts. The continuing revolution in 2DMs began with solitary GR studies. GR, whose composition is C-atoms organized hexagonally in a honeycomb pattern, has unique characteristics that distinguish it from graphite, the bulk form of graphite. Despite being the smallest substance ever discovered, GR is the toughest substance yet tested at one atom thick. It is also a superior electrical conductor to Cu. Although GR is considered a sophisticated substance, its origins extend back to nineteenth century. Benjamin Collins Brodie, an English chemist, discovered the layers in graphite in 1859, but it took Canadian theoretical

M. Ikram et al., *2D-Materials for Energy Harvesting and Storage Applications*,
Nanostructure Science and Technology,
https://doi.org/10.1007/978-3-030-96021-6_2

physicist Philip Russell Wallace until 1947 to coin the term GR. Wallace determined graphite's electrical characteristics by building layers of a hypothetical 2D substance termed GR. By the 1970s, scientists had discovered a method the strong interactions between these products makes identifying GR's physical characteristics are difficult. This changed in 2004, when Andre Geim and Kostya Novoselov, two physicists from the University of Manchester, were playing about in the lab one Friday evening tearing flakes off a piece of graphite using Scotch (sticky) tape. They eliminated fewer and smaller bits until after tens of rounds; they were able to generate Wallace's GR in a manner that allowed for the examination of its physical characteristics. This quickly exposed that these qualities are rather remarkable: GR is transparent and an excellent electricity conductor characteristics that are extremely unusual in the same material in addition to being extraordinarily strong, flexible, and light. Researchers have found a plethora of possible uses for this wonder substance, such as physical support for growing cells, an electrode material in batteries, a conductor in electronic circuits, and an additive for metals and alloys. Scientists began looking for similar layered compounds made of other elements after discovering GR. The study of 2DMs has exploded in the recent decade. In an ideal world, scientists would like to discover 2DMs with somewhat different and superior characteristics than GR. A new generation of atomic-scale electronic circuits might be built on semiconducting 2DMs. Comparatively, unlike GR, which conducts electricity constantly, semiconductors conduct electricity only when their electrons are sufficiently energized, usually by an applied voltage, causing them to jump across bandgap energy (E_g). The ones and zeroes that comprise digital information are created by turning the conductivity of semiconductors on and off. One strategy for developing semiconducting 2DMs is to attempt to synthesize GR using other elemental atoms. This resulted in large number of 2DMs formation of generally referred to as Xenes, due to their common suffix of "-ene." They comprise silicene, a Si-based compound; phosphorene, a phosphorus-based compound; stanene, a tin-based compound; germanene, a germanium-based compound; and borophene, a boron-based compound. These Xenes have the same honeycomb-like hexagonal structure as GR, but their atoms do not bind as tightly as C-atoms, resulting in a buckled surface of these Xenes only phosphorene and silicene contain bandgap, indicating that they are natural semiconductors. Like GR, germanene may get E_g by bonding molecules like H_2 to its surface. Borophene, like GR, is an efficient conductor, but stanene is a topological insulator means that the material's central region is an insulator yet its edges are highly conductive. A new age of computing known as spintronics may be born, in which digital information is collected in electron spins instead of electrical currents. Researchers are also looking at 2DMs made of molecules.

The suffix "-ene" refers to a class of possibly hundreds of distinct 2DMs called MXenes that mix ceramic and metal characteristics. MXenes are composed of C or N_2 with early TMs, which is a metal in the center of the periodic table, such as titanium. MXenes have been initially synthesized by chemically fragmenting Al from carbide known as a MAX Phase. MXenes, the same as GR, are strong and conduct well and are being investigated for a variety of possible uses as electrode materials in batteries and fuel cells. Another example is hexagonal boron nitride (h-BN), or white GR has the same hexagonal structure as GR but has alternating B and N atoms. Unlike GR, h-BN is an insulator, making it excellent for use with GR and other semiconducting 2DMs. This 2DM is extremely flexible and possesses a variety of unique characteristics, including fluorescence and resistance to oxidation. Scientists are investigating a variety of uses as nanofiller in electrical packaging, a laser emitter, and for oil spill cleanup. Aerogels composed of 2D h-BN NSs were shown to absorb up to 33 times their own weight in oil and organic solvents while remaining water-resistant. Several atomic layers make up other 2D compounds, including TMDCs. A TM atoms layer along with Mo or W is placed between layers of chalcogenide atoms, often S or Se are found below O_2 in the periodic table. Various fundamental element combinations can provide TMDCs with a range of characteristics, many of which were natural semiconductors. In addition, TMDCs can convert light to electricity and vice versa, which opens up the possibility of using them in photovoltaics and as light emitters in quantum cryptography. Due to the breadth of accessible 2DMs, research has shifted toward stacking several of these materials together to produce HS. These are kept together by weak vdW forces that help to keep the tension between the layers manageable. When several layers of GR are layered together, bulk GR is formed; however, when diverse 2DMs are stacked together, a substance is formed that does not exist in nature but may have some interesting features. For example, GR and tungsten disulfide, a TMDC, have been mixed to generate highly efficient solar panels. Once the semiconducting TMDC absorbs light, excited electrons jump into the GR layers and are carried away by an electric current. The same arrangement was been utilized in reverse to form a light-emitting diode (LED): sending an electrical current across GR allows a TMDC semiconductor to emit light, with various TMDCs producing light at different wavelengths. While GR was initially separated by physically peeling flakes from bulk graphite, a process known as mechanical exfoliation has been found not suited for industrial-scale production. Rather than that, researchers are resorting to procedures such as liquid exfoliation, as well as ion intercalation and exfoliation. Liquid exfoliation employs a solvent, such as water, to keep the GR layers apart after they were split by physical vibrations. Ions are injected between the layers during ion intercalation and exfoliation to weaken the vdW bonds prior to their disintegration by sound energy. Both of these methods, though, have drawbacks: liquid exfoliation yields are low, while intercalation reaction durations are lengthy. Furthermore, these exfoliation processes will not work with 2DMs that do not exist in bulk include silicene, germanene, and stanene. The only method to create these 2DMs also applies to GR that can be synthesized by exfoliation, which is from the bottom up via a chemical process. The most widely used approach is chemical vapor deposition (CVD), in which one or more gases containing the molecules that make up the 2DM are passed over a flat substrate is frequently coated with particles of a catalyst (iron). Following that, the gases react to produce the 2D substance on the substrate. This process is being utilized to create massive sheets of GR using methane as the C-source, however, achieving a uniform coating and separating the 2DM from the underlying substrate has made it challenging. 2DMs have progressed from the lab to the market as displayed in Fig. 2.1.

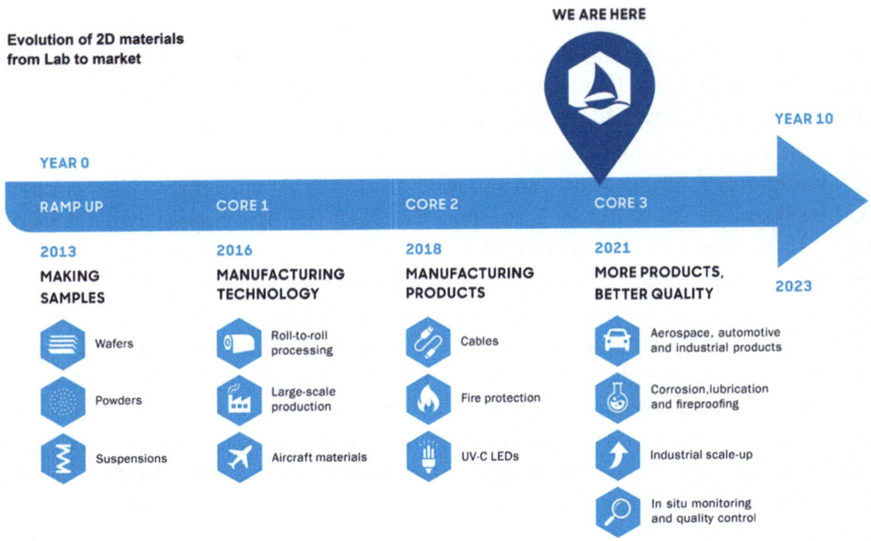

Fig. 2.1 Evolution of 2DMs from lab production to marketplace. (Reproduced with permission from [1], Figure courtesy "Graphene Flagship")

2.2 Recent Advances

In general, a lot of novel advancements for 2DMs were found or uncovered. To begin, the confinement of electrons in 2D ultrathin area, particularly for SL-2DMs, enhances their compelling electrical characteristics, making them suitable for basic research in condensed matter physics and electronics devices [2]. Second, their strong IP covalent bond and atomic thickness enable them to have exceptional mechanical strength, flexibility, and optical transparency, all of them is necessary for use in next-generation electronics [2–4]. Third, their enormous L-dimension, while retaining atomic thickness, confers an ultrahigh-SSA on them [5]. This is extremely advantageous for applications requiring a high degree of surface area as catalysis or supercapacitors [6, 7]. Fourth, the solution-based processibility of 2DM is critical for the production of high-quality freestanding thin films using straightforward techniques like vacuum filtering, spin coating, drop-casting, spray coating, and inkjet printing [8, 9]. Finally, the high exposure of surface atoms enables simple control of characteristics and functions by surface modification/functionalization, element doping, and/or defect/strain/phase engineering [10, 11]. Because electrons are relativistic particles related to GR's unique 2D framework and electronic band structure, they can be regarded as massless Dirac fermions [2, 12]. GR's unique electrical property enables it a desirable platform for fundamental condensed matter research since it allows detection of relativistic phenomena like the QHE and the Klein paradox to arise when GR has three layers or more [2, 13, 14]. Notably, GR has been shown to have a charge carrier mobility of up to ~ 10,000 $cm^2 \cdot V^{-1} \cdot s^{-1}$ at ambient temperature

[15, 16]. Despite this, one of GR's primary disadvantages is its lack of bandgap, which makes it unsuitable for high-performance low-power digital transistors [2, 17]. Ultrathin 2DMs include MoS_2, WS_2, $MoSe_2$, and WSe_2 have high E_g (~1–2 eV) but low carrier mobilities (a few to hundreds of $cm^2 \cdot V^{-1} \cdot s^{-1}$) [3, 4]. It has been proven that 2D-black phosphorus (BP) NSs have a broad E_g (0.3–2.1 eV) and carrier mobility of up to 1000 $cm^2 \cdot V^{-1} \cdot s^{-1}$ [18]. As a result, the low-dimensional materials described above are suitable channel materials for fabricating high-performance electrical devices and compensate for the weakness of GR [3, 4, 19].

Several simple fabricating techniques may turn 2D nanomaterials into free-standing thin films with good film quality, flexibility, and transparency (See Chap. 4). In addition to flexible and transparent electrical and optoelectronic devices, 2DMs may also be used as energy storage and conversion electrodes because of their outstanding film-forming ability. The exceptional film-forming ability and electron mobility of some 2DM, like rGO, MXenes, and 1T-MoS_2 are highly appealing as electrode materials for high-performance supercapacitors [7, 8, 20]. The freestanding thin films produced are employed as electrodes in supercapacitors due to their high capacitance ~900 Fcm^{-3} (for Ti_3C_2 NSs), good rate performance, and high cycle stability [20]. Cui and colleagues [21] recently proved that a hybrid electrode composed of GR and FL-BP Ns had a specific capacity of 2440 $mAhg^{-1}$ when utilized as a sodium-ion battery (SIB) electrode. Excellent film fabrication ability also demonstrates promise as active layers in functional devices as photovoltaics and memory devices where the active thin films quality is critical for the implementation of high-performance devices [9, 22]. With all atoms accessible for modification, low-dimensional materials provide an excellent platform for engineering their characteristics and features at the atomic level via surface modification/functionalization, element doping, and/or defect/vacancy/strain engineering [23]. Although neutral GR is unsuitable for oxygen reduction reactions, it may be doped with heteroatoms (N, S, P, and/or B) on its backbone to create N, S, P, and/or B-doped GR sheets that are extremely effective electrocatalysts [24]. Doping elements into 2DMs can dramatically modify their catalytic capabilities [24, 25]. Recent efforts to increase the number of defects on the backbone of the MoS_2 NSs clearly proved increased catalytic activity of 2D-MoS_2 toward electrocatalytic hydrogen evolution reaction (HER) [26, 27]. Sulfur (S) vacancies in the basal planes of MoS_2 Ns may also be used to enhance their electrocatalytic activity against HER [28, 29].

2.2.1 Research Highlights

2021

Ramesha et al. [30] developed cathode composite design using Si_3N_4–MoS_2 HS over multiwalled carbon nanotube (MWCNT) that can overcome the aforesaid problems. The S@Si_3N_4/MoS_2MWCNT, S cathode offers 1123 $mAhg^{-1}$ initial discharge capacity at 0.2 C rate. After 500 cycles at 1C rate, the cell retains 80.7% of its capacity with a 0.03% capacity fade each cycle. Even at 3C rate, the cell has 677 $mAhg^{-1}$ reversible capacities.

Hua et al. [31] reported that in potassium-ion batteries (KIBs), MoS_2 NS is coated over porous C polyhedrons made of ZIF-8. High electrical conductivity, N-doped C polyhedron structural stability, and MoS_2 layered crystal structure allow for good rate property and reversible capacity can be fixed at 375 $mAhg^{-1}$ after 100 cycles at 100 mAg^{-1}, 200 $mAhg^{-1}$ after 200 cycles at 500 mAg^{-1} in addition to excellent rate property.

Kim et al. [32] converted an N-MoS_2/PF precursor suspension in Si oil into a core–shell HS with a hierarchical core–shell structure via easy pyrolysis. Interlayer spacing is increased, as is electrical conductivity and mechanical stability, allowing for easy, highly reversible insertion and extraction of Na-ions by in situ N_2 doping in 2D-MoS_2 structure with C-inclusion. Its reversible capacity is 540.7 $mAhg^{-1}$, and its capacity retention is >100% after 200 cycles.

Zhang et al. [33] synthesized interlayer for LiPS stabilization using hierarchical porous C-nanonet flake coated with a small quantity of loosely dispersed MoS_2 for lithium-sulfur battery (LiSB). With the help of chemisorption and electrocatalytic actions of MoS_2, the soluble polysulfides intermediates are effectively transported out from the cell, resulting in a stable Li^+ transport pathway. A high initial capacity of 1413.8 $mAhg^{-1}$ at 0.1 C and a stable capacity of 897.5 $mAhg^{-1}$ at 0.5 C after 200 cycles are obtained, corresponding to a capacity decay rate of 0.063% per cycle.

Li et al. [34] introduced intercalation of GR into MoS_2 gallery leads in substantially larger MoS_2 interlayers (from 0.62 to 1.16 nm) and increased hydrophilicity (from 0.62 to 1.16 nm). Its flower-like structure enhances Zn-ion transport, electrolyte infiltration, and structural stability. A high rate capability (285.4 $mAhg^{-1}$ at 0.05 Ag^{-1} and 141.6 $mAhg^{-1}$ at 5 Ag^{-1}) and long-term cycling stability are therefore demonstrated in this new MoS_2/GR nanocomposite (88.2% capacity retention after 1800 cycles). Electrochemical studies and density functional theory (DFT) calculations support the improved Zn^{2+} migration kinetics and favorable pseudocapacitive characteristics.

2020

Huang et al. [35] reported interface enabler g-C_3N_4 was reported. As a result of the increased viscosity, lower surface tension of molten Li, and the in situ production of Li_3N at the interface, the researchers found that adding g-C_3N_4 into Li metal can not only transform the Li metal/garnet-type solid-state electrolyte (SSE) interface from point to intimate contact. An interfacial resistance of 11 Ωcm^2 and a critical current density (CCD) of 1500 μAcm^{-2} are obtained by symmetric Li-C_3N_4|SSE|Li-C_3N_4 cells. The identical symmetric cell design with clean Li metal electrodes has a significantly higher interfacial resistance (428 Ω cm^2) and a much lower CCD (50 μAcm^{-2}).

Wang et al. [36] performed surface of a commercial separator was coated with TM coordinated g-C_3N_4 with crystalline C (M–C_3N_4/C). To conclude, Ni–C_3N_4/C may firmly bind polysulfides, speed up polysulfide conversion rates, and facilitate electron/ion transfer. With good capacity retention of 89.4% after 300 cycles, the

LiSBs with Ni–C_3N_4/C modified separator can produce 999 $mAhg^{-1}$ at 0.5 Ag^{-1}. Even with 4.6 $mgcm^{-2}$ S loading and a 6 μLmg^{-1} electrolyte/S ratio, 1.0 Ag^{-1} yields a significant initial capacity of 724.6 $mAhg^{-1}$. With this new separator, researchers may study TM coordination as a dynamic controller for high-performance LiSBs and learn more about catalytic processes.

Ma et al. [37] reported unusual "stereoscopic" triangular pores, wide interlayer spacing, and flexible constitutional units that make the pores elastic and self-adaptable for the guest transmission were described for the first time a series of colyliform crystalline 2D covalent organic frameworks (COFs) with quasi-3D topologies. Their iodine adsorption rate (2.51 gh^{-1}) is quicker than that of conventional 2D COFs, with a maximum adsorption capacity (6.29 gg^{-1}) unmatched in the literature The QTD-COFs' high adsorption and irradiation stability enables them ideal for removing radioactive iodine from water quickly.

Samorì et al. [38] reported unusual usage of aza-Diels–Alder cycloaddition process to modify metal–organic framework (MOF)@COF-LZU1 to produce aza-MOFs@COFs hybrid porous materials with extended π-delocalization. As a proof of concept, the produced aza-MOFs@COFs are utilized as an electrode in supercapacitors with specific capacitance of 20.35 μFcm^{-2} and volumetric energy density of 1.16 Fcm^{-3}. The postsynthesis modification of MOFs@COFs hybrids implements rational design for the synthesis of functional porous materials and widens the potential uses of MOFs@COFs hybrid porous materials in energy storage applications.

Zhang et al. [39] developed a novel D_3h-symmetric tricyanomesitylene monomer that can be reticulated into a g-$C_{54}N_6$-COF through Knoevenagel condensation with D_{3h}-symmetric monomer 2,4,6-tris(4′-formyl-biphenyl-4-yl)-1,3,5-triazine. The octupolar conjugated properties of g-$C_{54}N_6$-COF were reflected in its rare solvatochromic effects in the ground and excited states, endowing it with more promising semiconducting behavior than g-$C_{52}N_6$-COF, including enhanced light-harvesting and excellent photoinduced charge generation and separation. Along with the matching energy level, g-$C_{54}N_6$-COF enabled photocatalytic water splitting in two-half reactions at an average O_2 evolution rate of 51.0 $\mu molh^{-1}$ g^{-} and an H_2 evolution rate of 2518.9 $\mu molh^{-1}$ g^{-1}.

Alsharaeh et al. [40] used microwave irradiation technique nanocomposites of Co_3O_4/rGO/h-BN as lithium-ion battery (LIB) anodes. As a consequence, incorporation of h-BN to Co_3O_4/rGO nanostructured but also increased the SSA and thermal stability. Co_3O_4/RGO/h-BN nanomaterials exhibited the largest SSA (191 m^2g^{-1}) suggesting the significant effects among h-BN and rGO. Furthermore, Co_3O_4 /rGO/h-BN also showed lowest internal resistance decrease to 150 °C with high cyclic stability (100% capacity retention) and the greatest specific capacity with reliable reversibility (100 cycles).

2019

Wang et al. [41] designed MoS_2-WS_2 films on FTO that can be vulcanized through CVD with MoO_3 and WO_3 as precursors by pulsed laser deposition. The electron transfer efficiency rate from FTO to the MoS_2 surface of FTO/WS_2/MoS_2 could be greater relative to efficiency from FTO to WS_2 surface of FTO/WS_2/MoS_2. The electrocatalysis reduction of I_3^- to I^- utilizing above-mentioned films as an electrode-counter in a dye-sensitized solar cell (DSSC) confirmed the supposition. The J_{sc} (14.16 $mAcm^{-2}$) of FTO/WS_2/MoS_2 is greater than FTO/MoS_2/WS_2 (11.72 $mAcm^{-2}$), suggesting that catalytic response of FTO/WS_2/MoS_2 has been superior compared to FTO/MoS_2/WS_2. The deposition sequence may be tuned to control the electron transport direction successfully.

Zhou et al. [42] prepared an organic oxygen battery using an organic metal membrane framework separator capable of conducting Li ions and separating other large species in the system. These batteries have excellent rate performance, extended cycle life, and low overpotential. Between the organic liquid anode and the ion-conducting separator, a solid electrolyte interface layer may be detected. A novel form of anode for LIBs is introduced here, as well as basic insights into biphenyl-based liquid anodes' use.

Qiao et al. [43] present a straightforward approach using single-step pyrolysis of bimetal-organic frameworks followed by a rapid in situ phosphorization procedure for the synthesis of cobalt phosphide anchored heteroatom (N, and P) codoped porous C (CoP/NP-HPC). In aqueous alkaline, the derived CoP/NP-HPC catalyst displays excellent bifunctional catalytic response to oxygen reduction reaction (ORR) and oxygen evolution reaction (OER) that is equivalent to or excellent than viable 20% Pt/C. Significantly, a rechargeable Zn-air battery using an electrocatalytic cathode of CoP/NP-HPC has a great power density of 186 $mWcm^{-2}$ and remarkable lasting charge or discharge cycle constancy, 20% Pt/C outperforming. The incorporated C hybrid catalyst with CoP-anchored heteroatom permits and encourages the economical and high-performance growth of bifunctional catalysts in metal-air batteries for ORR and OER.

Choi et al. [44] reported perovskite photodiode-solar cell nanosystems (PPSNs) by the use of two layers of bilayer (BL) MoS_2 and TETA-GR as an electron transport layer (ETL) and transparent conductive electrode (TCE), respectively. The rigid or flexible PPSNs display 0.42/0.40 AW^{-1} (R), 37.2/80.1 $pWHz^{-1/2}$ equivalent noise power, (1.1 × 1010/5.0 × 109 cm $Hz^{1/2}$ W^{-1}) definite detectivity with self-power operation, i.e., at a zero-bias photodiode mode is comparable to or exceeds previously reported values, along with 14.27/12.12% power conversion efficiency (PCE) in a photovoltaic mode. After 30 days, the PPSNs retain approximately 78% of their original R. These enhanced TCE properties, perhaps the best band offsets at the cathode/electron transport layer/active interface layers, and decreased recombination/charge carrier resistance achieved with TETA-GR TCE and BL-MoS_2 ETL all lead to these outstanding performance achievements.

Tan et al. [45] claimed as a superhygroscopic hydrogel, the solar cell and catalyst can be combined with the humidity digester, which can split the surrounding moisture into H_2 and O_2, producing the efficiency of the electrochemical cell. The hydrogel collects moisture from the surroundings. Moisture and water transfer to the catalyst vertical nanofilms of BTO and 2D MoS_2 are combined as catalysts: the cathode may increase electron transfer and attract H^+ to the surface of MoS_2 for dehydration, while water oxidation occurs at the anode. Using this method, humidity digester generates a constant current of 12.5 $mAcm^{-2}$ under ambient light when using this mechanism to control comparative moisture in an average-sized room with 60%.

Lin et al. [46] proposed thin films of MoS_2 were electrically bonded directly onto FTO using simple one-step potential reversal (PR) and potentiostatic (PS) techniques are used as electronic catalyst counter electrode (CE) for DSSC applications. The electrically integrated MoS_2-CE PR mode has superior electrocatalytic properties for reducing I_3^- to DSSC PS-MoS_2 mode. In a DSSC, PR-MoS_2 employed as a CE achieved a PCE of 8.77%, analogous to Pt-CE (9.01%). Meanwhile, backlighting studies have shown that PCE-DSSC using PR-MoS_2 is CE, reaching 4.82%, and is the same using Pt-CE (5.67%) is associated with the same transmission characteristics.

Sambur et al. [47] elucidated new carrier recombination and transfer pathways in MoS_2/WS_2 monolayer photoelectrochemical cells through spatially addressing the net carrier's assembly (photocurrent). They introduced a recombination process that depends on wavelength that relies on heterojunction (HJ) stacking and HS charge producing profile. Even though the liquid electrolyte and ITO electrode provide the best charge extraction ways by intimate e^- and h^+ selective contacts, charge transfer similar to the layers occurs over m-scale distances. As a result of these findings, researchers can better understand how composition variation affects HJ photoelectrode performance.

Yin et al. [48] produced an O_2-heteroatom-doped WS_2 (ED-O-WS_2) superstructure with a edges/defects-rich. The PCE of DSSCs with an ED-O-WS_2 CE was 10.36% (100 $mWcm^{-2}$) and 11.19% (under 40 $mWcm^{-2}$). These are the highest recorded PCE obtained through Pt-free CEs in I_3^-/I^- electrolytes for DSSCs. The micro/nanostructure and electrocatalysis process analysis reveals that ED-O-WS_2 displays metallic characteristics in electrolytes and that the presence of multiple edges and defects, and oxygen doping, contribute significantly to enhancement of WS_2's catalytic activity. Additionally, ED-O-WS_2 exhibits greater catalytic reversibility than Pt in I_3^-/I^- electrolytes.

Atwater et al. [49] showed the production of TMDC on 2D-organic and inorganic hybrid lead halide perovskites (2D-PVSKs) HS utilizing a solvent-free method results in unique optical conversions that are distinct to the HS and display significant PL. Additionally, a PL increase of two orders of magnitude is seen in comparison to WS_2 emission. Additionally, the incorporation of TMDC to 2D-PVSK enhances stability substantially when compared to 2D-PVSK alone. The enhanced emission is explained by altering the TMDC electronic structure via new interfacial relations among TMDC and 2D-PVSK materials, demonstrating the HS's potential for effective and steady optoelectronic devices.

Qiu et al. [50] make L-WSe_2 p–n junction was constructed using a laser scanning system. A laser-scanned region exhibits p-type behavior, whereas the adjacent area has an appropriate V_g and is electrically n-doped. This p-type doping is attributed to the laser-oxidized product WO_x. After laser scanning, WSe_2 exhibits a shift in its p-type property from ambipolar to unipolar. At the p–n junction, a strong photocurrent develops. As a result, a self-generated WSe_2 photodetector dependent upon this junction is possible, with an enormous photoswitching ratio of 10^6, a high R_{ph} of 800 mAW^{-1}, and a little photoresponse time that is stable and reproducible over time.

As a result, this approach of selective laser doping is potentially useful in near-future electronic devices.

Pataniya et al. [51] describe the use of a high-yield sonochemical exfoliation method to fabricate a flexible and high-performance photodetector. The WSe_2 nanodots have a $P6_3/mmc$ and a 1.63 eV indirect E_g. These techniques are utilized to confirm the exfoliation and layer number in WSe_2 nanodots. When tested for bias values ranging from 0 to 100 V, the device displays exceptional photoresponse with 0.68 s response time, R_{ph} of 17.78 mAW^{-1}, and detectivity of 5.86×10^{10} Jones at 5 V. Additionally, the paper-based detector was flexible and airtight. The photodetector performance was also tested from 390 to 880 nm. In short, the paper-based detector ushers in a new era of flexible and intelligent electronics.

Nie et al. [52] fabricated a lateral BL MoS_2 using a two-step CVD pathway. Raman, PL, and harmonic images demonstrate a clear distinction in the HS between MoS_2 and WS_2. HRTEM revealed that distinct boundaries are established by continuous bonds across the L-zigzag epithelium among MoS_2 and WS_2. The WS_2-MoS_2 photodetector based on the lateral BL HS demonstrates an extremely high R_{ph} and detectivity (6.72×10^3 AW^{-1} and 3.09×10^{13} Jones for 457 nm laser light) is orders of magnitude greater compared to single crystals of MoS_2 and WS_2. The lateral BL WS_2-MoS_2 HS is a potential candidate in future optoelectronic devices.

Warner et al. [53] demonstrated that ultrathin 2D vertically (V-) stacked type-II HS devices might be fabricated utilizing just CVD-grown materials, with GR serving as the top and bottom electrodes and MoS_2/WS_2 serving as the active semiconductor films in the center. Directing photovoltaic flow of current is not proved to be feasible for homo-BLs but only with hetero-BLs. They investigate the photovoltaic impacts of over 40 devices, allowing for statistical validation of efficiency and relative behavior. Due to the excellent charge transport among WS_2 and MoS_2 extraction to GR electrode, GR/TMDC-hetero-BL (MoS_2/WS_2)/GR) photovoltage generation is up to tenfold that of ML-TMDC devices when the same optical brightness power is used. External gate voltage (V_g) could be used to modify the band alignment, which in turn affects the photovoltaic effect in V-HS. These findings provide light on how to layer symmetry in vertically stacked GR/TMDC/GR ultrathin optoelectronics may be utilized to regulate the direction of electron flow during photoexcitation, therefore enabling tandem cell production.

Xu et al. [54] used a liquid-phase exfoliation (LPE) technique; various mixed ratios of WS_2/MoS_2 photoanodes are synthesized. The highest J_{sc} is observed to be eight times to WS_2 and four times to MoS_2 at a ratio of WS_2/MoS_2 (1:1). Using amperometric I-t, Mott–Schottky, and Nyquist measurements, the photoelectric enhancement can be analyzed which indicates low electrode–electrolyte interface resistance and active carrier separation in the photoelectrodes. Additionally, the carrier density difference between WS_2/MoS_2 suggests charge rearrangement during HS formation promotes charge transfer. The goal of this paper is to propose a simple and affordable approach for making WS_2/MoS_2 HJ-based photoanodes.

Lee et al. [55] offer a unified synthesis approach based on a solvothermal technique that is possible despite the use of a single common tetrathiometallate precursor. They investigate the structure–property relationships of TMDCs with respect to LSB performance using their capacity to function as polysulfide anchors in LiSB cathodes. In comparison to morphology, the TMDC phase has the greatest influence on cell performance in MoS_2 and WS_2. Furthermore, during long-standing cycle, 1T-phase basal-oriented polytypes retained the most ability, owing to their reduced charge transport resistances. Additionally, an enhanced areal capability ~4 $mAhcm^{-2}$ was proven in preparation for future applications of LiSBs. While more LiSB difficulties remain, the insights acquired through phase manufacturing and our morphology investigation led to improved sulfur-host cathode design.

Wang et al. [56] synthesized WS_2 NS and its suitability as a cathode material for a thermal mobile was examined as WS_2 is steady to 1200 °C. The open-circuit voltage (V_{oc}) vs anode is 1.43 v. The LIB/WS_2 single cell has a definite capacity of 334.7 $mAhg^{-1}$ with a J_{sc} of 150 $mAcm^{-2}$ and a cut-off voltage (1.0 V). W and Li_2S are the discharge's end products. As a possible cathode composite for long-lasting thermal batteries, WS_2 is now being investigated further.

Wu et al. [57] reported a self-assembled nanohoneycomb WS_2 was doped with GR utilizing an enhanced one-step hydrothermal method. As the morphology changes from microporous spheres nanowire to a GR-strengthened nanohoneycomb plane-nanostructure, GR is important in this hybrid structure because it promotes the transition. This anode allows Li/Na ion batteries to perform well electrochemically, with a large specific carrier capacity (953.1/522.3 $mAhg^{-1}$ at 0.1 Ag^{-1}), extended cycling life (over 350/200 cycles at 1 Ag^{-1}), and increase charge/discharge rates (~10/5 Ag^{-1}). They were interested in using this WS_2 composite anode in Li/Na batteries since it is easy to hydrothermally produce and performs well electrochemically.

Kulandaivel et al. [58] deposited 2D-NbS_2 directly on high-aspect-ratio Si-nanowires by CVD for HER electrocatalyst. After 20 h, the 2D-NbS_2 electrocatalyst exhibits an HER overpotential of 74 mV against RHE (reversible H_2 electrode) @ 1 $mAcm^{-2}$ in acidic circumstances. In the end, the better charge transfer properties at Si-NbS_2 interfaces resulted in a higher turn-on potential (from 0.06 to 0.34 V vs RHE) and a J_{sc} of −28 $mAcm^{-2}$ at 0 V versus RHE. The results show that 2D-NbS_2 electrocatalysts work synergistically to overcome Si's weak surface kinetics in solar water electrolysis. A novel type of photoelectrochemical cocatalyst for effective solar HER performance is revealed by NbS_2 promotes charge transfer while maintaining acid stability.

Angaiah et al. [59] found a single-step hydrothermal synthesis route yielded 2D $MoSe_2$-$Ni(OH)_2$). Owing to their synergistic impact, 2D-$MoSe_2$-$Ni(OH)_2$ nanohybrid exhibits superior electrochemical characteristics. At 1 Ag^{-1}, this 2D-2D hybrid electrode has a greater specific capacitance of 1175 Fg^{-1} than $Ni(OH)_2$ NS (933 Fg^{-1}) and a good rate ability of 85.6%. Additionally, with 5000 cycles at 1 Ag^{-1} current density, a 2D $MoSe_2$-$Ni(OH)_2$ nanohybrid-based asymmetric supercapacitor has a power density of 818 $Whkg^{-1}$, an energy density of 43 $Whkg^{-1}$, and an 85% capacitance retention. To demonstrate the great energy storage capacity of 2D $MoSe_2$-$Ni(OH)_2$ nanohybrid, two asymmetric supercapacitor devices charged for 1 min in sequence give power to an LED for more than 3 min.

Yang et al. [60] demonstrated that $MoSe_2$ NS may be used as electrode material in implanted microsupercapacitors via a laser-induced phase transition. $MoSe_2$ NS-based electrodes offer remarkable high-frequency responsiveness (~10^4 Hz) and cycle stability due to their metallic composition (~10^6 cycles) when incorporated in supercapacitors, with a power density ten times that of commercial aluminum electrolyte capacitors. Additionally, the fabrication techniques used to create the current device are entirely compatible with those used to manufacture system-in-package devices, which must adhere to stringent specifications regarding the size of embedded components. The current research marks a significant advancement in the use of electrolytic capacitors in-package and on-chip.

Angaiah et al. [61] describe a simple solvothermal technique for hybridizing 2D-$MoSe_2$ with 2D-GR. The exfoliated $MoSe_2$ NS are perpendicular to the GR NS surface. These $MoSe_2$ NS edges contain a high electrochemically active sites number, and GR sheets efficiently transport ions across the electrode–electrolyte interface. The specific ability is observed to be 945 Fg^{-1} with galvanostatic charge/discharge experiments. It has 75 Fg^{-1} specific capacitance (@1 Ag^{-1}), 0.8 $kWkg^{-1}$ power density, and 90% capacitance after 3000 cycles.

Zhang et al. [62] produced $MoSe_2$ sheet implanted on C nanofibers surfaces using a straightforward electro-spinning and selenization method. $MoSe_2$ / C-700 (selenization at 700 °C) keeps huge structure stability and enables K^+ intercalation/deintercalation, due to 1D nanofibers' structure stability and $MoSe_2$'s enhanced d-spacing. $MoSe_2$/C-700 has a reversible size of 316 $mAhg^{-1}$ at 100 mAg^{-1} after 100 cycles. It also possesses a specific ability of 81 $mAhg^{-1}$ after 100 cycles at 100 mAg^{-1} when employed as an electrode solid in nonaqueous K-based battery-supercapacitor hybrid devices.

2018

Wei et al. [63] fabricated synergetic composite WSe_2/ mesoporous C by employing nanocasting route, thereby offering an ideal structure owing to mesoporous C in the form of nanoreactors. Due to occupying synergetic effect for WSe_2 NSs as well as mesoporous C, the hybrid electrode (viz, WSe_2/ mesoporous C) presented high reversible capability, better efficiency rate, whereas superior long life cycles stability. For example, the capacity of 490.5 $mAhg^{-1}$ retains its existence with more than 600 cycles showing a current density 0.55 Ag^{-1}.

Gou et al. [64] prepared WSe_2-onion via manual CVD method. The capacity 205.5 $mAhg^{-1}$ at 10.5 Ag^{-1}vs sodium anion may proceed, regarding excellent sodium-ion-rate ability as in literature for WSe_2. An extraordinary capacitive addition showing rapid intercalation kinetics existed, attributing inherent properties as well as uniqueness for microstructure for WSe_2. Na-ion capacitors have supplied remarkable energy and/or power densities, such as optimal energy density of 123.10 $Whkg^{-1}$ whereas power density 14.15 $kWkg^{-1}$, thereby exhibiting for WSe_2 as a hopeful alternative toward Na-ion capacitors with the chance of Li-ion capacitors, thereafter shedding fresh light while 2DM in designing Na-ion capacitors occupying best energy/power density.

Pop et al. [65] evaluated a high-field-transport temperature depending function, very first time toward unique quality fabricated SL of MoS_2. The researchers mentioned typically fabricated device geometry, for example, on SiO_2 substrates normal heating may apparently lessen current channels upon high-field working conditions. However, varied ambient temperature evaluations, i.e., 100 K to 300 K, extract electron $V_{sat} = (3.45 \pm 0.45) \times 10^6\ cms^{-1}$ (i.e., in tri-atomic-thick semiconductor) thereby offering standard as compared with bulky as well as layered nanomaterials. Upon these results can pave paths from where saturation current flowing into SL-MoS_2 may increase (i.e., $1mA\mu m^{-1}$ at normal temperature) toward digital current paths associated with unique thermal management.

Yu et al. [66] prepared a sequence of MoS_2 NSs upon N-doped GR materials by employing solvothermal route. Moreover, at cut-off voltage associated with intercalation process, the ultrathin MoS_2 sheets composed of N-doped GR offer preferable cycling along with rate performance as compared with a variety of MoS_2 layered structures with N-doped GR, and also for various reported insertion nature anode materials. The kinetics analysis, as well as DFT calculations, is successfully employed for explaining the Na^+-sink trend, thereby leading importance toward surface-oriented pseudocapacitance effect at an optimum rate. Additionally, this literature presents a versatile fabrication as far as an investigation to frame electrode nanomaterials toward SIB occupying promising performance.

Roy et al. [67] evaluated nonvolatile memory devices that were sketched from PMMA embedded GR, as well as MoS_2 composites, which are followed as a charge trapping center. The high resistance ratio (i.e., $>10^4\ R_{off}/R_{on}$) has been evaluated even for ten days delay. The characteristics such as reliability as well as reproducibility owing to devices were investigated toward 10^4 cycles. Because of increasing interest in wide electronics, intentionally bending tests have been performed upon the well-designed devices on PET substrates. To test bendability characteristics upon a variety of bending diameters ranging 40–50 mm in 100 cycles were necessarily performed for showing mechanical robustness owing to these devices.

Lin et al. [68] fabricated photothermoelectric nanogenerator (PTENG) derived from MoS_2/PU photothermal film whereas Te/PEDOT thermoelectric toward collecting green IR light. In addition to photothermal layer into Te/PEDOT, a thermoelectric device, the PTENG may supply electrical output in the absence of spatial temperature gradient that is considered as necessary toward conventional thermoelectric devices. Under IR irradiated environment, wide temperature difference may be produced across the ends of device. Consequently, a potential difference is generated among two electrodes depending upon coupled photothermal effect as well as Seebeck effect. These PTENG's occupy substantial potential for applications for photothermoelectric energy devices to harvest wearable as well as implantable electronics.

Hu et al. [69] were inspired by Kirigami (an ancient Japan) because of drawing and art for paper cutting mechanisms. The said artists designed and framed on nanoscale basis (Kirigami architectures) owing to 2D-MoS_2 deposited on a substrate popular PDMS while adopting top-down fabrication route. Kirigami architectures have shown significant improvement in the form of reversible stretchability toward soft nature 2D-MoS_2 electronic devices caused by an increment from 0.76% to ~16%. Moreover, the increment into flexibility was extracted from composite that is multi-dimensionally deformed capability, originated from Kirigami structures comprising IP stretching as well as out-planes (OP) deformation. They further discovered refined fundamental relationship among electrical conductance associated with large strain into MoS_2 Kirigami structures employing experimental work along with definite elemental simulation. The above-mentioned approach has opened a novel route for fabricating flexible as well high stretchable properties for sensors and actuators while utilizing a variety of 2DMs.

Vashishta et al. [70] have studied lattice dynamics caused by electronic excitation toward an SL-$MoSe_2$ by following non-adiabatic quantum-molecular-dynamics simulations. Simulation outcomes have shown subpicosecond disorder relevant with lattice upon photoexcited electrons observed with Debye–Waller factor, whereas increment in disorder for higher density belonging to photogeneration of electron–hole pairing. Detailed study exhibited rapidly photoinduced lattice dynamics that are considered because of phonon mode softening, caused by arising of electronically Fermi-surface nesting. Such schematic study may facilitate optical material properties to functionalize TMDC layers, making enable for emerging applications like that of phase-change-memory as well as neuromorphic computational study.

Jamet et al. [71] presented vdW HJ owing to $MoSe_2$/FL-GR showing fine quality interfacial contact by employing cutting-edge surface techniques that are scaled with atomic to microscopic ranges. They have observed basic results of commensurability among $MoSe_2$ as well as some layers of GR lattices associated with bandgap generation for some layer GR. A large bandgap of 251 $_{meV}$ corresponds to interfacial charge transformation, resulting in electronic depletion connected with FL-GR. Moreover, commensurability among $MoSe_2$ as well as GR lattices along with bandgap creation clearly showed toward interlayer connection attributed to vdW interaction. Therefore, stacked 2DMs into vdW HJs open a new route for tailoring atomic as well as electronic properties owing to individual layers.

2.3 Classification of 2DMs

A solid knowledge of 2DMs structure comprised of single atom thick or single polyhedral-thick layer is required for reliable synthesis, via property understanding and application. 2DMs are therefore divided into three categories.

2.3.1 Layered Van Der Waals Solids

These materials are the most frequent and fastest way to exfoliate mono and FL-NSs due to strong IP covalent or ionic bonding, weak OP vdW or H_2 bonding. This is due to the weak vdW bonding (40–70 meV) and surface tension (60–90 mJm^{-2}) in typical solvents being used in graphite exfoliation [72, 73]. These Ns have a large surface area and aspect ratio after exfoliation. These exfoliated NSs are generally between few micrometers to 1 nm thick. A well-studied system for layering vdW solids, notably MoS_2, $MoSe_2$, and WS_2, are TMDCs. Presently, approximately, 40 TMDCs are published in the form of layered [74, 75]. Its exotic characteristics in terms of direct and indirect E_g, Dirac point, electrochemical and mechanical behavior have considered it a twenty-first century material with prospective applications in semiconductor [76], hydrogen generation [77], and thermoelectric devices [78]. In IT and 2H crystal structures (See Chap. 3), TMs occupy octahedrally (d0, d3, and some d1) and trigonal prismatic (d1 and d2) coordinates. These coordinates create a hexagonal honeycomb structure with their nearest neighbors within each layer. This group includes h-BN [79], SiC [80], V_2O_5 [81], and Sb_2Te_3 [82].

2.3.2 *Layered Ionic Solids*

Layered ionic compounds are a prominent family of layered 2DMs. Electrostatic forces sandwich a charged polyhedral layer between hydroxide or halide layers in these materials. Ion intercalation or ion-exchange liquid exfoliation works well for single or FL materials. By increasing the distance between the basal planes, and by exchanging parent ions with heavier ones, foreign molecules can be adsorbed. Exfoliated europium hydroxide is a typical layered ionic solid [83]. Examples of layered materials exfoliated from the ion-exchange technique include $La_{0.90}Eu_{0.05}Nb_2O_7$ [84], $K_{1.5}Eu_{0.5}Ta_3O_{10}$ [85], $KLnNb_3O_7$, and $RbLnTa_2O_7$ (R: Lanthanide Ion) [86], $KCa_2Nb_3O_{10}$ [87, 88], $Eu(OH)_{2.5}(DS)_{0.5}$ [83] and $Co_{2/3}Fe_{1/3}(OH)_2{}^{1/3+}$ [89]. The adsorption of solvent and bulkier exchanged ions causes the observed monolayer thickness to be greater than the crystallographic thickness.

2.3.3 *Surface-Assisted Non-layered Solids*

CVD and epitaxial growth are used to create this type of atomically thin layered materials on a substrate. Silicene, a GR counterpart, is the greatest example of surface-assisted non-layered solids. Si atoms were deposited with a low buckled hexagonal structure on Ag (111) [90], Ir (111) [91], and a thin layer of ZrB_2 on Si wafer [92]. Unlike GR, silicene's instability at ambient temperatures has restricted its use in electrical and optoelectronic applications. Al_2O_3/SiO_2 [93], Cu_2N/Cu(100) [94], MgO/Mo(001) [95], TiO_2 [96], and Ge/Ag(100) and Ag (111) NSs are produced for use in energy storage, FETs, and controlling the charge density of NPs [97–99]. Figure 2.2 demonstrates the atomic configurations in several types of 2DMs.

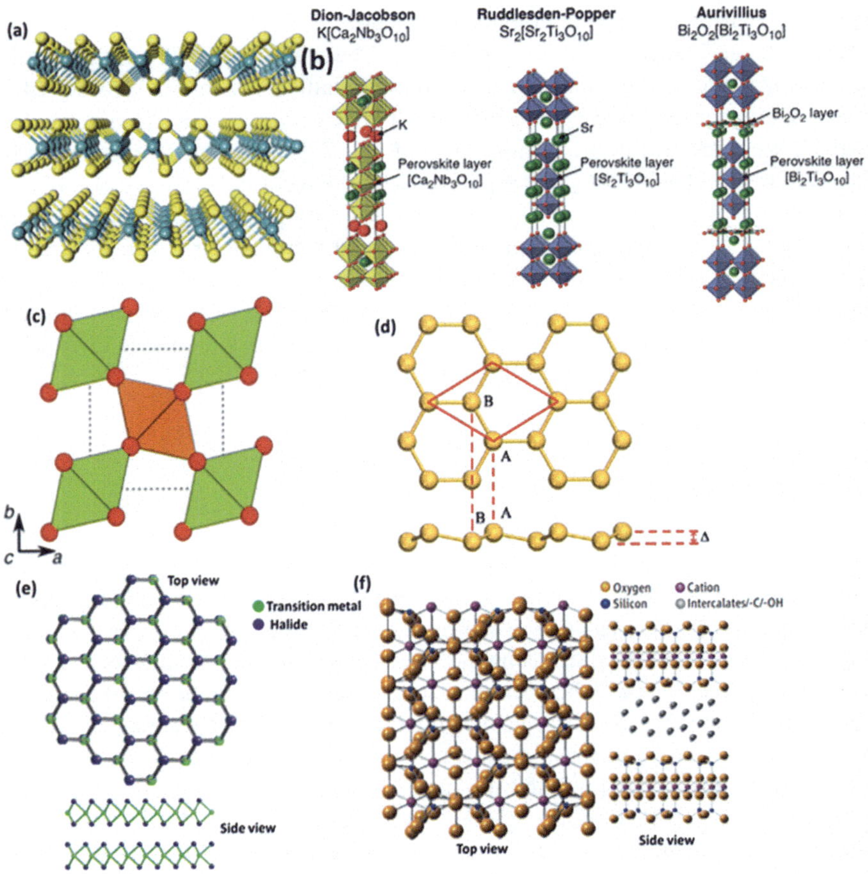

Fig. 2.2 Atomic structure of different classes of layered 2DMs: **a** MoS_2. (Reproduced with permission from [100]), **b** layered perovskites. (Reproduced with permission from [101]), **c** MnO_2 (Oxide layered materials). (Reproduced with permission from [102]), **d** silicene (surface-assisted non-layered solids). (Reproduced with permission from [103]), **e** $MoCl_2$ (transition metal dihalides), **f** vermiculite (layered silicates). (Reproduced with permission from [104])

References

1. The Graphene Flagship, European Research, Innovation, and Collaboration Link. https://www.mynewsdesk.com/graphene-flagship/images/evolution-of-graphene-production-from-lab-to-marketplace-2255249 (Visited at 23–08–2021)
2. Schedin F, Geim AK, Morozov SV, Hill EW, Blake P, Katsnelson MI, Novoselov KS (2007) Nat Mater 6:652
3. Fiori G, Bonaccorso F, Iannaccone G, Palacios T, Neumaier D, Seabaugh A, Banerjee SK, Colombo L (2014) Nat Nanotechnol 9:768
4. Chhowalla M, Jena D, Zhang H (2016) Nat Rev Mater 1:16052
5. Stoller MD, Park S, Zhu Y, An J, Ruoff RS (2008) Nano Lett 8:3498
6. Deng D, Novoselov KS, Fu Q, Zheng N, Tian Z, Bao X (2016) Nat Nanotechnol 11:218

7. Bonaccorso F, Colombo L, Yu G, Stoller M, Tozzini V, Ferrari AC, Ruoff RS, Pellegrini V (2015) Science 347:1246501
8. Mendoza-Sánchez B, Gogotsi Y (2016) Adv Mater 28:6104
9. Tan C, Liu Z, Huang W, Zhang H (2015) Chem Soc Rev 44:2615
10. Sun Y, Gao S, Xie Y (2014) Chem Soc Rev 43:530
11. Zhang X, Xie Y (2013) Chem Soc Rev 42:8187
12. Novoselov KS, Geim AK, Morozov SV, Jiang D, Katsnelson MI, Grigorieva IV, Dubonos SV, Firsov AA (2005) Nature 438:197
13. Abanin DA, Levitov LS (2007) Science 317:641
14. Williams JR, DiCarlo L, Marcus CM (2007) Science 317:638
15. Novoselov KS, Geim AK, Morozov SV, Jiang D, Zhang Y, Dubonos SV, Grigorieva IV, Firsov AA (2004) Science 306:666
16. Geim AK, Novoselov KS (2007) Nat Mater 6:183
17. Tan C et al (2017) Chem Rev 117:6225
18. Li L, Yu Y, Ye GJ, Ge Q, Ou X, Wu H, Feng D, Chen XH, Zhang Y (2014) Nat Nanotechnol 9:372
19. Krishnan U, Kaur M, Singh K, Kumar M, Kumar A (2019) Superlattices Microstruct 128:274
20. Ghidiu M, Lukatskaya MR, Zhao M-Q, Gogotsi Y, Barsoum MW (2014) Nature 516:78
21. Sun J, Lee H-W, Pasta M, Yuan H, Zheng G, Sun Y, Li Y, Cui Y (2015) Nat Nanotechnol 10:980
22. Bertolazzi S, Brivio J, Kis A (2011) ACS Nano 5:9703
23. Guo Y, Xu K, Wu C, Zhao J, Xie Y (2015) Chem Soc Rev 44:637
24. Wang X, Sun G, Routh P, Kim D-H, Huang W, Chen P (2014) Chem Soc Rev 43:7067
25. Deng J et al (2015) Energy Environ Sci 8:1594
26. Xie J, Zhang H, Li S, Wang R, Sun X, Zhou M, Zhou J, Lou XW, Xie Y (2013) Adv Mater 25:5807
27. Ouyang Y, Ling C, Chen Q, Wang Z, Shi L, Wang J (2016) Chem Mater 28:4390
28. Li H et al (2016) Nat Mater 15:48
29. Li H, Du M, Mleczko MJ, Koh AL, Nishi Y, Pop E, Bard AJ, Zheng X (2016) J Am Chem Soc 138:5123
30. Angamuthu G, Bosubabu D, Ramesha K, Rengarajan V (2021) Appl Mater Today 22: 100916
31. Jiang Q, Wang L, Chen J, Huang Z, Yang H-J, Zhou T, Hu J (2021) J Alloys Compounds 855: 157505
32. Lim H, Yu S, Choi W, Kim S-O (2021) ACS Nano 15:7409
33. Gao K et al (2021) Energy Fuels 35:10303
34. Li S et al (2021) Adv Mater 33:2007480
35. Huang Y et al (2020) Angew Chem Int Ed 59:3699
36. Chen M et al (2020) Chem Eng J 385: 123905
37. Guo X et al (2020) Angew Chem Int Ed 59:22697
38. Peng H, Raya J, Richard F, Baaziz W, Ersen O, Ciesielski A, Samorì P (2020) Angew Chem Int Ed 59:19602
39. Xu J, Yang C, Bi S, Wang W, He Y, Wu D, Liang Q, Wang X, Zhang F (2020) Angew Chem Int Ed 59:23845
40. Mussa Y, Ahmed F, Arsalan M, Alsharaeh E (2020) Sci Rep 10:1882
41. Liu Y, Yin J, Zhou Y, Sun L, Yue W, Sun Y, Wang Y (2019) ChemElectroChem 6:2737
42. Deng H, Chang Z, Qiu F, Qiao Y, Yang H, He P, Zhou H (2020) Adv Energy Mater 10:1903953
43. Wang Y, Wu M, Li J, Huang H, Qiao J (2020) J Mater Chem A 8:19043
44. Shin DH, Ko JS, Kang SK, Choi S-H (2020) ACS Appl Mater Interfaces 12:4586
45. Yang L, Loh L, Nandakumar DK, Lu W, Gao M, Wee XLC, Zeng K, Bosman M, Tan SC (2020) Adv Mater 32:2000971
46. Chang C-Y, Anuratha KS, Lin Y-H, Xiao Y, Hasin P, Lin J-Y (2020) Sol Energy 206:163
47. Wang L, Tahir M, Chen H, Sambur JB (2019) Nano Lett 19:9084
48. Yin J et al (2019) Chemsuschem 12:795
49. Yang A et al (2019) Nano Lett 19:4852

50. Chen J et al (2019) ACS Appl Mater Interfaces 11:43330
51. Pataniya P, Zankat CK, Tannarana M, Sumesh CK, Narayan S, Solanki GK, Patel KD, Pathak VM, Jha PK (2019) ACS Appl Nano Mater 2:2758
52. Ye K et al (2019) Adv Optical Mater 7:1900815
53. Zhou Y, Xu W, Sheng Y, Huang H, Zhang Q, Hou L, Shautsova V, Warner JH (2019) ACS Appl Mater Interfaces 11:2234
54. Lu C, Ma J, Si K, Xu X, Quan C, He C, Xu X (2019) Physica Status Solidi (a) 216: 1900544
55. Eng AYS, Cheong JL, Lee SS (2019) Appl Mater Today 16:529
56. Guo S, Guo H, Wang X, Zhu Y, Yang M, Zhang Q, Chu Y, Wang J (2019) Mater Lett 249:81
57. Song Y, Liao J, Chen C, Yang J, Chen J, Gong F, Wang S, Xu Z, Wu M (2019) Carbon 142:697
58. Gnanasekar P, Periyanagounder D, Varadhan P, He J-H, Kulandaivel J (2019) ACS Appl Mater Interfaces 11:44179
59. Kirubasankar B, Palanisamy P, Arunachalam S, Murugadoss V, Angaiah S (2019) Chem Eng J 355:881
60. Jiang Z et al (2019) Adv Func Mater 29:1807116
61. Kirubasankar B, Vijayan S, Angaiah S (2019) Sustain Energy Fuels 3:467
62. Shen Q, Jiang P, He H, Chen C, Liu Y, Zhang M (2019) Nanoscale 11:13511
63. Wang J, Chen L, Zeng L, Wei Q, Wei M (2018) ACS Sustain Chem Eng 6:4688
64. Wang Y et al (2018) J Mater Chem A 6:21605
65. Smithe KKH, English CD, Suryavanshi SV, Pop E (2018) Nano Lett 18:4516
66. Xu X, Zhao R, Ai W, Chen B, Du H, Wu L, Zhang H, Huang W, Yu T (2018) Adv Mater 30:1800658
67. Bhattacharjee S, Das U, Sarkar PK, Roy A (2018) Org Electron 58:145
68. He M, Lin Y-J, Chiu C-M, Yang W, Zhang B, Yun D, Xie Y, Lin Z-H (2018) Nano Energy 49:588
69. Zheng W et al (2018) Chem Mater 30:6063
70. Bassman L, Krishnamoorthy A, Kumazoe H, Misawa M, Shimojo F, Kalia RK, Nakano A, Vashishta P (2018) Nano Lett 18:4653
71. Dau MT et al (2018) ACS Nano 12:2319
72. Coleman JN et al (2011) Science 331:568
73. Butler SZ et al (2013) ACS Nano 7:2898
74. Wilson JA, Yoffe AD (1969) Adv Phys 18:193
75. Gupta A, Sakthivel T, Seal S (2015) Prog Mater Sci 73:44
76. Wang QH, Kalantar-Zadeh K, Kis A, Coleman JN, Strano MS (2012) Nat Nanotechnol 7:699
77. Li Y, Wang H, Xie L, Liang Y, Hong G, Dai H (2011) J Am Chem Soc 133:7296
78. Poudel B et al (2008) Science 320:634
79. Taha-Tijerina J, Narayanan TN, Gao G, Rohde M, Tsentalovich DA, Pasquali M, Ajayan PM (2012) ACS Nano 6:1214
80. Lin SS (2012) J Phys Chem C 116:3951
81. Yang S, Gong Y, Liu Z, Zhan L, Hashim DP, Ma L, Vajtai R, Ajayan PM (2013) Nano Lett 13:1596
82. Kong D, Dang W, Cha JJ, Li H, Meister S, Peng H, Liu Z, Cui Y (2010) Nano Lett 10:2245
83. Hu L, Ma R, Ozawa TC, Sasaki T (2010) Chemistry – Asian J 5: 248
84. Ozawa TC, Fukuda K, Akatsuka K, Ebina Y, Sasaki T (2007) Chem Mater 19:6575
85. Ozawa TC, Fukuda K, Akatsuka K, Ebina Y, Sasaki T, Kurashima K, Kosuda K (2008) J Phys Chem C 112:17115
86. Ida S, Ogata C, Eguchi M, Youngblood WJ, Mallouk TE, Matsumoto Y (2008) J Am Chem Soc 130:7052
87. Ebina Y, Sasaki T, Harada M, Watanabe M (2002) Chem Mater 14:4390
88. Osada M, Akatsuka K, Ebina Y, Funakubo H, Ono K, Takada K, Sasaki T (2010) ACS Nano 4:5225
89. Ma R, Liu Z, Takada K, Iyi N, Bando Y, Sasaki T (2007) J Am Chem Soc 129:5257
90. Lalmi B, Oughaddou H, Enriquez H, Kara A, Vizzini S, Ealet B, Aufray B (2010) Appl Phys Lett 97: 223109

91. Meng L et al (2013) Nano Lett 13:685
92. Fleurence A, Friedlein R, Ozaki T, Kawai H, Wang Y, Yamada-Takamura Y (2012) Phys Rev Lett 108: 245501
93. Jung H, Park J, Oh I-K, Choi T, Lee S, Hong J, Lee T, Kim S-H, Kim H (2014) ACS Appl Mater Interfaces 6:2764
94. Ruggiero CD, Badal M, Choi T, Gohlke D, Stroud D, Gupta JA (2011) Phys Rev B 83: 245430
95. Sterrer M, Risse T, Martinez Pozzoni U, Giordano L, Heyde M, Rust H-P, Pacchioni G, Freund H-J (2007) Phys Rev Lett 98: 096107
96. Hao B, Yan Y, Wang X, Chen G (2013) ACS Appl Mater Interfaces 5:6285
97. Oughaddou H, Aufray B, Bibérian JP, Hoarau JY (1999) Surf Sci 429:320
98. Golias E, Xenogiannopoulou E, Tsoutsou D, Tsipas P, Giamini SA, Dimoulas A (2013) Phys Rev B 88: 075403
99. Oughaddou H, Mayne A, Aufray B, Bibérian JP, Le Lay G, Ealet B, Dujardin G, Kara A (2007) J Nanosci Nanotechnol 7:3189
100. Wang H et al (2013) Proc Natl Acad Sci 110:19701
101. Osada M, Sasaki T (2018) Dalton Trans 47:2841
102. Sai Gautam G, Carter EA (2018) Phys Rev Mater 2: 095401
103. Ni Z, Liu Q, Tang K, Zheng J, Zhou J, Qin R, Gao Z, Yu D, Lu J (2012) Nano Lett 12:113
104. Nicolosi V, Chhowalla M, Kanatzidis MG, Strano MS, Coleman JN (2013) Science 340:1226419

Chapter 3
Composition and Materials Chemistry

The composition and structural chemistry of several materials vary but are mainly categorized into two classes: non-layered and layered materials. At present, our concern is a layered material, inside each layer of layered materials, IP atoms are linked by strong chemical bonding with vertical tacking via vdW forces to establish bulk crystals. Graphite is a famous layered material being formed by GR's weakly stacked layers. Apart from graphite, some other emerging materials are also evolved such as h-BN, TMDCs, g-C_3N_4, BP, transition metal oxides (TMOs), and layered double hydroxides (LDHs). These materials are survived into a variety of crystal phases depending on atomic arrangement, configuration, and stacking sequence of layers. The properties and functionalities of materials are greatly influenced by these crystal phases and some properties could be dramatically enhanced by tuning of crystal phases. This chapter will primarily surround the compositions, atomic structures, and structural chemistries of each material and also address how crystal phases affect the applications.

3.1 Graphene

Recently, 2D GR has garnered considerable interest due to its numerous interesting characteristics and ultrathin structural structure. The graphene has high carrier mobility (200,000 $cm^2 \cdot V^{-1} \cdot s^{-1}$), high heat conductivity (~5300 $Wm^{-1} \cdot K$), and exceptional optical transparency (97.7%) are suitable for a variety of uses (see Chap. 6) [1–3]. All C-atoms in GR are chemically manageable; hence, covalent chemistry facilitates in tailoring of physical characteristics of pure GR as a result, stronger, more appealing, materials can be achieved [4, 5]. GR is constructed two-dimensionally using an unlimited number of benzene rings combined by an infinite number of benzene rings in a 2D architect. The chemically grown GR demonstrated

M. Ikram et al., *2D-Materials for Energy Harvesting and Storage Applications*, Nanostructure Science and Technology,
https://doi.org/10.1007/978-3-030-96021-6_3

optimal reactivity attributed to the presence of defects; on the other hand, mechanically exfoliated GR has been predicted to show similar inertness when the defects have been removed [6, 7]. Consequently, high temperature and energy reagents have been typically expected to solve the energy barrier of GR toward covalent modification. Indeed, handful of chemical reactions with poor reaction efficiency has been established to date for GR chemistry.

3.1.1 Structural Chemistry

Carbon is a 6th periodic element having electronic configuration $1s^2 2s^2 2P_x^1 2P_y^1 2P_z^0$ as shown in Fig. 3.1b. For simplicity, energy level $2p_z$ is maintained without an electron, even though it is equal to $2p_x$ and $2p_y$. The C-atom has six electrons around its nucleus, four of which are valence electrons, as shown in Fig. 3.1a. These electrons valence shells of C-atom can hybridize in three different ways (sp, sp^2, and sp^3), among these sp^2-hybrids formations as shown in Fig. 3.1c. Accordingly, C-atoms that share sp^2 electrons with their three neighbors form a honeycomb network termed monolayer GR. A GR crystal's unit cell, denoted in Fig. 3.1d by a purple parallelogram, comprises two C-atoms, and a_1 and a_2 unit cell vectors have the same lattice constant (2.46 Å). Electron delocalization and resonance determine stability of planar ring. On the GR layer, sp^2 hybridization of two neighboring C-atoms layer (see Fig. 3.1e), π-bond is formed by $2p_z$ orbitals that are perpendicular to planar structure, whereas hybridized orbital, sp^2 ($2s$, $2p_x$, and $2p_y$) form an IP-σ bond. The ensuing covalent connection is even stronger than diamond's sp^3 hybridized C–C bonds due to its low interatomic length of 1.42 (1 TPa Young's modulus and 130.5 GPa intrinsic

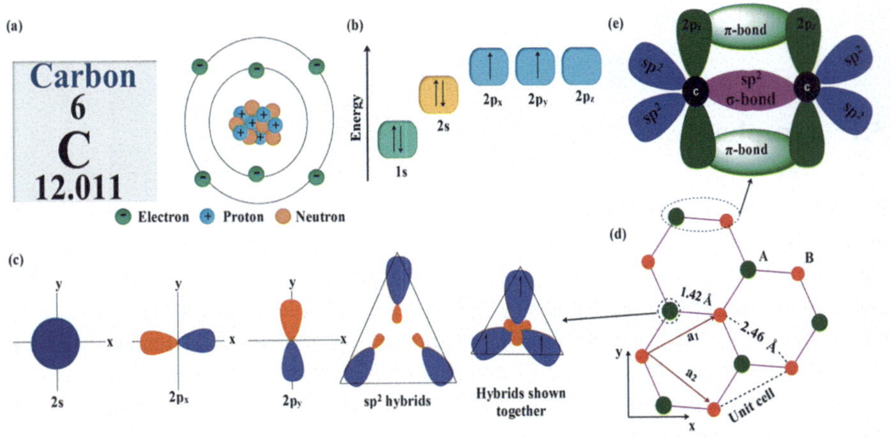

Fig. 3.1 **a** C-atom structure, **b** C-atoms' outer electron energy levels, **c** sp^2 hybrids formation, **d** GR crystal lattice, wherein A and B are atoms of C from distinct sublattices, a_1 and a_2 are vectors of unit cell, **e** sp^2 hybridization creates σ and π-bond

tensile strength [8]). The CB and VB with zero E_g are formed by the half-filled band that lets electrons freely flow in monolayer GR. Also, in BL and ML-GR, π-bonds serve as a vdW contact between GR layers [9].

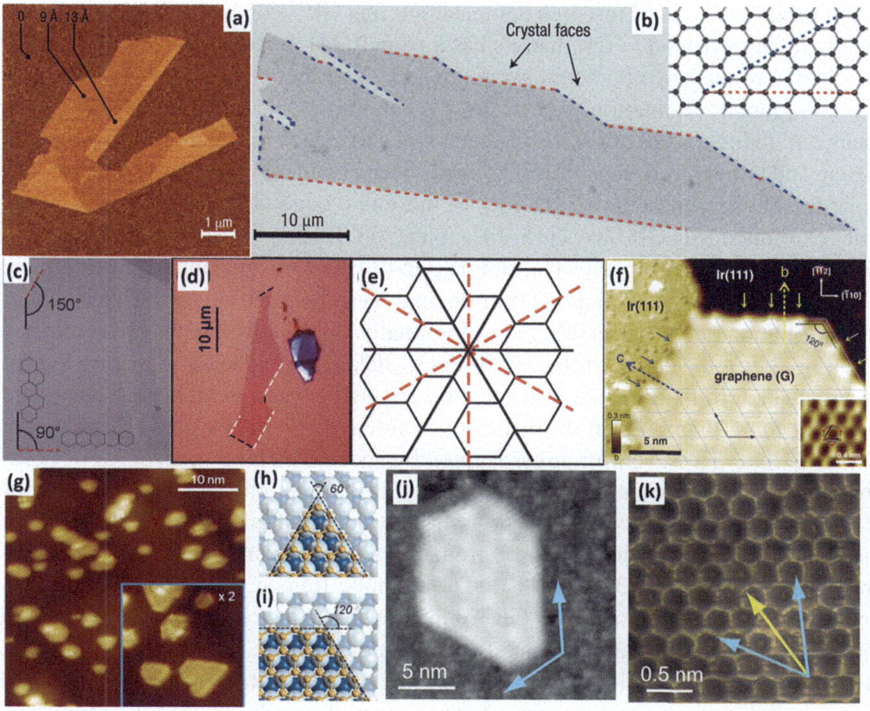

Fig. 3.2 One-atom-thick single crystals: **a** GR visualized by AFM. (Reproduced with permission from [25]), **b** inset: Blue and red lines on an SEM image of a big GR crystal show zigzag and armchair edges. (Reproduced with permission from [21]), **c** typical micromechanical cleavage GR flake The edge types are zigzag (dashed lines) and armchair (solid lines), **d** an optical image of an SL-GR flake produced via micromechanical cleavage. For each sort of edge on this specific flake, a dashed line black and white has been used to emphasize it, **e** GR honeycomb crystal lattice schematic with two potential crystallographic orientations: armchair solid lines and zigzag dashed lines. (Reproduced with permission from [13]), **f** CC-STM image of GR edges on Ir(111), with Ir substrate crystallographic directions shown top right. (Reproduced with permission from [26]), **g** STM image of GR structures on Co substrate, **h, i** triangular and hexagonal corners for zigzag-edged GR structures on Co (0001). (Reproduced with permission from [27]), **j** STM image and **k** STM GR island topography on 6H-SiC(0001) substrate. Here are the two lowest energy edge directions: zigzag (yellow arrow) and armchair (blue arrows). (Reproduced with permission from [15])

3.1.2 Edge Alignment in Graphene

According to spectroscopic measurements and electron transfer theory, monolayer GR sheets have double the chemical reactivity at their edges as at their basal planes [10]. Additionally, STM research demonstrates that GR's edges can display a greater electrical DOS around Fermi than its basal planes [11]. Localized electron distribution is determined by edge configurations [12], and therefore, the choice of GR's crystallographic orientation is critical for regulating its electrical characteristics in localized states. In GR, the two primary types of edges are recognized including zigzag and armchair. Extensive research has been conducted in recent years to determine the physics and relative stability of these two edge orientations for mechanically exfoliated ML-GR flakes [13, 14], epitaxial GR islands [15, 16], CVD-derived GR grains [17], and graphene nanoribbon (GNRs) [18, 19] are connected with them. Breaking is thought to occur along the principal crystallographic axes during mechanical exfoliation of GR flakes [20, 21]. Additionally, due to GR's hexagonal symmetry (see Fig. 3.2e), the edges of the resulting GR flakes are expected to feature zigzag or armchair terminations. As a result, the angles between neighboring edges are typically more than 30% in mechanically exfoliated GR flakes [13, 21], as shown in Fig. 3.2b, c. However, for edges formed by specific etching procedures or holes created by electron beam irradiation of GR, zigzag directions appear to be superior [18, 22]. The GR flakes' edges directions may be identified using HR-STM which can visualize GR crystal lattice up to angstrom, allowing to clearly distinguish armchair and zigzag edges (Fig. 3.2a, d) [13]. The G-mode Raman spectroscopy may also be utilized to identify monolayer GR's edge direction since the polar behavior of G modes armchair at zigzag-dominated edges is unique [14]. Xu et al. [23] explored GR edges thermal stability and dynamics using polarized Raman spectroscopy and discovered that zigzag and armchair edges appeared unstable at 200 °C. Moreover, GR flake edges showed a high proportion of zigzag terminations below 400 °C [24] but are dominated by armchair and rebuilt zigzag terminations beyond 600 °C.

Due to their tiny size and potential for new electrical characteristics, GR islands are of significant scientific interest (quantum confinement) [19]. The underlying substrates have a substantial impact on the edge orientations of epitaxial GR islands. For example, GR nanoislands produced epitaxially on metal surfaces (Ir(111) that is monocrystalline [16, 27] Co(0001) [27, 28], and Pt(111) [29]) have a high proportion of zigzag ends, but those formed on 6H-SiC(0001) have a high proportion of armchair edges [15]. The CC-STM image of GR edges on Ir(111) is shown in Fig. 3.2f, which symbolizes the gray mesh representing the GR moiré superstructure and the substrate step edges and the zigzag edges on the substrate surface are represented by blue and yellow arrows, respectively. At low temperatures, TEM studies show that GR structures have an on-top registry with respect to the Co(0001) surface [28]. The on-top registry reveals that on a pristine Co surface, GR's straight edges usually end in zigzags. On a Co(0001) substrate, comparable GR can generate islands with triangular or hexagonal corners because of fixed edge chirality, as shown in Fig. 3.2g–i [27]. In the first instance, neighboring atoms share a similar registry, but in the next situation, neighboring edges have polarized conformations. As shown in Fig. 3.2j, the edges of GR island are associated with directions (lattice vector) of the underlying substrate of SiC (note the inset's red arrows). Figure 3.2k depicts an AR-STM image of the GR island's step edges, armchair, and zigzag orientations represented by blue and yellow arrows, respectively. Arrows in Fig. 3.2j, k, determined the completely encircled GR island by armchair edges. As a material's size decreases, its surface free energy grows and surface interactions lead to signify the characteristics. Nanomaterials are especially associated with this effect have an average size of 1–100 nm. Similarly, the GNRs electrical structure with an ultrathin width of <10 nm is tremendously sensitive to the orientation of their edges. The production of GNRs can be accomplished through epitaxial growth on SiC substrates using templates [30]. On structured substrates, CVD growth is possible (steps [31], twins [32], and trenches [33]), or directly partial growth on Ge(001) [34]. AFM shows that GNR's lattice has zigzag and armchair edge orientations [35] and STM [34]. An HRTEM image of overlapping zigzag–armchair edges has been edited in Fig. 3.3. Typically, GNRs with a larger proportion of zigzag edges have a lower energy gap relative to a ribbon with a high proportion of armchair edges of comparable width [19]. According to tight-binding theory, metallic characteristics are always associated with zigzag GNRs, whereas armchair GNRs is linked with either one metallic or semiconducting, (contingent with the width). For instance, DFT demonstrates that

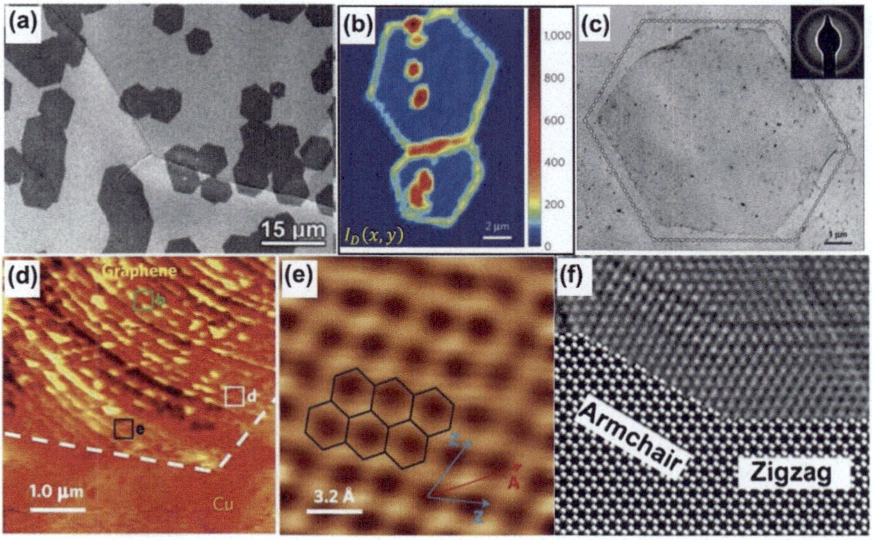

Fig. 3.3 **a** SEM micrograph of annealed Cu foil after GR growth, **b** D band intensity map for two coalesced GR grains, **c** montage of bright-field TEM (80 kV) images showing a hexagonal GR grain with its distinctive SAED pattern inset, **d** STM topography near a GR grain on Cu, **e** Fig. 6d shows an atomic-resolution STM image of the black square. Z: zigzag, A: armchair (Reproduced with permission from [17]), **f** modified HRTEM image of overlapping zigzag–armchair edges; (Reproduced with permission from [18])

when the width of semiconducting armchair GNRs shrinks, their energy E_g rise [36]. Additionally, studies demonstrate this characteristic of armchair GNRs [37]. Thus, the GNRs electronic structure may be controlled by modifying the local structure of the edge [38], modifying the GNR configuration (width and crystallographic orientation) [39, 40], and manipulating e^-–e^- and spin–orbit interactions [41, 42]. Dutta, Pati [43], and Yazyev et al. [44] provided more extensive information on the design of GNRs' electrical characteristics.

3.1.3 Band Engineering

As Wallace predicted in 1947 [45], GR possesses an exceptional band structure because of the unusual 2D framework of binary sp^2 hybridized C-atoms in a single unit cell (Fig. 3.4a, b). The sp^2 C-atoms offer π and π* orbitals, which together create the VB and CB. Since perpendicular π and π* orbitals may not intersect, these two bands are enabled to come into contact at just six places, designated Dirac points, making GR a zero E_g semimetal. At the Fermi level, the linear energy–momentum dispersion is demonstrated via Dirac-cone band structures would be extremely beneficial for electron transport. In GR, the linear dispersion demonstrates that in the relativistic Dirac equation, massless fermions do not exist. Indeed, GR is a

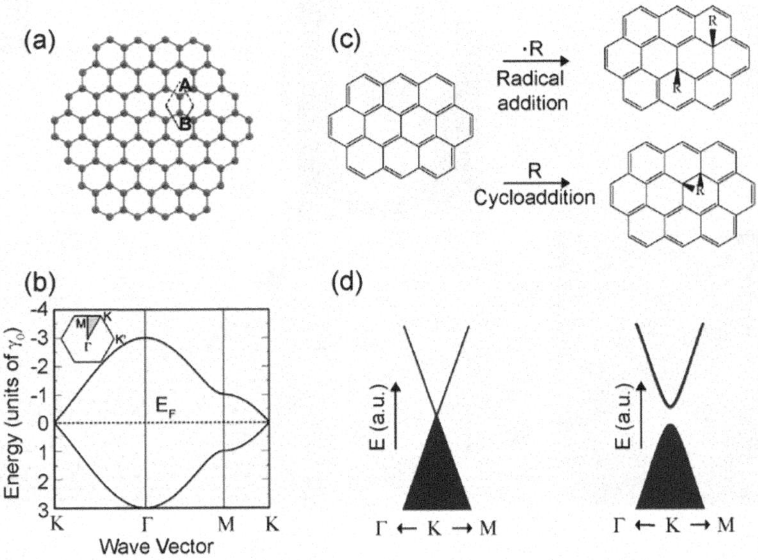

Fig. 3.4 **a** Hexagonal and honeycomb-like GR containing A and B (i.e., two Cs) in a unit cell, **b** energy momentum dispersed GR. (Reproduced with permission from [52]), **c** schematic covalently bonded GR, **d** band structure modification owing SL-GR close to Brillouin zone K point with **left** and **right** chemically modified. (Reproduced with permission from [7])

great 2D system for studying the half-inter QHE and other intriguing Dirac fermion phenomena [46, 47]. Additionally, In GR, electrons have received extensive interest in electronic applications like radio-frequency transistors since they are massless and move at vF $\approx 10^6$ m/s (velocity of light/300) [48, 49], and ultrafascinating. In this sense, it is extremely desirable to introduce an E_g into GR by band structure engineering. GR's zero E_g structure is caused by the symmetry between two C-atoms in a unit cell. To open an E_g, the potentials of the two atoms must be different [48]. Although a large E_g ($\geq$0.5 eV) and a high on/off current ratio (10^2-10^5) [7, 50] have been acquired in GNRs with a width of 2–3 nm, their use was already limited due to the challenges associated with experimental control of scalable manufacturing, ribbon width, and finally structure of edge (Fig. 3.4c, d). By establishing a single bond with exterior groups, transformation of sp^2–sp^3 hybridization of C-atoms has occurred via covalent chemistry. As noted before, several benzene-related processes that involve the dissociation of –C=C– double bonds, such as the addition of free radicals, offer themselves to GR modification [51].

Thereby, tailored GR sustained on a SiO_2/Si substrate acted like a granulated metal with an E_g of 0.1 eV for charge mobility [5]. At ambient temperature, the suspended GR sample has been observed to act as a semiconductor (80 meV gap), contrary to expectations that it would be changed on both sides by aryl groups [5]. The production of high-energy free radicals using light irradiation can promote chemically inert GR addition process. From the standpoint of device manufacturing, light is predominantly simple-to-use, and pattern-able tool to utilize. Brus et al. [53] invented photoinduced GR chemistry, starting with benzoyl peroxide; they used photoassisted aryl radical addition to put aryl groups onto the GR scaffold. The photochlorination reaction has been prompted via interacting benzene and Cl_2 in the existence of light [4, 54]. A significant increase in electrical conductivity occurs when GR is visible to Cl_2 gas under UV light. Once a threshold concentration of Cl radicals was reached, an addition reaction occurred, creating chlorinated GR with ~8 atom % Cl coverage. As shown in Fig. 3.5a, b, the photochlorination reaction was quite effective, where saturation coverage was achieved in a matter of minutes. The transport gap of the photochlorinated GR derivative was about ~45 mV (Fig. 3.5c). Using di tert-butyl peroxide (DTBP) as the methyl radical source, a similar method worked on photomethylation GR [55]. The methyl groups might be eliminated from methylated GR by heating to 380 °C. Surprisingly, by addition of methyl, the process frequently begins near the GR edge, suggesting a novel route for fabricating GR nanoribbons by simply narrowing the electrically conductive pure GR regions (Fig. 3.5d). Further effective use of GR photochemistry has been the photocatalytic oxidation of GR using TiO_2 (Fig. 3.5e) [56]. Under UV irradiation and sufficient humidity, extremely reactive ·OH radicals could be generated. Hydroxyl radicals are capable of converting GR to GO, an electrically insulating substance.

Cycloaddition reactions as the Diels–Alder (DA) reaction were previously used to break the double bonds (–C=C–) [57, 58]. The HOMO and LUMO orbitals of GR cross at the Dirac argument and operate as a parallel to DA orbitals, according to frontier molecular orbital theory. Thus, at the Dirac point, GR's orbitals are twofold degeneracy and half-filled allows it to function as a "D" or "A" in the reaction by

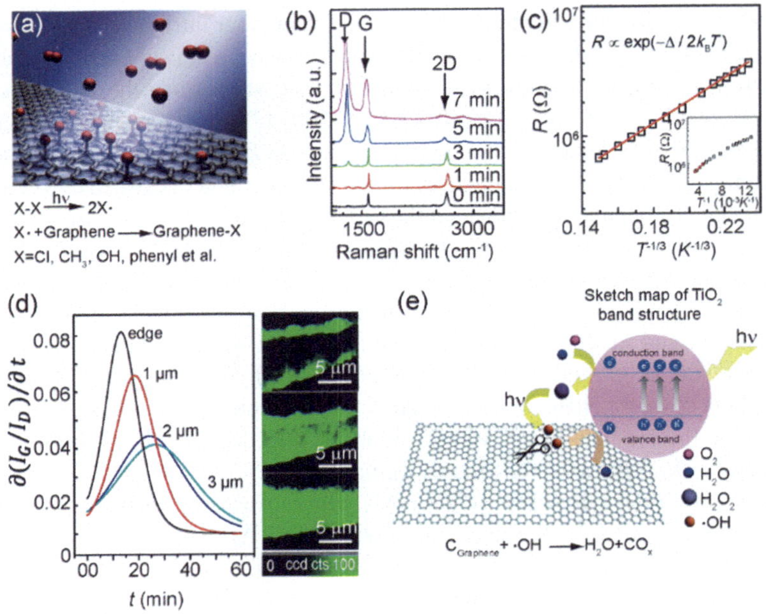

Fig. 3.5 **a** Schematic demonstration for GR photochemistry depending upon photogeneration of high-energy free-radical, **b** Raman spectra indicating GR over photochlorination, **c** temperature parameter for electrical resistance owing to photochlorination of GR. (Reproduced with permission from [4]), **d** time measurement for photomethylation representing GR edges chemical reactions. (Reproduced with permission from [55]), **e** schematic pictorial regarding TiO_2 dependent toward photocatalytic oxidized GR, representing hydroxyl radicals as chemical scissors of GR tailoring. (Reproduced with permission from [56])

comparing its antisymmetric (A) or symmetric (S) orbitals to those of its DA partner. Haddon and coworkers achieved the DA reaction between GR and tetracyanoethylene using GR as a diene [57]. In addition, GR has been used as a dienophile in its DA reaction with methylanthracene, illustrating the conversion of two adjacent sp^2 C-atoms to 1, 2-sp^3 hybridized complexes at the same time [57]. Because GR is a molecule with all of its atoms available its reactivity is extremely reliant on its surroundings. Indeed, the energy levels HOMO and LUMO of GR might be simply controlled via chemical doping or the E_{ext} effect, resulting in a change in reactivity. Strano et al. [59] established that the GR doping level has a considerable impact on the electron transport between GR and diazonium salt. Due to the increased Fermi level, the n-type-doped GR displayed increased reactivity. They discovered that CVD-produced GR supported on SiO_2 had a greater reactivity than GR supported on modified SiO_2 substrates or BN due to moderate doping of charge impurities on SiO_2's substrate [59]. Additionally, due to decreased electronic delocalization, sp^2 hybridized C linkages with greater curvature, including such CNTs and C_{60}, displayed increased reactivity. Ruoff et al. [60] found that the curved portion, generally the GR wrinkle, had greater reactivity than the flat area [60]. Compressing or stretching GR alters the length of –C=C–, therefore altering its reactivity. Stretching stress is being applied effectively into GR supported on a PDMS substrate, significantly increasing response rate (up to tenfold) [6]. However, compression stress is anticipated to increase reactivity more than stretching stress had not been reported.

3.2 Transition Metal Dichalcogenides

2DMs and advances in their synthetic processes have unlocked Pandora's Box of new layered nanomaterials designated as 2D TMDCs, in addition to the bandgap-free GR. These sophisticated 2D semiconducting materials seem to be currently gained the attention of numerous researchers, as seen by the publication record. Around 40 distinct TMDC combinations were investigated in the published work [61], these fortunate materials with an extensive range of applications for years; however, their magnificent performance in diverse technological frontiers has been surprisingly taken into consideration. Further, vdW gaps among the neighboring layers, sheet-like structure, and SSA mark the family of TMDCs as model species for sensing, energy storage, as well as capacitive storage (including supercapacitors and batteries), [62–64]. IP and OP vibrational modes of single-atom-thick sheets with different TMDCs, such as vertically layered HS (see Chap. 5), permit the discovery of novel functionalities and characteristics that are not possible otherwise [62, 65]. Numerous innovative electrical or optoelectronic devices including photodetectors, flexible electronics, and tunneling transistors may be produced by utilizing the peculiar features of these vdW HS including such band tuning, tunneling transport, and strong interlayer coupling [62, 66–69], different 2DMs' electronic band structures (Fig. 3.7a). $MoSe_2 > WS_2 > MoSe_2 > WSe_2 > WTe_2 > MoTe_2$ has a high (~1.4–2.3) hole to mobility of electron ratio in MX_2 monolayers [70].

3.2.1 Structural Chemistry

Generally, first-row TMDCs (Fig. 3.6) have garnered the greatest attention because of their enormous elemental abundance, although heavier TMs impose extraordinary characteristics on the chalcogenides structure [71, 72]. Moreover, the propensity of heavier elements to form 2D structures is pillared by vdW interactions and first row of TMDCs toward the development of crystal as 3D proportion [73, 74]. In general, these layered materials have an MX_2 stoichiometry, where M denotes TMs and X denotes Chalcogen. Remarkably, M is sandwiched among the layers structure of 2 X atoms, typically being ~6–7 Å layers deep and containing hexagonal symmetry [72, 75].

TMDCs exist in various structural proportions caused by the contacts of the few coordination spheres of TM (Fig. 3.7d); the most frequently faced polymorphs are recognized as hexagonal, rhombohedral, and lastly trigonal denoted by the letters 3R,

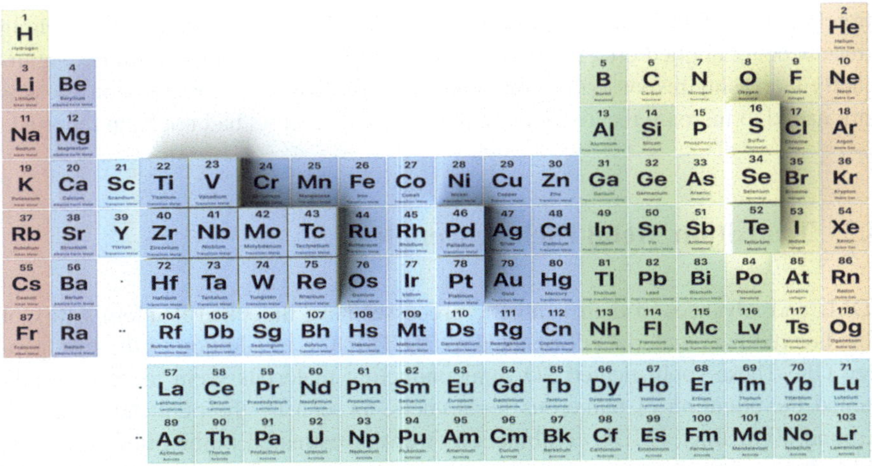

Fig. 3.6 TMDCs periodic table demonstration, where light blue colors indicate TMs (M) whereas others as chalcogens. (Reproduced with permission from [76])

2H, and 1T, respectively (Fig. 3.7c) [77, 78]. These are typically limited to chemical elements belonging to groups IVB (Ti, Zr, and Hf), VB (V, Nb, and Ta), VIB (Mo, and W), VIIB (Tc and Re), and X (Pt and Pd) and chalcogens (S, Se, and Te); their behavior varies from semiconductor to superconductor; for instance, the semiconductor group contains MoS_2 and WS_2 are classified as semimetals. WTe_2 and $TiSe_2$ are possible possibilities, but VSe_2 and NbS_2 are regarded true metals in TMDCs, and $NbSe_2$ and TaS_2 are proficient for contributing superconductivity at very low values (0.8 K) [79, 80]. The quantum size effect, surface imperfections, and nanoscale thickness all contribute to TMDCs having unique adjustable optical and physical characteristics. For example, although bulk semiconducting TMDCs in the 1T phase have an indirect E_g, monolayers have a direct optical and electronic E_g, resulting in increased PL [71, 80]. The confinement of photon transit, heat generation, and charge carrier mobility in 2D planes dramatically alter electrical and optical characteristics [70, 71, 81]. Among their numerous unmatched properties, their adaptability E_g works in tandem with their strong PL and high excitonic binding energy photovoltaic, LEDs, photoelectric modulators, as well as photodetectors and phototransistors [62, 75, 82]. As an example, the distinctive properties of MoS_2 include a direct E_g (1.8 eV, see Fig. 3.7b), high charge mobility (~700 cm^2V^{-1} s^{-1}), a high current on/off ratio (10^7–10^8), a high optical absorption in the visible range (~10^7 m^{-1}), and a huge rising of PL phenomenon from direct E_g in a monolayer; as a result, it is extensively investigated for electronics and optoelectronics [71, 83]. By adding an extra degree of freedom to electrons in crystals, they get a pair of in-equivalent valleys in the k-space electronic structure. This unprecedented degree of charge carrier freedom is referred to in novel physics as "valleytronics" since it is established by the polarization of the PL MoS_2 monolayer and valley Hall effect, a crucial effect for valley-based electronic engineering [80, 84].

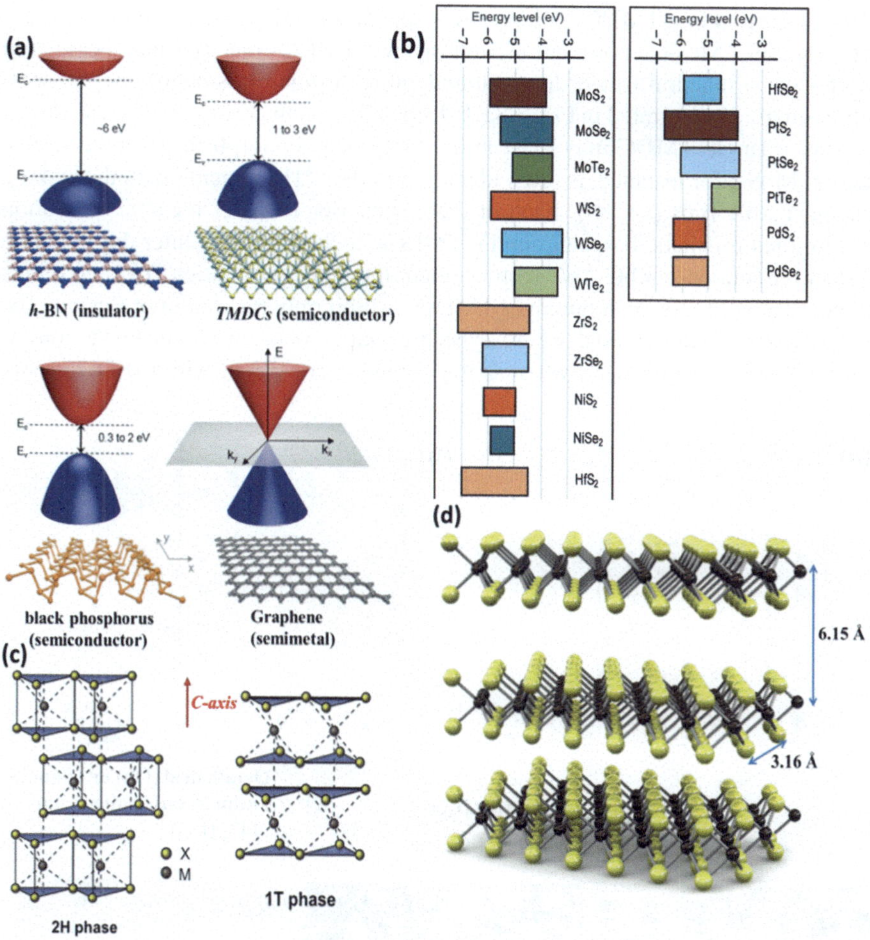

Fig. 3.7 **a** Schematic band structure for 2DMs. (Reproduced under the terms of CC-BY 4.0 license [85], https://creativecommons.org/licenses/by/4.0/), **b** energy level diagrams indicating various TMDCs. (Reproduced with permission from [62]), **c** top view picture for variety of polytypes into 2H, two-layered repeating mechanism, whereas unit showing trigonal prismatic coordination as well as 1T display, where one layer/repeat unit forming octahedral coordination owing to TMDCs, **d** three-dimensional schematic representation of a typical MX_2 structure, with the chalcogen atoms (X) in yellow and the metal atoms (M) in gray. (Reproduced with permission from [86]), **d** 3D illustration showing typical MX_2 structure containing chalcogens with yellow atoms whereas metals for gray atoms. (Reproduced with permission from [87])

TMDCs require four electrons supplied by metal atoms, whose oxidation states are +4 and −2, respectively. Owing to lack of dangling bonds, the layers become more stable against environmental reactions. The detected values are approximately 15–20% more than those reported for elemental TM solids, indicating that the d-orbital in the corresponding TMDCs compound is spatially overlapping and partially energetic [88, 89]. Metals coordination with layered TMDCs material may be seen in two forms: octahedral (often distorted and called trigonal prismatic) and trigonal-antiprismatic as illustrated in Fig. 3.8a, b. Depending on the history of TMDC development, a single TMDC may exist in a variety of polymorph or polytype forms. Natural MoS_2, for instance, is often identified in the "2H segment" with the staking sequence AbA BaB [88, 89]. Each of these scenarios involves metal coordination on a trigonal prismatic basis. Group IV TMDCs, including TiS_2, enter the 1T phase in stacking sequence AbC AbC with octahedral metal coordination. To keep things simple, we will focus on monolayer TMDCs exhibit only two polymorphisms. The 1st phase is prismatic trigonal, whereas the 2nd phase is octahedral. The first is consistent with D_3h point clusters, whereas the second is consistent with the D_{3d} group.

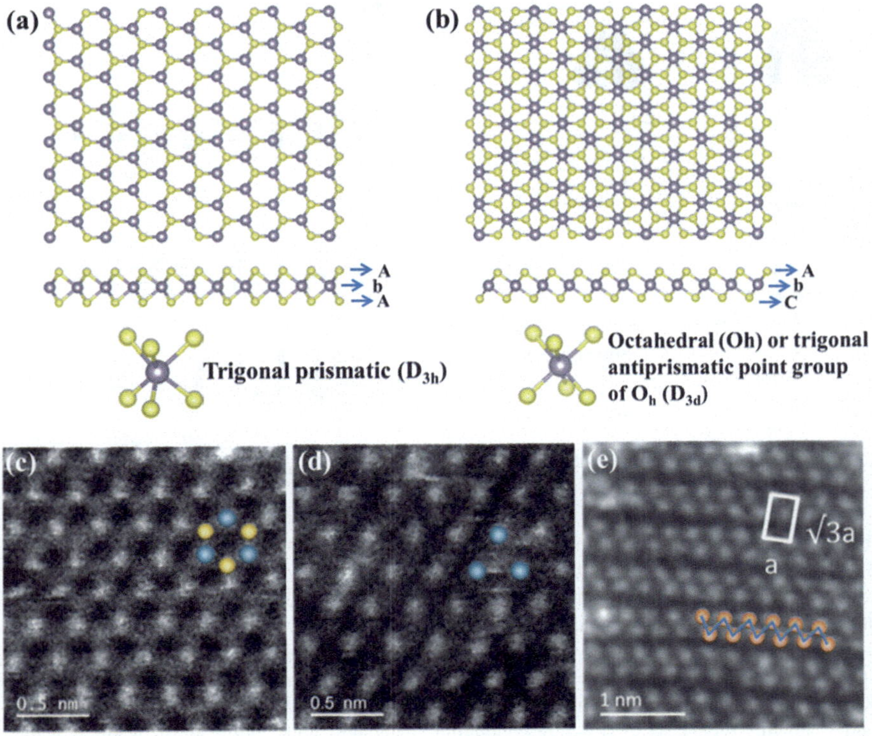

Fig. 3.8 SL-demonstration for **a** trigonal prismatic as well as **b** octahedral arrangements showing cut labels with AbA as well as AbC stacking series, further upper chalcogen, and metal with low atoms. **c, d** STM for monolayered MoS_2 with the dissimilarity of 1H **c**,1T **d**, **e** clustered zigzag chain owing to W atoms caused by Jahn–Teller distortion relevant with monolayer WS_2. The clustering of W atoms identifies orange balls comprising $\sqrt{3}a \times$ nature unit cell having superstructure and representing a rectangle. (**a, b,** and **e** Reproduced with permission from [88], and **c, d** Reproduced with permission from [90])

Table 3.1 Electronic properties of different layered TMDCs (Reproduced from [88])

Group	M	X	Properties
4	Ti, Hf, Zr	S, Se, Te	Semiconducting ($E_g = 0.2 \sim 2$ eV). Diamagnetic narrow band metals ($\rho \sim 10^{-4}$ Ω cm) or semimetals. Superconducting. Charge density wave (DCW). Paramagnetic, antiferromagnetic, or diamagnetic
5	Va, Nb, Ta	S, Se, Te	
6	Mo, W	S, Se, Te	Sulfides and selenides are semiconducting ($E_g \sim 1$ eV). Tellurides are semimetallic ($\rho \sim 10^{-3}$ Ω cm). Diamagnetic
7	Tc, Re	S, Se, Te	Small-gap semiconductors. Diamagnetic
10	Pd, Pt	S, Se, Te	Sulfides and selenides are semiconducting ($E_g = 0.4$ eV) and diamagnetic. Tellurides are metallic and paramagnetic. $PdTe_2$ is superconducting

ρ = in-plane electrical resistivity

Monolayers are what they are called, 1H (or D_3h)- and 1T (or D_3d)- MX_2 polymorphs, respectively. As shown in Fig. 3.8c, d can be used to differentiate between crystal phases. Recent research suggests that coherent interfaces between the two segments domains may arise through 1H and 1T segment lattice matching of MoS_2 and WS_2 [90]. Moreover, clustering can occur as a result of distortion, as illustrated in Fig. 3.8e with zigzag chain-like patterns.

The TM's coordinating atmosphere and the d-electron are vital in the electronic structure of TMDCs leads to exposing electrical and magnetic properties (Table 3.1). The electrical properties of TMDCs vary from IV group to X group species, as demonstrated by the non-bonding d bands being filled in Table 3.1. TMDCs display metallic conductivity after their orbitals are partly filled, as in the case of 2H-$NbSe_2$ and 1T-ReS_2. When all orbitals are completely occupied in 1T-HfS_2, 2H-MoS_2, and 1T-PtS_2, the material is classified as a semiconductor. In comparison to metallic atoms, the chalcogen atom's electrical structure is relatively unaltered. Nonetheless, a tendency may be observed with increasing chalcogen atomic numbers, such as a decrease in E_g and an increase in the corresponding d-orbital. For instance, the E_g of 2H-MoS_2, 2H-$MoSe_2$, and 2H-$MoTe_2$ gradually decrease from 1.3 to 1.0 eV [88, 91]. At first, the desired segment selected by TMDCs was determined by the TM's d-electron count. While TMDCs belonging to group IV have an octahedral structure, Group V members have trigonal and octahedral prismatic segments (d_1). Trigonal prismatic geometry correlates to TMDC group VI (d_2), whereas deformed octahedral structure belongs to TMDC group VII (d_2) (d_3). Additionally, the TMDCs' group X (d_6) is completely octahedral in structure. The 1T-MoS_2 (Fig. 3.8d) and 1T-WS_2 interfaces exhibit metallic properties, whereas the 1H-1T interfaces demonstrate unique electrical HJ via chemically homogeneous layers. In terms of phase transition, due to the instability of their electronic structure, distinct TMDCs cause periodic distortions in crystallographic lattices under particular conditions [92]. Generally, the charge density wave (CDW) phases are stable at low temperatures, like 40 K for $NbSe_2$ and 120 K for $TaSe_2$. At ambient temperature, this type of CDW lattice distortion was seen for some intercalated TMDCs. This type of distortion is thought to be caused by "Jahn–Teller instability." This instability occurs as a result of the splitting of partially filled degenerate orbits results in a decrease in free energy [93]. At normal temperature, the formation of superlattices of metal atoms by chain clusterization must also be seen in Fig. 3.8e [90]. In contrary to a hexagonal arrangement with an "a a" unit cell ("a" denotes the lattice parameter), a $\sqrt{3}a \times a$ (or $2a \times a$), $\sqrt{3}a \times \sqrt{3}a$, or $2\sqrt{3}a \times 2\sqrt{3}a$ superlatives, the atoms are moved out of their equilibrium locations to create these. Surprisingly, even without intercalants, warped phase can be metastable [94]. Such distorted TMDCs should exhibit considerably different characteristics than their undistorted counterparts.

3.3 Mxenes

This sparked fresh study on existing 2DMs such TMDCs and h-BN in addition to the discovery of several novel 2DMs [95, 96]. However, several of these materials have gained popularity due to their exceptional characteristics and practical applications. MXenes are a new type of 2DM made up of TM carbides and nitride carbides. MXene ($Ti_3C_2T_x$) was found at Drexel University in 2011 [97], with no past knowledge of such 2D compounds' stability. Ever since over 30 MXene compositions were published (blue in Fig. 3.9), and dozens more were investigated computationally (gray in Fig. 3.9). When two TMs are combined in an MXene structure, a unique property of MXenes is revealed. Along with the expected formation of solid solutions, including such (Ti, Nb)CT_x (green in Fig. 3.9), TMs can create ordered structures in a single 2D-MXene flake, either by forming atomic sandwiches of TM planes or by forming ordered structures in a single 2D-MXene flake (for $n \geq 2$), such as $Mo_2TiC_2T_x$ or through the formation of IP ($n = 1$) ordered structures, like $(Mo_{2/3}Y_{1/3})_2CT_x$. While ordered MXenes were initially synthesized in 2014 and published in 2015 [98], several additional compounds in this subfamily of MXenes have been produced (marked in red in Fig. 3.9).

In 2D flake of MXene, $n + 1$ ($n = 1$–3) layers of early TMs (M, elements in yellow in Fig. 3.10a) are interspersed with n layers of C or N (X, elements in gray in Fig. 3.10a), with the general formula $M_{n+1}X_nT_x$. T_x in the formula denotes surface terminations as O, OH, F, and/or Cl (yellow in Fig. 3.10b) that are bound to the outer M layers. In Fig. 3.10a, elements in blue indicate the elements with the components reported in the MAX phases underlined, the letter "A" originates from the previous American periodic table terminology. MXenes (Fig. 3.10a), MXenes contain intercalated metal ions as well as their precursor MAX phases (Fig. 3.10b) provide illustrative examples of basic chemistry, showing the elements' use as building blocks for nanomaterials.

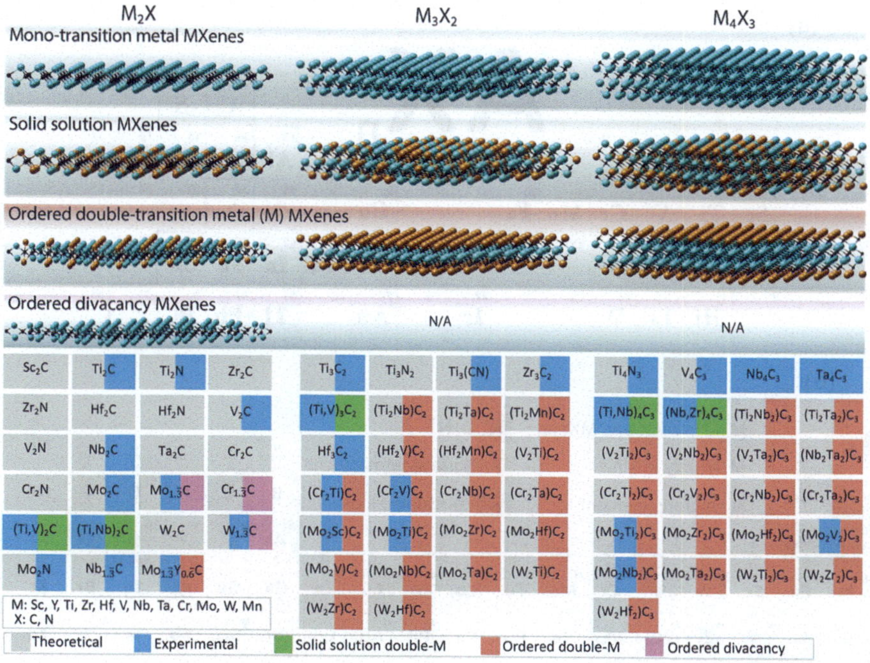

Fig. 3.9 MXene well-defined order with top row showing mono-M, MXenes whereas second-row exhibit double-M in solid solution. The double-M solid solution composition marks with green. The third row identifies orderly double-M MXenes, while composition identifies with red. The fourth row exhibits orderly divacancy structure, clearly reported toward M_2C MXenes containing $M_{1.33}C$ compositions caused by ~33 atom % showing vacancy in every M layer, whereas composition is highlighted with pink, and color-coding is also mentioned in the figure. (Reproduced with permission from [99])

3.3.1 Crystal Structure

Modeling seems to have been important in elucidating the structure and characteristics of MXenes from the beginning. Indeed, the initial structure of ML-MXenes has been predicted using a DFT simulation namely, stacked OH-terminated Ti_3C_2 layers [97, 101]. With the exception of mix of hydroxyl and fluorine terminations in the geometry optimized MXene structure, the calculated c value was close to the experimental XRD data. Because incomplete or mixed F, OH, and O terminations are possible in the real world, the scenario could be much more complicated (XPS verified the existence of OH and O) [97, 102]. Complicating matters further is the significant possibility of water molecules' presence in interlayer space [103]; particularly in V_2C and Nb_2C, when the c values after etching are quite high (see Table 3.2).

DFT calculations [106, 107] revealed two powerfully advantageous locations for T in $Ti_3C_2T_2$, leading to the formation of two different configurations: I and II (Fig. 3.11). On other side, T groups in configuration I is on lie right above hollow

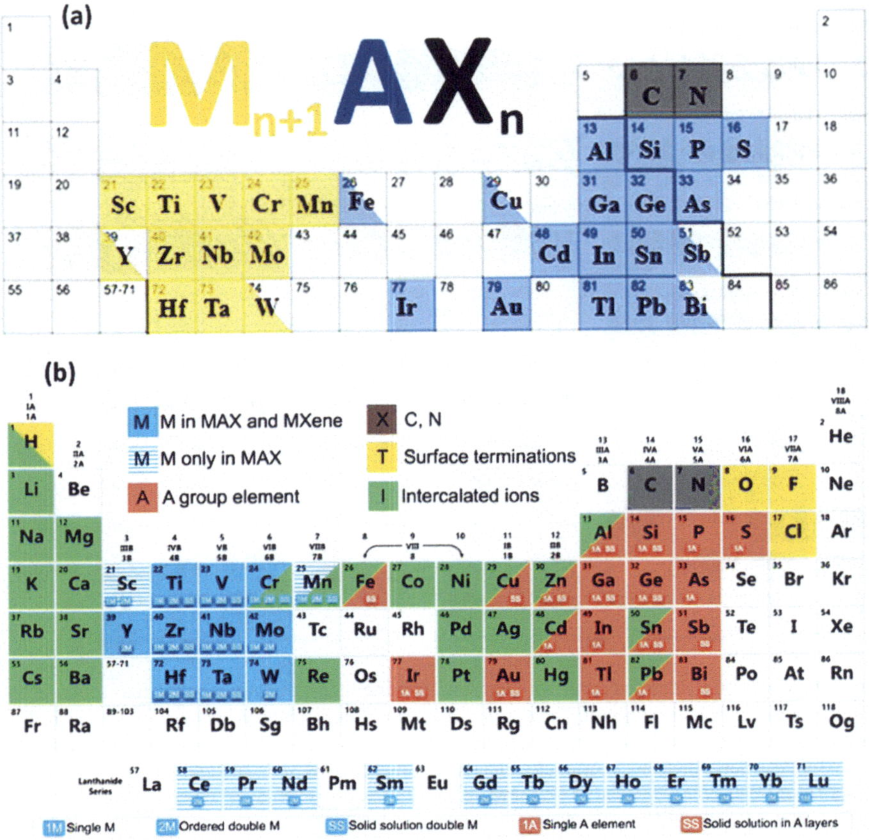

Fig. 3.10 **a** Periodic table owing to M, A, and X elements, identifying yellow, blue, and gray colors as reported by MAX phases, half-filled squared shape denoting M (yellow) and/or A (blue) as reported by solid solutions or orderly quarter MAX phases. **b** Elements for MAX phases and those with blue barred background have been reported in MXene showing MAX phases, while MXenes synthesis is still incomplete. The elements decorated by red background indicate "A" elements for MAX phases etched by for making MXenes. The cations with green background indicate intercalation into MXenes. The bottom legend, 1 M, 1A shows production probability for single and pure TM, whereas "A" element MXene with MAX phase. MAX phasing of a solid solution tells the presence of solid solutions into transition metal atomic-plane (blue) and/or "A" element-plane (red), whereas 2 M shows production probability for ordering double-TM (MAX phase) and/or MXene with IP as well as OP). (Reproduced with permission from [100])

positions that were formed via adjacent C-atoms, or the Ti(2) atoms on both sides of the Ti_3C_2 layers. Furthermore, configuration II corresponds to T groups that reside above C-atoms through both sides of the Ti_3C_2 layers. A mixed structure was also proposed, designated as configuration III, in which sheet has one side in configuration I and another in II. Different writers have employed a variety of different notations to denote these combinations. For instance, in one research, a configuration designated I was renamed "model II" or "configuration A" in another, and so

Table 3.2 Process conditions and c-lattice parameters for MXene synthesis from MAX phases (Reproduced with permission from [101])

MAX Structure	MAX	MXene	Etching conditions		*c*-lattice parameter, A		Ref
			HF Conc.%	Time, h	MAX	MXene	
211	Ti_2AlC	Ti_2CT_x	10	10	13.6	15.04	[104]
	V_2AlC	V_2CT_x	50	8	13.13	23.96	[103]
				90		19.73	
	Nb_2AlC	Nb_2CT_x	50	90	13.38	22.34	[103]
	$(Ti_{0.5},Nb_{0.5})_2AlC$	$(Ti_{0.5},Nb_{0.5})_2CT_x$	50	28	13.79	14.88	[104]
312	Ti_3AlC_2	$Ti_3C_2T_x$	50	2	18.42	20.51	[97, 104]
			40	20	18.62	20.89	[105]
	$(V_{0.5},Cr_{0.5})_3AlC_2$	$(V_{0.5},Cr_{0.5})_3C_2T_x$	50	69	17.73	24.26	[104]
	Ti_3AlCN	Ti_3CNT_x	30	18	18.41	22.28	[104]
413	Ta_4AlC_3	$Ta_4C_3T_x$	50	72	24.08	30.34	[104]
	Nb_4AlC_3	$Nb_4C_3T_x$	50	90	24.19	30.47	[103]

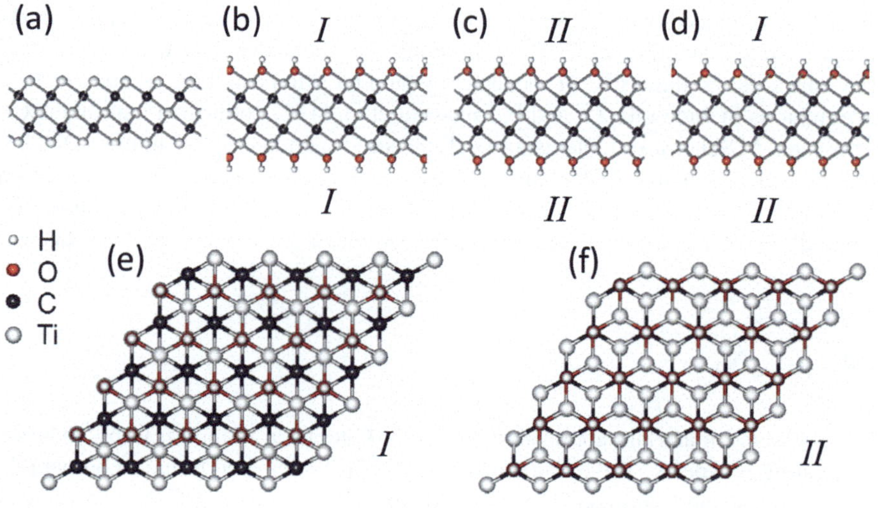

Fig. 3.11 Functionalized MXenes along with a variety of surface atoms showing side view for **a** simple Ti_3C_2, **b** and I-$Ti_3C_2(OH)_2$, **c** whereas II-$Ti_2C(OH)_2$ as well as **d** and III-$Ti_3C_2(OH)_2$, **e, f** top view indicating I-$Ti_3C_2(OH)_2$ along with II-$Ti_3C_2(OH)_2$. But configuration III being a mixture of I, II is without top view. (Reproduced with permission from [106])

on. The structural stability of different $Ti_3C_2F_2$ and $Ti_3C_2(OH)_2$ combinations was observed to decline in the order I > III > II when their respective DFT total energies were compared. This means that the F and OH groups both prefer arrangement I. The steric repulsion between T groups and C-atoms is responsible for the least stable structural configuration II [107]. T terminations bonded to and slightly above Ti(1) atoms of Ti_3C_2, another possible configuration, has been found to be avoided due to its instability and frequent transformation into configurations I–III with optimized geometry [107]. The MXenes produced thus far may have mixed functional groups (like OH, O, and F) on their surfaces that might change their structures. These configurations, as well as their order of stability, have been verified for a larger range of T groups as H, O, OH, and F, further DFT studies for a variety of MXenes with varied n values and elemental compositions are being conducted [108, 109], like solid solution (carbonitrides) [110].

The relative total energy disparities are unaffected by layer thickness, between configurations I and II or III, with the greatest disparities seen for $M_{n+1}X_nO_2$ compared to $M_{n+1}X_nH_2$, $M_{n+1}X_nF_2$, and $M_{n+1}X_n(OH)_2$ when X = C or N. Furthermore, the energy differences between I–III $M_{n+1}N_nT_2$ and I–III $M_{n+1}C_nT_2$ are less compared to those between I–III $M_{n+1}C_nT_2$. No consensus exists on the thermodynamic stability of fully vs. partially terminated structures ($M_{n+1}X_nT_x$, $0 < x < 2$), or mixed structures with different Ts, which remains unresolved [108]. The same publication's phonon dispersion curves for Ti_2CO_2 do not contain any imaginary frequencies, demonstrating the stability of structure. Gan et al. [111] discovered that oxygen adsorbs on the surface of Ti_2C, generating super- and per-oxo species, which then dissociate to form Ti_2CO_x. The surface resists additional O_2 formation once Ti_2CO_2 is produced, even at 550 °C. An important factor in catalytic applications is the expected stability of Ti_2CO_2. It should be noted that this work used ideal Ti_2C surfaces (no flaws like Ti vacancies) would have impacted the results. Surface terminations include OH, O, and F. Enyashin et al. [112] confirmed that methoxy-terminated MXenes are stable using DFT, density functional tight-binding (DFTB), and molecular dynamics calculations. These results indicate that MXenes could be useful as catalysts in some processes, such as esterification. Until now, the majority, if not all, of DFT computations were performed on single, isolated MXene sheets.

3.4 g-C_3N_4

To understand materials application in which C-atom is required, C_3N_4 exhibits stimulating prospects. The diamond-like β-C_3N_4 is anticipated to have comparable compressibility and hardness [113, 114]. At ambient conditions, g-C_3N_4 is perceived as highest stable allotrope; however, numerous published reports are attempting to synthesize various modifications of this material. Kroke and Schwarz [115] and Fuess et al. [116] have discussed the history of chemical techniques to various C_3N_4 derivatives. A polymeric derivative initially prepared by Berzelius and called "melon" (named in1834 by Liebig), considered as earliest synthetic polymers [117]. Tunable optoelectronic capabilities and the fact that it is a metal-free semiconductor have been cited as advantages [118, 119]; since it shows better performance, along with being environmentally good [120, 121], it was applied in several fields including photocatalysis, [122] sensing [123, 124] and photovoltaics [125]. Photocatalytic g-C_3N_4-based materials eliminate environmental contaminants synergistically via adsorption and photocatalysis and thus have a wide range of applications in pollutant degradation [126]. Likewise, in photovoltaics, this material's narrow E_g (about 2.7 eV) offers for more light absorption resulting in a higher rate of photo charge production and therefore more electrons accessible in the CB to drive photoreactions [127].

3.4.1 Crystal Structure

Teter and Hemley foretold five distinct kinds of CN in 1996: α-C_3N_4, β-C_3N_4, c-C_3N_4, p-C_3N_4, and g-C_3N_4 (Fig. 3.12). Except for g-C_3N_4, the CNs are all very hard materials based on their crystal structure [128, 129]. Determining the shape and structure of g-C_3N_4 is therefore much simpler. Due to this, the study of g-C_3N_4 had become a prominent topic, and different g-C_3N_4 was reported. The C and N atoms in g-C_3N_4 are both sp^2 hybridized. They are linked together through bonds to create a hexagonal structure. Triazine is the name given to this hexa-atomic ring. Each triazine ring is joined by a C–N connection to a smaller unit. It is widely believed that g-C_3N_4 can have one of two chemical structures: g-C_3N_4 with a triazine ring (C_3N_3) belonging to the R3m space group, or g-C_6N_7 with a tri-s-triazine ring (C_6N_7) belongs to P6m2 space group [130].

In g-C_3N_4, the N atom at the end connects each triazine ring, resulting in an endlessly extended planar grid structure. The illustrational graph of g-C_3N_4

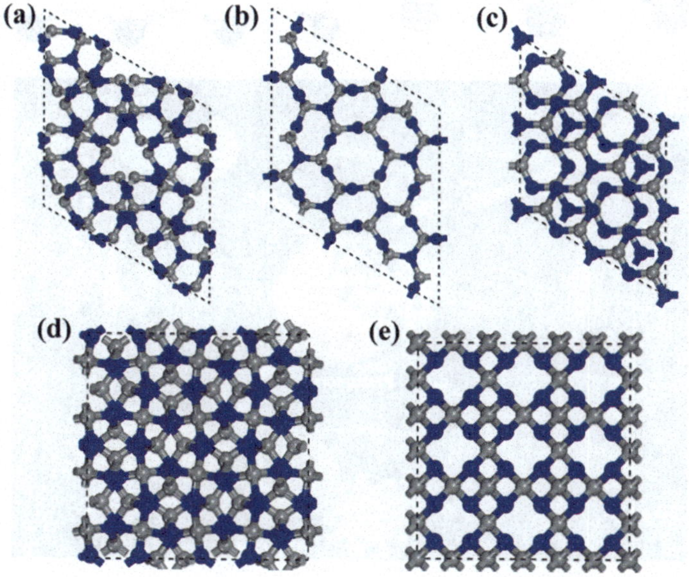

Fig. 3.12 Structure indicating **a** α-C_3N_4, **b** and β-C_3N_4, **c** whereas g-C_3N_4, **d** cubic-C_3N_4 as well as, **e** pseudocubic-C_3N_4 pictorial for [001] axis with gray-sphere C-atoms as well as blue-sphere for N atoms. (Reproduced with permission from [129])

(Fig. 3.13a) indicates that the length of the C–N bond in the g-C_3N_4 ring is 0.1315 nm and the angle of the C–N–N bond is 116.5 °C. Outside the g-C_3N_4 ring, the length of the C–N bond is 0.1444 nm, and angle between the C–N–C bonds is similarly 116.5 °C. In Fig. 3.13b, the C–N bond in the ring of g-C_3N_4 has a length of 0.1316 nm and an angle of 116.6 °C between the C–N–C bonds. Outside the g-C_3N_4 ring, the length of the C–N bond is 0.1442 nm, and the angle between the C–N–C bonds is similarly 120.0 °C [130]. The g-C_3N_4 ring structure with the tri-s-triazine ring is claimed to be thermodynamically highest stable compared to triazine structure [131]. The major tectonic units responsible for the formation of the allotropes of g-C_3N_4 are the triazine and tri-s-triazine ring structures, represented in Fig. 3.13a, b. The C and N atoms in these two kinds of g-C_3N_4 exhibit sp^2 hybridization allow for the formation of strongly conjugated π bonds with lone electron pairs in p_z orbitals. As a consequence, a conjugated system that is substantially delocalized. g-C_3N_4 has an absorption edge of around 460 nm and an E_g ~2.7 eV, having the CB at 1.13 eV and

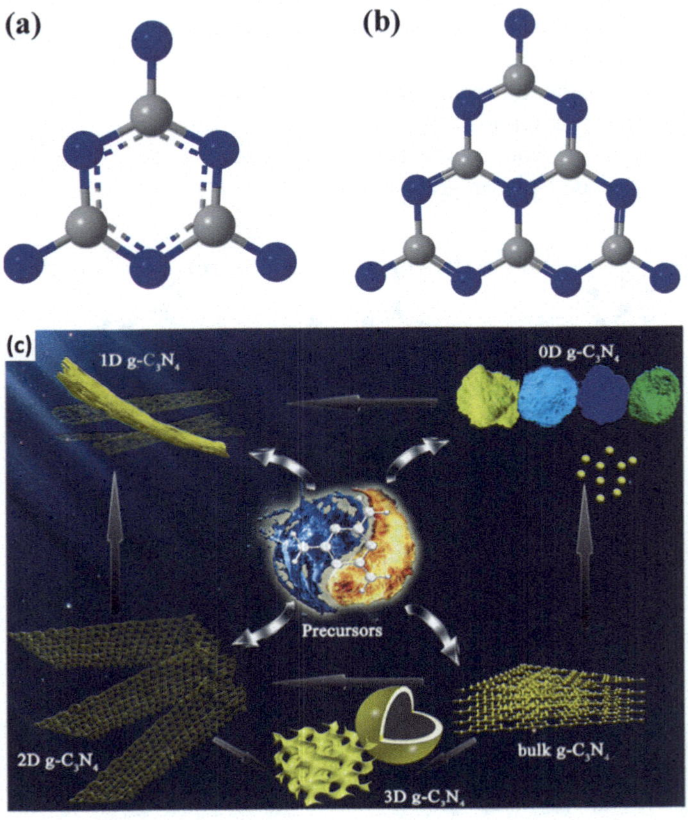

Fig. 3.13 Demonstration for **a** Triazine-, **b** tri-s-triazine-pattern owing to g-C_3N_4 allotropes, whereas **c** configuration of transitions lying between bulk, 0D structure, 1D structure, 2D structure as well as 3D-g-C_3N_4. (Reproduced with permission from [129])

VB at +1.74 eV [129, 132] Due to the contribution of the low-coordinated nitrogen atoms to CB and VB, the major bonds produced have a low degree of delocalization that is prominent explanations for extraordinary recombination rates of e^-h^+ for bulk $g-C_3N_4$. Thermal breakdown of C precursors with a high concentration of nitrogen is frequently utilized procedures for $g-C_3N_4$ growth. Consequently, resulting $g-C_3N_4$ is often bulky as well as nonporous, with strong vdWs interactions between the $g-C_3N_4$ layers. Consequently, bulk $g-C_3N_4$ has a low surface area specified due to uneven layer stacking, severely limiting its use in catalytic processes because of constrained interaction surface with reactants. Significantly, calcined $g-C_3N_4$ has a poor polymerization degree, a low crystallinity, and a high number of surface defects, all of which impair its performance and limit its extensive applicability in a variety of disciplines including sensing, adsorption and separation, and catalysis. Another critical element is the dimension of $g-C_3N_4$, which may be adjusted by adjusting the synthesis parameter. Dimensional control is critical for the performance of these nanostructures toward environmental, as well as energy applications, influence pore structure, separation rate for e^-h^+, optical characteristics, and SSA of $g-C_3N_4$ [133]. The transitions between bulk, 0D, 1D, 2D, and 3D $g-C_3N_4$ are schematically shown in Fig. 3.13c [129]. The $g-C_3N_4$ has a 2D layered structure similar to GR. The 3-s-triazine rings are linked inside each layer by an endlessly extended network-like planar structure, whereas the interlayer contact occurs via vdW forces. As a result of the success of GR stripping, 2D $g-C_3N_4$ NS with a higher precise surface area may also be produced by stripping bulk $g-C_3N_4$. Additionally, by altering the amount and arrangement of structural units, 0D and 1D $g-C_3N_4$ may be synthesized. For instance, one can synthesize 1D $g-C_3N_4$ using the template growth technique. 0D, 1D, 2D, and bulk $g-C_3N_4$ may be synthesized using a variety of techniques. Bulk $g-C_3N_4$ may be transformed to a 0D, 1D, 2D, or 3D structure by the use of external force and solution. Meanwhile, 0D $g-C_3N_4$ may be joined to form 1D $g-C_3N_4$, and 2D $g-C_3N_4$ can be transformed to a 1D nanostructure by specific treatment. Thus, it is critical to create $g-C_3N_4$ with a range of dimensionalities in order to maximize its application potential [134].

3.5 Covalent Organic Frameworks

Because of the immense design space offered by atomically accurate spatial assembly of molecular organic building components, COFs have recently gained prominence. COFs are crystalline, porous networks formed by reversible covalent bond formation from organic linkers [135–137]. The size, symmetry, and organic linkers' connectivity dictate geometry of resultant framework. COFs unlike regular linear polymers allow for two- and three-dimensional component positioning control. By controlling the network's chemical and physical properties, very regular structures may be created. The specified nanoscale channels and gaps of the COF scaffolds develop the wide and defined interface that supports catalysis and sensing applications, whereas the positive environment favors molecule storage, release, and separation. Likewise, the regularity and interconnectivity of the organic units make COFs excellent candidates for applications requiring charge carrier transfer, including optoelectronics and electrochemical energy storage. COFs provide an alternative to "traditional" inorganic porous crystalline materials including such zeolites [138, 139] and hybrid MOFs [140, 141]. Porous aromatic frameworks [142] and polymers with intrinsic microporosity [143] are examples of porous organic compounds without long-range order. Instead, the huge design area accessible to COF buildings is mirrored in large structures number produced too far, even though this study subject only began in 2005.

3.5.1 Design Principle

The topology design diagram aims to indicate the direction in which covalent bonds will form and to govern the polymer backbone development. To guarantee that each covalent connection has a unique orientation, the monomers must have relatively stiff backbones with a discrete geometry of reactive sites [144]. The block model, as depicted in Fig. 3.14, simplifies and illustrates this design principle. Each monomer has its own form that correlates to comparative positions of reactive sites. The spatial orientation and relative locations determined by covalent connections of the following units of monomers; by repeating this rule in each connection, the chain development directions are rigorously constrained to match the predesigned topology diagram. To build and control the primary-order structure, this topology map is used in conjunction with monomer geometry matching. As illustrated in Fig. 3.14, the use of planar monomers combination restricts the development of polymer backbones into 2D atomic layers with specified topologies [135, 137, 145]. During polymer chain development, on the *x*–*y* planes, periodic ordered open polygon spaces occur. These 2D covalent polymers have distinct nanopores as well as the extended polygon-based basic order structure [146, 147]. The stiff skeleton of the 2D covalent polymer also provides the system orientation of spatial lattice and crystallinity. There are several organic monomers with rigid backbones that are required for support, and they frequently contain π-systems [148]. This basic characteristic allows for more investigation of noncovalent interactions to build well-defined high-order structures. The stiff 2D covalent polymers encourage the development of layered structures in which interlayer interactions drive and control the relative positions of adjacent layers. To optimize the stacked structure's attractive energy, 2D covalent polymers overlap each monomer unit. The z-direction alignment of 2D covalent polymers defines the high-order structure of 2D COFs. Thus, a distinguishing property of 2D COFs is their ability to generate not just completely ordered π-arrays but also 1D open channel. Obviously, the monomer structure predetermines both the main and high-order structures of 2D COFs.

Formation of 2D COFs can be done using regular or irregular (isotropic and anisotropic topologies) polygon skeletons, respectively. Each example has a highly organized lattice structure and a distinct pore. The structural diversity of 2D COFs

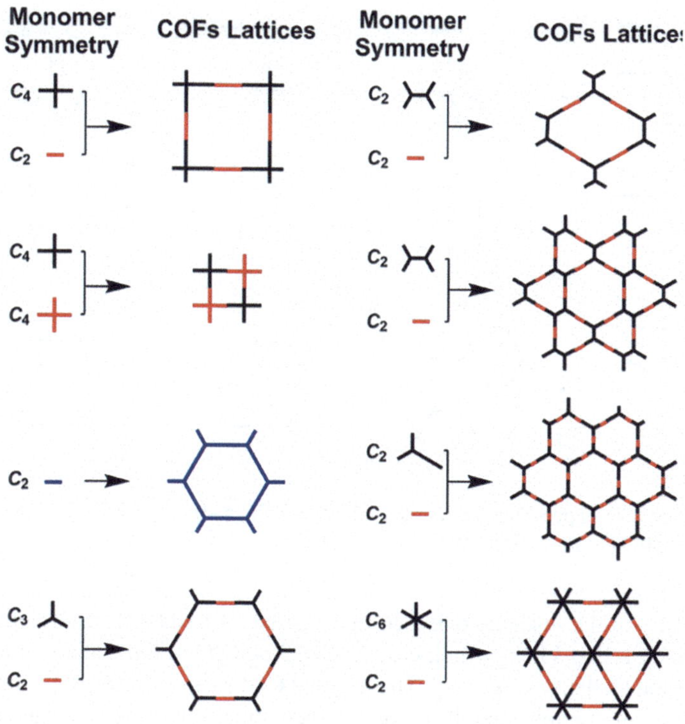

Fig. 3.14 Topological pictorial view indicating structure of 2D COFs. (Reproduced with permission from [144])

is founded on topological variety. Each topology is supported by a large number of COFs comprised of a variety of building pieces and linkers.

3.5.1.1 Symmetric Topologies

The topology diagram offers for the generation of COFs with varying skeletons and pores resulting in a variety of structural orderings of the frameworks [137, 145, 149]. For example, the combination of [C_3 + C_2] generates hexagonal 2D COFs [150, 151], whereas the combinations of [C_2 + C_2 + C_2] and [C_3 + C_3] also generate hexagonal COFs with different pore size and π-ordering. For example, the COFs formed by the [C_2 + C_2 + C_2] and [C_4 + C_4] combinations have micropores, but the COFs formed by the [C_3 + C_2], [C_3 + C_3], and [C_4 + C_2] combinations often contain mesopores. Correspondingly, the topology diagram regulates the skeletons'-density. The invention of the [C_6 + C_2] topology diagram allows for the creation of microporous COFs with the highest π density and smallest pore size [152]. As illustrated in Fig. 3.15, the resulting COFs have triangular pores with a pore size of only $L/\sqrt{3}$, which is approximately $1/\sqrt{3}$ and $1/\sqrt{3}$ the size of tetragonal and hexagonal COFs,

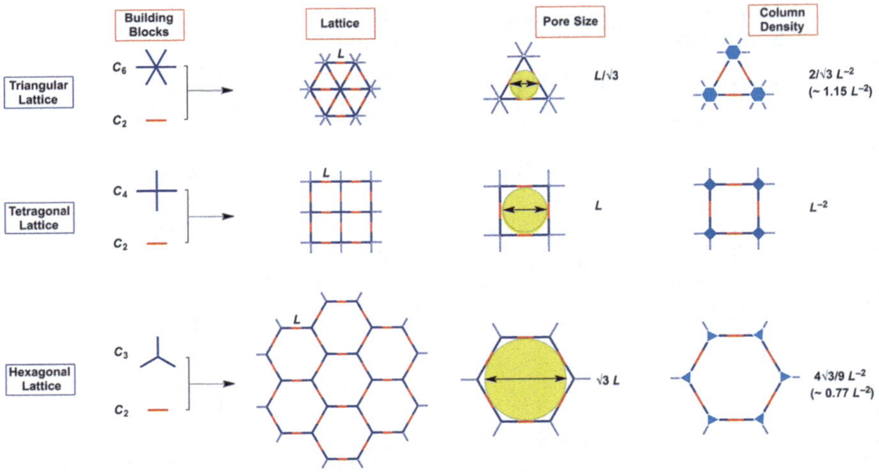

Fig. 3.15 Pore size dependence effect as well as column π-density owing to 2D COFs over topologies. (Reproduced with permission from [144])

respectively, and the distance between two vertices is fixed at L. Subsequently, the π-unit density is $2/\sqrt{3}\ L^{-2}$, which is about $2/\sqrt{3}$ and 2/3 of the density of tetragonal and hexagonal COFs, respectively. The trigonal COFs can display molecular separation and semiconducting capabilities due to their small micropore size and high-unit density. Hexa-substituted C_3-symmetric vertices can also be used to provide triangular lattices with dual-pore architectures [152]. A kagome COF with one central dodecagonal and six peripheral triangular pores was produced by constructing a C_3-symmetric macrocyclic as a knot (Fig. 3.14). Another intriguing topology diagram is $[C_{2\,+}C_2]$ results in two distinct skeletons and pore architectures. The rhombic skeleton contains a single kind of pore, whereas the kagome skeleton has a core hexagonal mesopore and six trigonal micropores on the perimeter. The topology is regulated by the knot and linker structures; the knot unit's bulkiness and interlayer interactions are most likely dominant [153]. Most C_2-symmetric knots with large systems and a preference for strong π-π interactions tend to build rhombic polygons, although other C_2-symmetric knots show to build kagome-type architecture. The weak interlayer π-π contacts in the kagome polygon may be accounted for by docking effect created by the linker portions' twisted structures [154].

3.5.1.2 Asymmetric Topologies

In frameworks topology diagrams, a knot and a linker are alternately connected in the [1 + 1] two-component system to form regular polygons and tiling structures that are isotropic. The traditional [1 + 1] combination of two-component requirements for COFs, on the other hand, severely restricts the diversity of COFs. For instance, only ten distinct COFs may be created and generated for a single knot and ten linkers. This [1 + 1] combination-based diversity constraint is circumvented by incorporating a multicomponent idea into the design of COFs. The many-component method is based on the employment of two or three distinct linkers to create COF skeletons (Fig. 3.16a, b) [155]. For instance, one C_3-symmetric knot and ten C_2-symmetric linkers may yield 210 distinct hexagonal COFs. The three-component [1 + 2] methods generate two distinct hexagonal COFs, based on the two linkers' molar ratios at 1/2 or 2/1 [1 + 3] produces hexagonal COFs with three distinct pairs of edges. The presence of multicomponent in bones significantly increases structural complexity. In two-component [1 + 1] systems, the pore changes from a regular hexagonal to an irregular hexagonal, and the lattice changes from isotopic to anisotropic tiling. Indeed, ten different C_2-symmetric linkers were synthesized, each with different lengths, systems, and redox activity. For instance, the [1 + 2] scheme was used to synthesize 48 three-component COFs, and the [1 + 3] scheme was used to synthesize eight four-component COFs. [156]. Assuming phase integrity of multicomponent COFs, these COFs have distinct lattice structures and only one type of pores. This approach may be used to combine donor and acceptor units

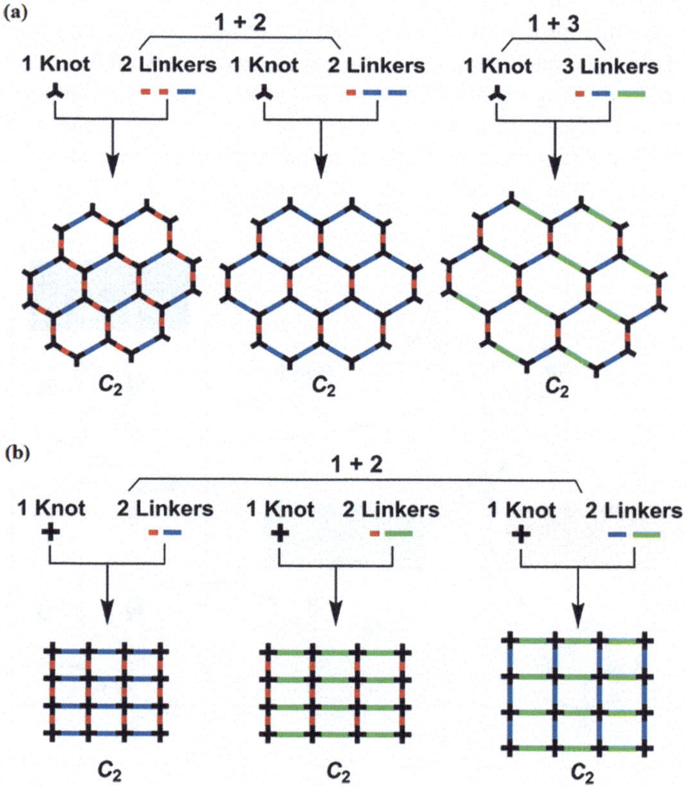

Fig. 3.16 Multicomponent designed view owing to **a** hexagonal and, **b** tetragonal COFs. (Reproduced with permission from [144])

with different components. The acceptor and donor π-columns arrays in these COFs may induce intercolumnar interactions, resulting in remarkable conductivity which is not a straightforward summation of the [1 + 1] COF counterparts. Remarkably, this multicomponent method works with tetragonal skeletons as well as hexagonal COFs (Fig. 3.16b). In this example, the three-component [1 + 2] technique enhances the number of tetragonal COFs by using one knot and two linkers and results in anisotropic tiling and uniquely shaped pores (Fig. 3.16b). By using phthalocyanine units as a single knot and coupling them with ten distinct linkers [157], tetragonal COFs may be generated in the two sets of edges that are asymmetrically tiled [156].

3.6 Metal–organic Framework

MOFs are a sort of porous material that is new to science, have modular in nature (see Fig. 3.17 for structural understanding) enables great structural variety and a broad range of options for tailoring the characteristics of materials. MOFs are synthesized using a variety of techniques, including microwave, ultrasonic, mechanochemical, and electrochemical processes. The repeatability of the synthesis and postsynthesis handling of the generated samples is critical in this context. MOFs are typically functionalized further to enable them more suited for a particular application [158]. The ability to modify the structure, topology, porosity, and composition of elements (the Al:Si ratio and substitution of TM atoms in tetrahedral locations isomorphous) of zeolites has made them more popular in recent years [159, 160] and their results

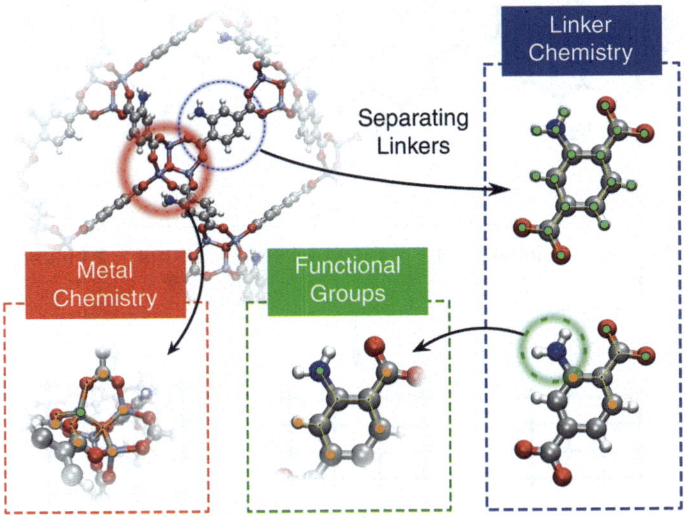

Fig. 3.17 Description of the three domains of MOF chemistry (Reproduced under the terms of CC BY 4.0 license [177], https://creativecommons.org/licenses/by/4.0/)

[161–163] the aptest materials for a spectrum of uses: gas separation and adsorption [164, 165], catalysis [166, 167], and photocatalysis [168]. However, MOFs outperform zeolites in a number of ways [169–172]; a distinguishing property of MOFs is their high surface area. Organic linkers connect secondary building units in the MOF. Secondary building blocks are metal ions/clusters, often coordinated by O_2 or N_2 atoms, but rarely by F or other nonmetal atoms. The MOFs differ significantly from zeolites in several important ways [173]. The main distinction between MOFs and zeolites is their structural diversity and lack of topological constraints on the development of porous structures. New MOF structures are synthesized every year, proving their variety and generating attention in possible applications [174]. Whereas, zeolites are comprised of tetrahedral fragments, MOFs' inorganic secondary building units could be as simple as one metal atom or complex metal atom. Combining organic and inorganic structural components can result in the formation of materials with unique characteristics. In one scenario, MOFs exhibit a high degree of strength and a large number of pores (50% of total volume or more). Conversely, by selecting the initial building units, other important parameters such as pore size (in order to raise the pore diameter to 98 Å) [175], density (to reduce to 0.126 gcm^{-3}) [176], and SSA (up to 1000–10,000 m^2g^{-1}) can be varied [141], enabling the generation of materials with precisely controlled physicochemical characteristics. MOFs are well suited for usage in gas separation, ion exchange processes, and catalysis, because of their porous structure, higher SSA, stability, and capacity to be functionalized; zeolites have been used in the past. MOFs have exceeded zeolites in terms of pore size and specific volume and hence may be employed as selected catalysts in a variety of processes, even those involving large molecules that cannot pass through zeolite pores. Despite this, MOFs remain less thermally stable than zeolites due to their structural characteristics.

3.6.1 Structural Chemistry

Interestingly, MOFs are composed of inorganic and organic molecules carboxylates/anions such as sulfonate, phosphate, and heterocyclic compounds make up organic units (birding/linkers ligands). (Fig. 3.18) [178]. Metal ions/clusters are the inorganic units denoted by the abbreviation secondary-building units (SBUs). The coordination geometry, coordination number, and nature of functional groups all contribute to its geometry. Numerous SBU geometries with varying numbers of extension points were found in MOF structures, including the octahedron (six points), the trigonal prism (six points), the square paddlewheel (four points), and the triangle (three points) (Fig. 3.19). In theory, a bridging ligand interacts with a metal ion that possesses numerous vacant or labile sites (ditopic, tritopic, tetratopic, or multitop linkers). Both SBU connections and organic ligand linkers contribute to the final framework architecture of MOF. Depending on the system under consideration, polymeric structures that are indefinitely stretched or oligomeric structures that are discretely closed may form (Fig. 3.19). Organic linkers and metal-containing units

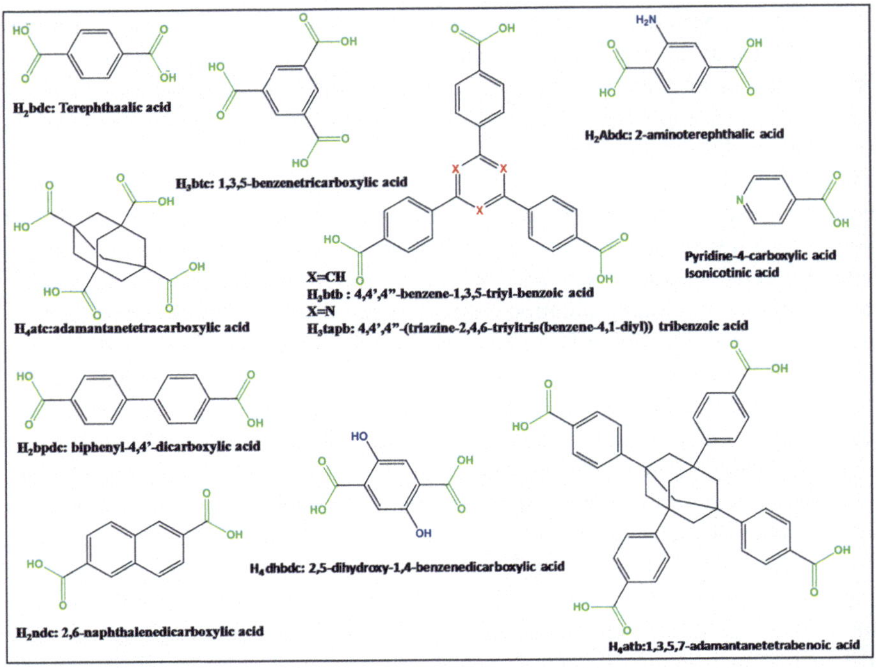

Fig. 3.18 Illustration representing organic ligands along with carboxylic functionality utilized toward MOFs fabrication. (Reproduced under the terms of CC BY 3.0 license [178], https://creativecommons.org/licenses/by/3.0/)

may be changed to create a range of MOFs with customized properties for specific purposes [140].

The development of interpenetrating structures may occur in MOFs with vast gaps. As a result, it is critical to prevent interpenetration by selecting organic linkers with care. Metal centers and legends that are organic are carefully chosen and also by changing their synthesis conditions, pore size and spatial cavity layout may be adjusted. Their high porosity enables their use in gaseous molecule adsorption and separation, catalysis, optics, sensing applications, microelectronics, drug delivery, and bioreactors are some of the technologies that have been developed among others [179–181]. MOFs feature pores as tiny as 2 nm in diameter, allowing for the passage of small molecules. However, big molecules are rarely allowed to pass through the pore holes (proteins and enzymes). The pore size has been increased to that of a mesopore (2–50 nm) and the size of crystal was down to nanometers. Surface modification is possible because to the large pore aperture, despite affecting MOF porosity, and allows encapsulation of huge MOFs molecule. Furthermore, MOFs can be synthesized using solvothermal, ionothermal, diffusion, microwave, and template-guided methods, among others [140, 141]. Combining MOFs with functional NPs is an intriguing and major breakthrough in the industry; as a result of innovative nanocomposite materials with unrivaled properties and performance are produced.

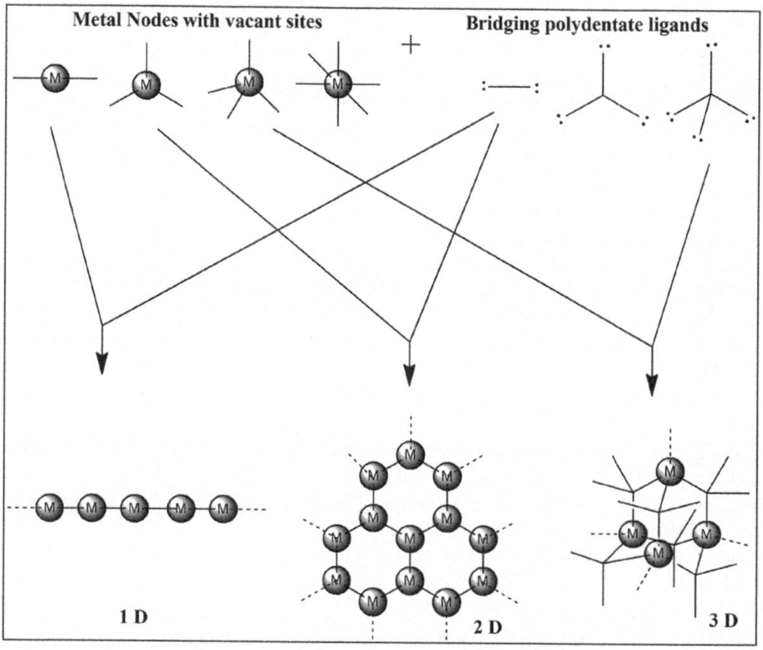

Fig. 3.19 MOFs caused by a variety of metal nodes as well as bridging ligands. (Reproduced under the terms of CC BY 3.0 license [178], https://creativecommons.org/licenses/by/3.0/)

In terms of structural and chemical variety, biodegradability, and capacity as well as high loading, nanoMOFs outperform traditional nanomedicines. The ultimate characteristics are determined by the content, morphology, and size of the particles. These materials are available in two forms: crystalline and amorphous. MOFs as porous soft crystals can demonstrate flexibility of framework even in the absence of guests or without adsorption and desorption [178, 182].

References

1. Huang X, Yin Z, Wu S, Qi X, He Q, Zhang Q, Yan Q, Boey F, Zhang H (2011) Small 7:1876
2. He Q, Wu S, Yin Z, Zhang H (2012) Chem Sci 3:1764
3. Huang X, Qi X, Boey F, Zhang H (2012) Chem Soc Rev 41:666
4. Li B, Zhou L, Wu D, Peng H, Yan K, Zhou Y, Liu Z (2011) ACS Nano 5:5957
5. Zhang H, Bekyarova E, Huang J-W, Zhao Z, Bao W, Wang F, Haddon RC, Lau CN (2011) Nano Lett 11:4047
6. Bissett MA, Tsuji M, Ago H (2013) J Phys Chem C 117:3152
7. Liao L, Peng H, Liu Z (2014) J Am Chem Soc 136:12194
8. Lee C, Wei X, Kysar JW, Hone J (2008) Science 321:385
9. Yang G, Li L, Lee WB, Ng MC (2018) Sci Technol Adv Mater 19:613
10. Sharma R, Baik JH, Perera CJ, Strano MS (2010) Nano Lett 10:398
11. Klusek Z et al (2005) Appl Surf Sci 252:1221

12. Bellunato A, Arjmandi-Tash H, Cesa Y, Schneider GF (2016) Chem Phys Chem 17:785
13. Neubeck S, You YM, Ni ZH, Blake P, Shen ZX, Geim AK, Novoselov KS (2010) Appl Phys Lett 97:053110
14. Cong C, Yu T, Wang H (2010) ACS Nano 4:3175
15. Rutter GM, Guisinger NP, Crain JN, First PN, Stroscio (2010) Phys Rev B 81:245408
16. N'Diaye AT, Engler M, Busse C, Wall D, Buckanie N, Meyer zu Heringdorf F-J, van Gastel R, Poelsema B, Michely T (2009) New J Phys 11:023006
17. Yu Q et al (2011) Nat Mater 10:443
18. Jia X et al (2009) Science 323:1701
19. Ritter KA, Lyding JW (2009) Nat Mater 8:235
20. Sen D, Novoselov KS, Reis PM, Buehler MJ (2010) Small 6:1108
21. Geim AK, Novoselov KS (2007) Nat Mater 6:183
22. Krauss B, Nemes-Incze P, Skakalova V, Biro LP, Klitzing Kv, Smet JH (2010) Nano Lett 10:4544
23. Xu YN, Zhan D, Liu L, Suo H, Ni ZH, Nguyen TT, Zhao C, Shen ZX (2011) ACS Nano 5:147
24. Hyun C, Yun J, Cho WJ, Myung CW, Park J, Lee G, Lee Z, Kim K, Kim KS (2015) ACS Nano 9:4669
25. Novoselov KS, Jiang D, Schedin F, Booth TJ, Khotkevich VV, Morozov SV, Geim AK (2005) Proc Natl Acad Sci USA 102:10451
26. Phark S-H, Borme J, Vanegas AL, Corbetta M, Sander D, Kirschner J (2012) Phys Rev B 86:045442
27. Prezzi D et al (2014) ACS Nano 8:5765
28. Eom D et al (2009) Nano Lett 9:2844
29. Yamamoto M, Obata S, Saiki K (2010) Surf Interface Anal 42:1637
30. Sprinkle M, Ruan M, Hu Y, Hankinson J, Rubio-Roy M, Zhang B, Wu X, Berger C, de Heer WA (2010) Nat Nanotechnol 5:727
31. Ago H, Ito Y, Tsuji M, Ikeda K-I (2012) Nanoscale 4:5178
32. Hayashi K, Sato S, Ikeda M, Kaneta C, Yokoyama N (2012) J Am Chem Soc 134:12492
33. Ago H, Tanaka I, Ogawa Y, Yunus RM, Tsuji M, Hibino H (2013) ACS Nano 7:10825
34. Jacobberger RM et al (2015) Nat Commun 6:8006
35. Kawai S, Nakatsuka S, Hatakeyama T, Pawlak R, Meier T, Tracey J, Meyer E, Foster AS (2018) Sci Adv 4:eaar7181
36. Barone V, Hod O, Scuseria GE (2006) Nano Lett 6:2748
37. Han MY, Özyilmaz B, Zhang Y, Kim P (2007) Phys Rev Lett 98:206805
38. Kan E-J, Li Z, Yang J, Hou JG (2008) J Am Chem Soc 130:4224
39. Wang ZF, Li Q, Zheng H, Ren H, Su H, Shi QW, Chen J (2007) Phys Rev B 75:113406
40. He HY, Zhang Y, Pan BC (2010) J Appl Phys 107:114322
41. Raza H, Kan EC (2008) Phys Rev B 77:245434
42. Dalosto SD, Levine ZH (2008) J Phys Chem C 112:8196
43. Dutta S, Pati SK (2010) J Mater Chem 20:8207
44. Yazyev OV (2013) Acc Chem Res 46:2319
45. Wallace PR (1947) Phys Rev 71:622
46. Zhang Y, Tan Y-W, Stormer HL, Kim P (2005) Nature 438:201
47. Novoselov KS et al (2007) Science 315:1379
48. Lin YM, Dimitrakopoulos C, Jenkins KA, Farmer DB, Chiu HY, Grill A, Avouris P (2010) Science 327:662
49. Liao L et al (2010) Nature 467:305
50. Li X, Wang X, Zhang L, Lee S, Dai H (2008) Science 319:1229
51. Elias DC et al (2009) Science 323:610
52. Avouris P (2010) Nano Lett 10:4285
53. Liu H, Ryu S, Chen Z, Steigerwald ML, Nuckolls C, Brus LE (2009) J Am Chem Soc 131:17099
54. Zhou L et al (2013) Small 9:2485

55. Liao L, Song Z, Zhou Y, Wang H, Xie Q, Peng H, Liu Z (2013) Small 9:1348
56. Zhang L et al (2011) J Am Chem Soc 133:2706
57. Sarkar S, Bekyarova E, Niyogi S, Haddon RC (2011) J Am Chem Soc 133:3324
58. Sarkar S, Bekyarova E, Haddon RC (2012) Acc Chem Res 45:673
59. Wang QH et al (2012) Nat Chem 4:724
60. Wu Q et al (2013) Chem Commun 49:677
61. Gupta A, Sakthivel T, Seal S (2015) Prog Mater Sci 73:44
62. Kalantar-zadeh K, Ou JZ, Daeneke T, Strano MS, Pumera M, Gras SL (2015) Adv Func Mater 25:5086
63. Choudhary N, Patel MD, Park J, Sirota B, Choi W (2016) J Mater Res 31:824
64. Panda PK, Grigoriev A, Mishra YK, Ahuja R (2020) Nanoscale Adv 2:70
65. Lin S, Li X, Wang P, Xu Z, Zhang S, Zhong H, Wu Z, Xu W, Chen H (2015) Sci Rep 5:15103
66. Samadi M, Sarikhani N, Zirak M, Zhang H, Zhang HL, Moshfegh AZ (2018) Nanoscale Horiz 3:90
67. Huang HH, Fan X, Singh DJ, Zheng WT (2020) Nanoscale 12:1247
68. Wang H, Liu F, Fu W, Fang Z, Zhou W, Liu Z (2014) Nanoscale 6:12250
69. Rhodes D, Chae SH, Ribeiro-Palau R, Hone J (2019) Nat Mater 18:541
70. Rawat A, Jena N, Dimple D, De Sarkar A (2018) J Mater Chem A 6:8693
71. Ahmadi M, Zabihi O, Jeon S, Yoonessi M, Dasari A, Ramakrishna S, Naebe M (2020) J Mater Chem A 8:845
72. Parvez K (2019) in Biomedical Applications of Graphene and 2D Nanomaterials, edited by M. Nurunnabi, and J. R. McCarthy (Elsevier, 2019), pp 1
73. Di Bartolomeo A (2020) Nanomaterials (Basel) 10
74. Eftekhari A (2017) J Mater Chem A 5:18299
75. Zhou X, Sun H, Bai X (2020) Frontiers in Bioengineering and Biotechnology 8:236
76. Raza A, Hassan JZ, Ikram M, Ali S, Farooq U, Khan Q, Maqbool M (2021) Adv Mater Interfaces 8:2002205
77. Manzeli S, Ovchinnikov D, Pasquier D, Yazyev OV, Kis A (2017) Nature Rev Mater 2
78. Chen F, Su W, Ding S, Fu L (2020) Appl Surf Sci 505
79. Yoshida M, Ye J, Nishizaki T, Kobayashi N, Iwasa Y (2016) Appl Phys Lett 108:202602
80. Lv R, Robinson JA, Schaak RE, Sun D, Sun Y, Mallouk TE, Terrones M (2015) Acc Chem Res 48:56
81. Tedstone AA, Lewis DJ, O'Brien P (2016) Chem Mater 28:1965
82. Greben K, Arora S, Harats MG, Bolotin KI (2020) Nano Lett 20:2544
83. Li X, Zhu H (2015) J Materiomics 1:33
84. Mak KF, He K, Shan J, Heinz TF (2012) Nat Nanotechnol 7:494
85. Lee JY, Shin J-H, Lee G-H, Lee C-H (2016) Nanomaterials 6
86. Wang QH, Kalantar-Zadeh K, Kis A, Coleman JN, Strano MS (2012) Nat Nanotechnol 7:699
87. Radisavljevic B, Radenovic A, Brivio J, Giacometti V, Kis A (2011) Nat Nanotechnol 6:147
88. Chhowalla M, Shin HS, Eda G, Li L-J, Loh KP, Zhang H (2013) Nat Chem 5:263
89. Yumei J, Baoze L, Xukun Z, Fangping O, Jian S, Yu Z (2020) Nanophotonics 9:1675
90. Eda G, Fujita T, Yamaguchi H, Voiry D, Chen M, Chhowalla M (2012) ACS Nano 6:7311
91. Wilson JA, Yoffe AD (1969) Adv Phys 18:193
92. Kuc A, Zibouche N, Heine T (2011) Phys Rev B 83:245213
93. Castro Neto AH (2001) Phys Rev Lett 86:4382
94. Heising J, Kanatzidis MG (1999) J Am Chem Soc 121:11720
95. Zhang H (2015) ACS Nano 9:9451
96. Gogotsi Y, Anasori B (2019) ACS Nano 13:8491
97. Naguib M, Kurtoglu M, Presser V, Lu J, Niu J, Heon M, Hultman L, Gogotsi Y, Barsoum MW (2011) Adv Mater 23:4248
98. Anasori B, Xie Y, Beidaghi M, Lu J, Hosler BC, Hultman L, Kent PRC, Gogotsi Y, Barsoum MW (2015) ACS Nano 9:9507
99. Anasori B, Lukatskaya MR, Gogotsi Y (2017) Nat Rev Mater 2:16098

100. Anasori B, Gogotsi Y (2019). In: Anasori B, Gogotsi Y (eds) 2D metal carbides and nitrides (MXenes): structure, properties and applications. Springer International Publishing, Cham, p 3
101. Naguib M, Mochalin VN, Barsoum MW, Gogotsi Y (2014) Adv Mater 26:992
102. Mashtalir O, Naguib M, Mochalin VN, Dall'Agnese Y, Heon M, Barsoum MW, Gogotsi Y (2013) Nat Commun 4:1716
103. Naguib M, Halim J, Lu J, Cook KM, Hultman L, Gogotsi Y, Barsoum MW (2013) J Am Chem Soc 135:15966
104. Naguib M, Mashtalir O, Carle J, Presser V, Lu J, Hultman L, Gogotsi Y, Barsoum MW (2012) ACS Nano 6:1322
105. Chang F, Li C, Yang J, Tang H, Xue M (2013) Mater Lett 109:295
106. Enyashin AN, Ivanovskii AL (2012) Comput Theor Chem 989:27
107. Tang Q, Zhou Z, Shen P (2012) J Am Chem Soc 134:16909
108. Khazaei M, Arai M, Sasaki T, Chung C-Y, Venkataramanan NS, Estili M, Sakka Y, Kawazoe Y (2013) Adv Func Mater 23:2185
109. Xie Y, Kent PRC (2013) Phys Rev B 87:235441
110. Enyashin AN, Ivanovskii AL (2013) J Solid State Chem 207:42
111. Gan L-Y, Huang D, Schwingenschlögl U (2013) J Mater Chem A 1:13672
112. Enyashin AN, Ivanovskii AL (2013) J Phys Chem C 117:13637
113. Thomas A, Fischer A, Goettmann F, Antonietti M, Müller J-O, Schlögl R, Carlsson JM (2008) J Mater Chem 18:4893
114. Liu AY, Cohen ML (1989) Science 245:841
115. Kroke E, Schwarz M (2004) Coord Chem Rev 248:493
116. Horvath-Bordon E, Kroke E, Svoboda I, Fuess H, Riedel R (2005) New J Chem 29:693
117. Liebig J (1834) Annalen der Pharmacie 10:1
118. Mousavi M, Habibi-Yangjeh A, Pouran SR (2018) J Mater Sci: Mater Electron 29:1719
119. Li Y-Y, Ma S-F, Zhou B-X, Huang W-Q, Fan X, Li X, Li K, Huang G-F (2019) J Phys D: Appl Phys 52:105502
120. Li J et al (2019) Angew Chem Int Ed 58:1985
121. Xu H, Yan J, She X, Xu L, Xia J, Xu Y, Song Y, Huang L, Li H (2014) Nanoscale 6:1406
122. Hong Hak C, Ching Sim L, Leong KH, Chin YH, Saravanan P (2018) In: IOP conference series: materials science and engineering 409:012008
123. Hatamie A, Jalilian P, Rezvani E, Kakavand A, Simchi A (2019) Measurement 134:679
124. Alizadeh T, Rafiei F (2019) Mater Chem Phys 227:176
125. Afshari M, Dinari M, Momeni MM (2018) Ultrason Sonochem 42:631
126. Li K, Su F-Y, Zhang W-D (2016) Appl Surf Sci 375:110
127. Liao G, Gong Y, Zhang L, Gao H, Yang G-J, Fang B (2019) Energy Environ Sci 12:2080
128. Zhang J-S, Wang B, Wang X-C (2013) Acta Phys Chim Sin 29:1865
129. Hao Q, Jia G, Wei W, Vinu A, Wang Y, Arandiyan H, Ni B-J (2020) Nano Res 13:18
130. Molina B, Sansores LE (1999) Mod Phys Lett B 13:193
131. Kroke E, Schwarz M, Horath-Bordon E, Kroll P, Noll B, Norman AD (2002) New J Chem 26:508
132. Hao Q, Xie CA, Huang Y, Chen D, Liu Y, Wei W, Ni B-J (2020) Chin J Catal 41:249
133. Zhao Z, Sun Y, Dong F (2015) Nanoscale 7:15
134. Zhu J, Xiao P, Li H, Carabineiro SAC (2014) ACS Appl Mater Interfaces 6:16449
135. Lohse MS, Bein T (2018) Adv Func Mater 28:1705553
136. Diercks CS, Yaghi OM (2017) Science 355:eaal1585
137. Huang N, Wang P, Jiang D (2016) Nat Rev Mater 1:16068
138. Davis ME (2014) Chem Mater 26:239
139. Li Y, Yu J (2014) Chem Rev 114:7268
140. Lu W et al (2014) Chem Soc Rev 43:5561
141. Furukawa H, Cordova KE, O'Keeffe M, Yaghi OM (2013) Science 341:1230444
142. Ben T, Qiu S (2013) Cryst Eng Comm 15:17
143. Ramimoghadam D, Gray EM, Webb CJ (2016) Int J Hydrogen Energy 41:16944

144. Geng K et al (2020) Chem Rev 120:8814
145. Dalapati S, Addicoat M, Jin S, Sakurai T, Gao J, Xu H, Irle S, Seki S, Jiang D (2015) Nat Commun 6:7786
146. Colson JW, Dichtel WR (2013) Nat Chem 5:453
147. Feng X, Ding X, Jiang D (2012) Chem Soc Rev 41:6010
148. Kouwer PHJ et al (2013) Nature 493:651
149. Chen X, Geng K, Liu R, Tan KT, Gong Y, Li Z, Tao S, Jiang Q, Jiang D (2020) Angew Chem Int Ed 59:5050
150. Côté AP, El-Kaderi HM, Furukawa H, Hunt JR, Yaghi OM (2007) J Am Chem Soc 129:12914
151. Wan S, Guo J, Kim J, Ihee H, Jiang D (2008) Angew Chem Int Ed 47:8826
152. Xu S-Q, Zhan T-G, Wen Q, Pang Z-F, Zhao X (2016) ACS Macro Lett 5:99
153. Zhou T-Y, Xu S-Q, Wen Q, Pang Z-F, Zhao X (2014) J Am Chem Soc 136:15885
154. Ascherl L et al (2016) Nat Chem 8:310
155. Pang Z-F, Xu S-Q, Zhou T-Y, Liang R-R, Zhan T-G, Zhao X (2016) J Am Chem Soc 138:4710
156. Huang N, Zhai L, Coupry DE, Addicoat MA, Okushita K, Nishimura K, Heine T, Jiang D (2016) Nat Commun 7:12325
157. Lin G, Ding H, Chen R, Peng Z, Wang B, Wang C (2017) J Am Chem Soc 139:8705
158. Butova VV, Soldatov MA, Guda AA, Lomachenko KA, Lamberti C (2016) Russ Chem Rev 85:280
159. Kokotailo GT, Lawton SL, Olson DH, Meier WM (1978) Nature 272:437
160. Davis ME, Lobo RF (1992) Chem Mater 4:756
161. Anderson MW, Terasaki O, Ohsuna T, Philippou A, MacKay SP, Ferreira A, Rocha J, Lidin S (1994) Nature 367:347
162. Morris RE, Bu X (2010) Nat Chem 2:353
163. Nemeth L, Bare SR (2014) In: Jentoft FC (ed) Advances in catalysis. Academic Press, pp 1
164. Groen JC, Peffer LAA, Pérez-Ramírez J (2003) Microporous Mesoporous Mater 60:1
165. Zecchina A, Bordiga S, Vitillo JG, Ricchiardi G, Lamberti C, Spoto G, Bjørgen M, Lillerud KP (2005) J Am Chem Soc 127:6361
166. Pérez-Ramírez J, Christensen CH, Egeblad K, Christensen CH, Groen JC (2008) Chem Soc Rev 37:2530
167. Haw JF, Song W, Marcus DM, Nicholas JB (2003) Acc Chem Res 36:317
168. Llabrés i Xamena FX, Calza P, Lamberti C, Prestipino C, Damin A, Bordiga S, Pelizzetti E, Zecchina A (2003) J Am Chem Soc 125:2264
169. Perry Iv JJ, Perman JA, Zaworotko MJ (2009) Chem Soc Rev 38:1400
170. Furukawa H, Cordova KE, O'Keeffe M, Yaghi OM (2013) Science 341:1230444
171. Colón YJ, Snurr RQ (2014) Chem Soc Rev 43:5735
172. Schneemann A, Bon V, Schwedler I, Senkovska I, Kaskel S, Fischer RA (2014) Chem Soc Rev 43:6062
173. Férey G (2001) Chem Mater 13:3084
174. Long JR, Yaghi OM (2009) Chem Soc Rev 38:1213
175. Deng H et al (2012) Science 336:1018
176. Furukawa H, Go YB, Ko N, Park YK, Uribe-Romo FJ, Kim J, O'Keeffe M, Yaghi OM (2011) Inorg Chem 50:9147
177. Moosavi SM, Nandy A, Jablonka KM, Ongari D, Janet JP, Boyd PG, Lee Y, Smit B, Kulik HJ (2020) Nat Commun 11:4068
178. Sharmin E, Zafar F (2016) In: IntechOpen, London, p 1. https://doi.org/10.5772/64797
179. Shi Y, Yang A-F, Cao C-S, Zhao B (2019) Coord Chem Rev 390:50
180. Canivet J, Fateeva A, Guo Y, Coasne B, Farrusseng D (2014) Chem Soc Rev 43:5594
181. Czaja AU, Trukhan N, Müller U (2009) Chem Soc Rev 38:1284
182. Tamames-Tabar C, Cunha D, Imbuluzqueta E, Ragon F, Serre C, Blanco-Prieto MJ, Horcajada P (2014) J Mater Chem B 2:262

Chapter 4
Synthetic Protocols

Achieving the appropriate size, structure, crystal phase, thickness, defect, and surface characteristic of 2DMs is important for future research for their prospective applications. The number of effective synthetic techniques currently developed in response to their appealing characteristics and potential uses has kept an upward growth trajectory. So far, the methods that have been elucidated include micromechanical cleavage, ion intercalation, oxidation-based techniques, wet-chemical (WC) routes, 2D-templated syntheses, and CVD are denoted either as top-down or bottom-up. Among these methods, the most famous mechanical cleavage and force-aided liquid exfoliation, top-down approaches, rely on exfoliating 2D thin-layer crystals from their bulk layered crystal parents. Bear in mind that top-down techniques apply only to materials consisting of layered compounds in their bulk crystals. In contrast, CVD growth and WC synthesis are bottom-up techniques based on chemical interactions between particular precursors under predetermined experimental conditions. Bottom-up approaches are more adaptive than top-down techniques and apply to the fabrication of any type of 2DMs. This chapter will discuss many established synthetic approaches for fabricating 2DM. Each technique is thoroughly discussed, including its benefits and drawbacks, as well as personal observations.

4.1 Micromechanical Cleavages

Micromechanical cleavage or mechanical cleavage is the most prevalent technique that involves the isolation of single or FL nanostructures from their bulk counterparts. This procedure is particularly appropriate for research works when a better quality layer; for instance, transistor prototyping is needed [1–3].

M. Ikram et al., *2D-Materials for Energy Harvesting and Storage Applications*,
Nanostructure Science and Technology,
https://doi.org/10.1007/978-3-030-96021-6_4

4.1.1 Scotch Tape Method

Firstly introduced by R.F. Frindt in 1966 as the "scotch tape method" for thin-layered composites, this approach resulted in the separation of MoS_2 flakes around 10 nm thick [4]. More recently, the lamellar crystals were known to have been further improved by Geim and Novoselov [5], and since then numerous 2DMs have been synthesized in feature of micromechanical cleavage. Various prepared materials such as MoS_2 (Fig. 4.1a), $MoSe_2$ (Fig. 4.1b), WS_2 (Fig. 4.1c), WSe_2 (Fig. 4.1d), $TaSe_2$ (Fig. 4.1e), and ReS_2 (Fig. 4.1f) show different number of layers in Fig. 4.1. This method offers crystalline samples suitable for electronic and optical devices on microscale and relies on adhesiveness of a lamellar crystal among two pieces of adhesive tape until peeling the tape separate to cleave the crystal. This process repeating on newly peeled flakes results in single or FL-NSs can be moved on a substrate to characterize and fabrication of devices. This method has often been

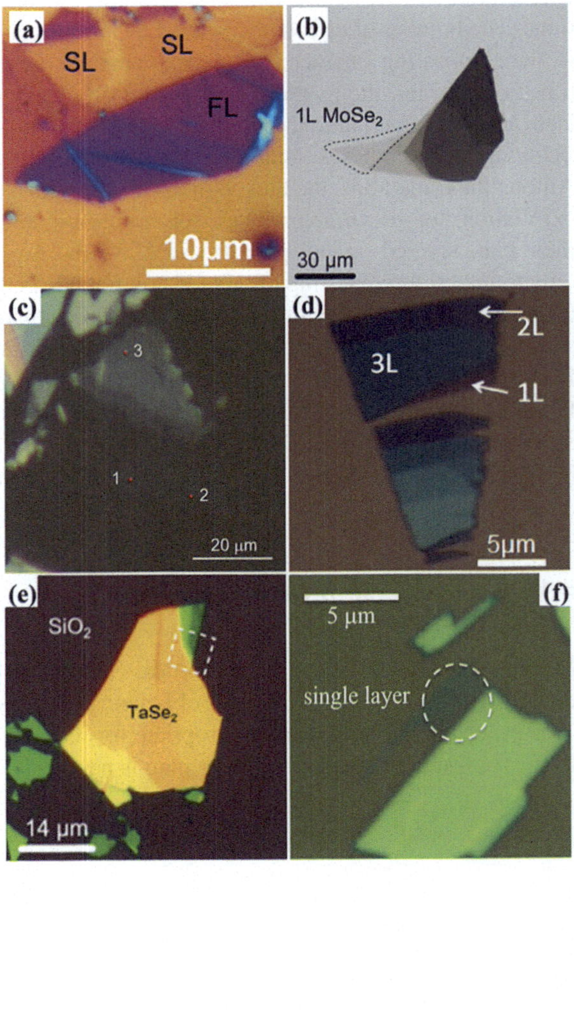

Fig. 4.1 Micromechanical exfoliation images of **a** MoS_2 (SL and FL) on SiO_2 substrate. Reproduced with permission from [8], **b** $MoSe_2$ (SL) on PDMS substrate. Reproduced under the terms of CC BY 4.0 license [9], https://creativecommons.org/licenses/by/4.0/, **c** SL and FL nanosheets of WS_2. Reproduced under the terms of CC BY 3.0 license [10], https://creativecommons.org/licenses/by/3.0/, **d** WSe_2 film exhibiting 1L, 2L, and 3L over Al_2O_3/Si substrate. Reproduced with permission from [11], **e** $TaSe_2$ nanoflakes over 300 nm SiO_2/Si substrate. Reproduced with permission from [12], **f** SL-ReS_2 flakes over 300 nm SiO_2/Si substrates. Reproduced with permission from [13]

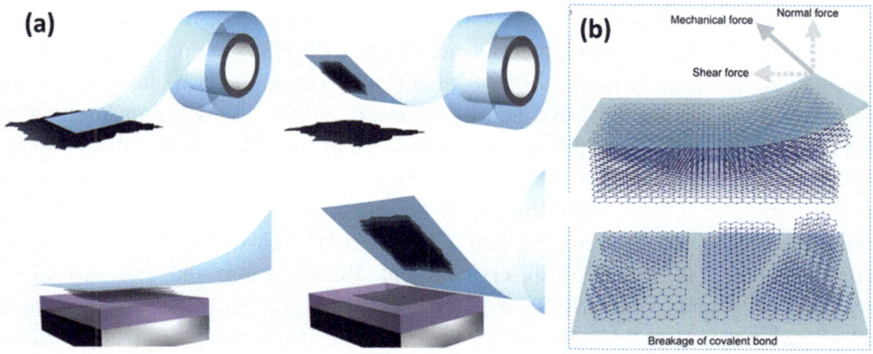

Fig. 4.2 **a** Adhesive tape cleavage of 2D crystal. Reproduced with permission from [14], **b** mechanical force exerted could be resolved into two parts, normal force and shear force, breakage of covalent bond occurred during mechanical exfoliation process. Reproduced with permission from [16]

employed in making NS of lamellar crystals, comprising several low-dimensional materials with their parent lamellar crystals are available [6, 7].

Scotch tape approach involves major steps as fascinated in Fig. 4.2a Beginning from a single crystal with an area of 1 cm^{2} × 1 cm^2, the typical thinning process includes sequential peeling of structure layers through a regular adhesive tape (e.g., 3 M scotch tape). Firstly, certain primary cleavage stages are taken to attain a fresh surface of TMDCs crystals, and then an adhesive side of tape is placed on this fresh basal plane. The thickness of a layer about 1 mm is isolated from bulk crystal when the tape is pulled off, then another piece of tape is put over the second layer to make a sandwich (tape/TMDC/tape). Repeatedly, folding and unfolding the tape until the layers on its surface are openly visible to the human eye, and layers can be separated into two thin layers. In addition, the adhesive force between tape and crystal basal plane is extensively larger than the weaker vdW forces among the layers enables separation of the thin flakes (Fig. 4.2b). In the second stage, the tape should be sticked to Si substrate usually covered with (90 or 285 nm) thick SiO_2 layer and rubbed gently with a soft object (i.e., dry sponge); eventually, peel away the sticky tape leaves behind TMDC sheets on substrate [14–16].

4.1.2 Viscoelastic Stamps

Castellanos-Gomez established "viscoelastic stamps" to be employed for exfoliating lamellar crystals that achieve the same result as scotch tape [16]. The viscoelastic transfer using PDMS stamps is and most favorable route to achieve deterministic placement of 2DMs along with their fabrication of vdW HS (see Fig. 4.3a, b) [17]. It depends on the viscoelasticity of PDMS that respond as short-term elastic solid but a long-term viscous fluid. Consequently, the implication is PDMS films kinetically show controlled adhesion: the quick peeling off of an object from film surface, higher adhesive force will be experienced. Hence, faster peel-off rates may apply for deposition of 2DMs on PDMS, and slower rates are used to release it on desired substrate. MoS_2 [16], $NbSe_2$ [16], and $TaSe_2$ [12] NSs have been produced using repeated peeling of TMDCs bulk crystals with PDMS stamp and then moved to any other substrate, e.g., Si/SiO_2 or glass, and monolayer sheet is recleaved onto untouched of stamp could be shifted to substrate. The viscoelastic stamp approach is preferred over the scotch tape method since it prevents adhesive traces from adhering to the sheet surface is important for clean NSs.

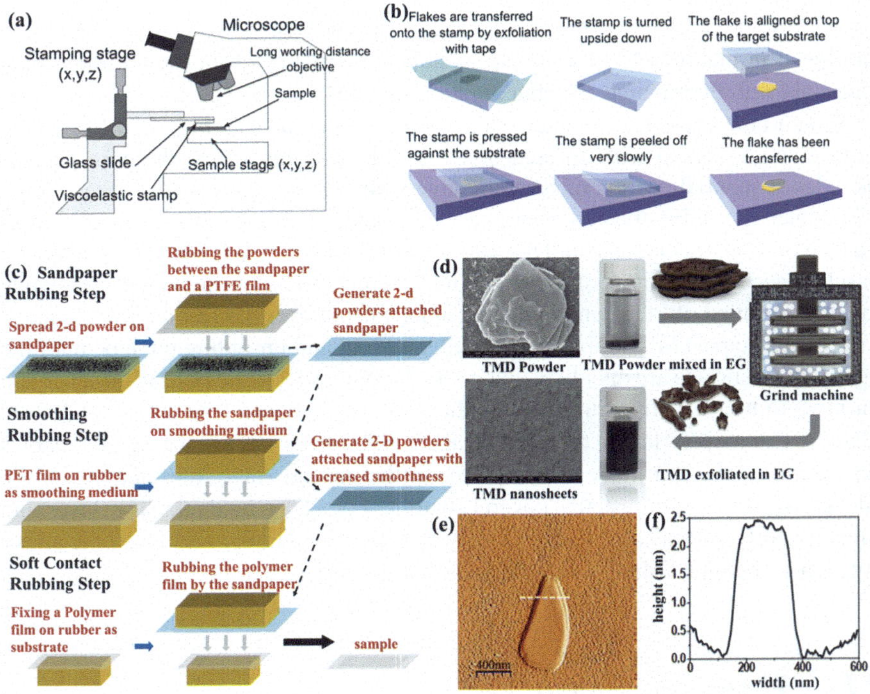

Fig. 4.3 Transfer setup procedure of viscoelastic stamp: **a** Schematic experimental setup owing to overall dry transfers, **b** forming stages for viscoelastic stamp, observed transfer atomic-thin flakes over targeted substrate. Reproduced with permission from [17], **c** sandpapers containing micro mechanically cleavage procedure owing to sandpaper soft rubbing, fine smoothing, and soft-contact-rubbing. Reproduced with permission from [19], **d** wet-grinding micromechanical exfoliation, **e** AFM photograph for WS_2, **f** height data for WS_2 showing thickness 2–2.5 nm toward NS through wet-grinding mechanism. Reproduced with permission from [20]

4.1.3 *Sandpaper-Assisted Exfoliation*

Soon after, the scotch-type technique was published, Geim and Novoselov also provide a simple rubbing method of lamellar crystal including $NbSe_2$ and MoS_2, among related examples on targeted substrates that afford shear forces essential to extract sheets from their parent crystals [6]. The invention of the basic "Chalkboard" method contributed to attempts to significantly increase in production of sheets. By this improvement, a "sandpaper-assisted" rubbing technique was established that consumes fine-grain friction of sandpaper on TMDCs powder to produce shear forces mandatory for exfoliation (Fig. 4.3c). Utilization of sandpapers initially facilitates to production of greater than 100 layers of thick NSs [18]. The mechanism was improved to produce single and FL-NSs by smooth-rubbing thicker material into a film, and then sandpaper-assisting thinning as elucidated in Fig. 4.3c. The freshly developed 2D sheets can be moved to solid or flexible substrate for more examination (Fig. 4.3b).

4.1.4 *Electrostatic-Assisted Exfoliation*

Apart from afore-mentioned processes, another favorable method of micromechanical exfoliation is "electrostatic-assisted exfoliation." Figure 4.4a shows a schematic demonstration of exfoliation process of TMDCs; initially, on the surface of highly oriented pyrolytic graphite (HOPG) disk, nano- and micrometer-sized features are assembled using lithographic techniques that followed by RIE as described in Fig. 4.4a. This designed HOPG disk acts as template and is contacted with a SiO_2/Si substrate, an electrostatic attractive force generated from voltage between HOPG and Si surface that severs between prepatterned graphitic surface and Si substrate (see Fig. 4.4b). Further, HOPG template is perpendicularly pulled away from substrate; the electrostatic force isolates prepatterned ML samples and connects them to surface of SiO_2. The thin screening deepness of HOPG (<0.5 nm) ensures that electrostatic force exploits only the outermost layer of the sample during every exfoliation cycle. An extremely localized friction force will adjust the morphology or flatness of sample outer layers to render them smooth, on substrate, but the rest of layers do not affect their bulk counterparts [21]. Gacem and coworkers [22] attempted micromechanical exfoliation of MoS_2, $NbSe_2$, and TiS_2 using an extensively related, electrostatic-assisted exfoliation method. The electric field exposure at higher pressure and temperature on a glass substrate and TMDC crystal endorses the (Na_2O) decomposition to comparatively mobile Na^+ and static O_2 ions in the substrate. By transferring Na^+ cations yield (−ve) space charge, an electrostatic field could be induced inside the lamellar crystal. Among the crystal surface and substrate, many attractions enable TMDC layers to be cleavaged that are properly attached to glass. In certain instances, it is not feasible to separate a monolayer using this technique, but attached flakes may be more thinned using previously mentioned micromechanical methods. This method has been found to develop considerably larger NS; when disabling the vdW

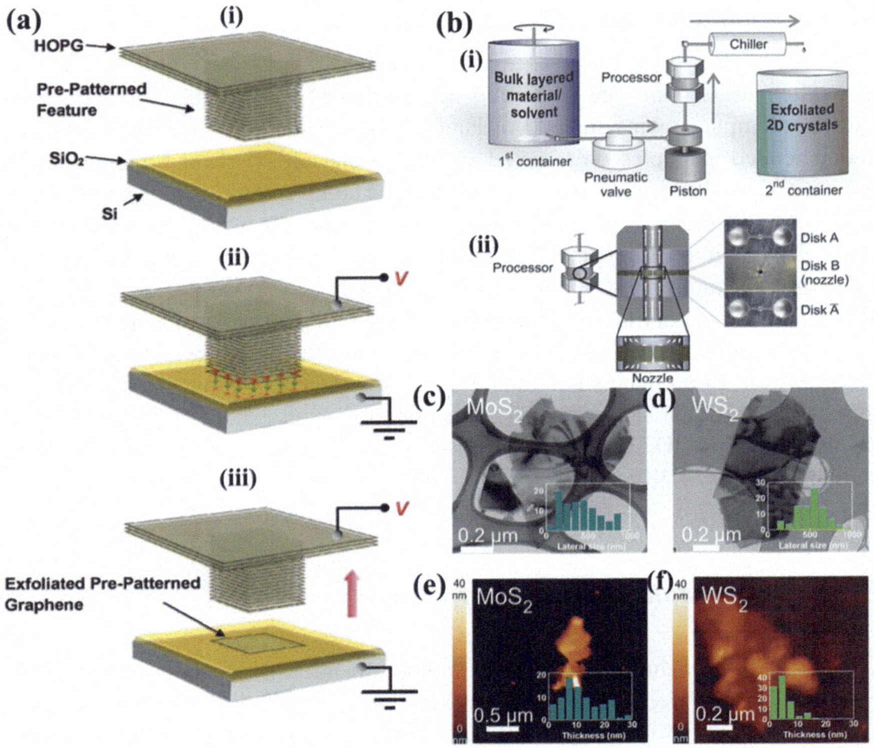

Fig. 4.4 **a** Electrostatic force exfoliated demonstration, **(i)** HOPG template setup **(ii)** voltage between HOPG and Si substrate **(iii)** exfoliation exhibiting SL and/or FL when HOPG is perpendicular to substrate. Reproduced with permission from [21], **b** wet-jet milling arrangement setup; **(i)** arrows showing solvent stream, **(ii)** close-up step of processor. Zoomed space of image **(ii)** indicating network of disk arrangement, white arrows indicating solvent flow on right of zoom space gives hole top view as well as channels over disks. **c, d** TEM view of exfoliation of MoS_2 as well as WS_2 flakes via wet-jet milling method, whereas statistical data with inset images, **e, f** AFM photographs for exfoliated MoS_2 as well as WS_2 flakes along with thickness distributions as given by insets. Reproduced with permission from [27]

interlayer forces, the micromechanical forces are more evenly used across the crystal surface. The process of NS tearing is then repressed, and thin flakes with larger lateral dimensions can be attained [23]. The primeval quality of TMDCs sheet produced is demonstrated by maximum recorded values of μ_e and I_{ON}/I_{OFF} in electronic devices. Therefore, various research groups are most commonly deploying micromechanical approaches to analyze fundamental properties of these materials or for assembling proof-concept devices [11, 24–26].

4.1.5 Wet-Grinding Technique

To enhance the mechanical exfoliations scalability, a "wet-grinding technique" has been established to uphold significantly scalable micromechanical exfoliation of TMDCs crystals through forming MoS_2 aqueous dispersions, WS_2 [20] and $NbSe_2$ [28]. The wet-grinding process allows TMDC crystals to delaminate and to produce NS in an aqueous medium by generating shear and compression forces. Komatsu and colleagues [29] suggested that this method can achieve large production of MoS_2 and WS_2 sheets with uncertain dimensions. It is obvious, this method will not yield monolayers; in MoS_2, 3–5 numbers of layers are produced averagely with lateral dimension of 500 nm–1 μm as described in Fig. 4.3d, e with height profile in Fig. 4.3f [20]. Exfoliated $NbSe_2$ [28] and WS_2 [20] 2DNS were also of smaller dimensions of cross section than for their molybdenum analog.

4.1.6 Wet-Jet Milling Method

Castillo et al. [27] produced monolayers of MoS_2 and WS_2 using large scalable micromechanical method considered as "wet-jet milling method." The wet-jet method allows pure and better quality production of 2D crystal dispersions on big scale which pave their broad manipulation in various commercial applications such as fabrication of anodes in LIB, and production of conductive inks, etc. The exfoliation of MoS_2 and WS_2 using this approach yields flakes having comparable lateral diminutions (~500 nm) with thickness of ~3 nm (Fig. 4.4c–f). Moreover, obtained TMDC NSs are favorable for several applications where a huge amount of substance is required for commercial application, procedure as explained step by step in Fig. 4.4b [27].

4.1.7 Liquid Exfoliations

The micromechanical cleavage approach demonstrated that exerting mechanical force on bulk layered structures effectively exfoliates them into SL or FL-2DM. Using this as a model, bulk layered structures can likewise be exfoliated into 2D sheets in the aqueous by applying appropriate mechanical forces to the dispersed bulk layered structures. With this in mind, many techniques for high-yield and huge-scale liquid exfoliation of ML bulk crystals have thus been devised. Exfoliation methods in liquids may be classified into two broad groups: sonication-assisted liquid exfoliation (ALE) and shear force-ALE. These groups are based on mechanical forces.

4.1.7.1 Sonication-Assisted Liquid Exfoliation

Sonication is the utmost broadly mechanical force that is used to exfoliate stacked bulk crystals into 2D NSs. Typically, bulk layered structures are spread in a particular solvent N-methyl-pyrrolidone (NMP) and then subjected to sonication at a predetermined time (Fig. 4.5) [30–32]. Centrifugation is then used to purify the suspension following sonication, yielding NS suspension. The underlying concept is that sonication may cause liquid cavitation resulting in solution bubbles. When these bubbles end, the shock waves and microjets travel through them, dispersing the layered bulk crystals into a solution. In this scenario, intense tensile stress is applied to the layered bulk crystals, resulting in their exfoliation into thin sheets. Matching the layered bulk crystal surface energy with the solvent solution is critical for effective exfoliation. Developed by Coleman's group [33] in 2008 for the exfoliation of graphite into GR, this process is relatively straightforward and successful, requiring no sophisticated instruments or costly chemicals. It paves the path for low-cost, high-yield liquid-phase GR generation. The concentration of GR suspension in their initial study was 0.01 mgmL^{-1} which is relatively low for future uses. Subsequently, it was demonstrated that a nonpolar solvent, ortho-dichlorobenzene, is reasonably effective for producing homogenous GR suspensions [34]. Despite their great efficacy,

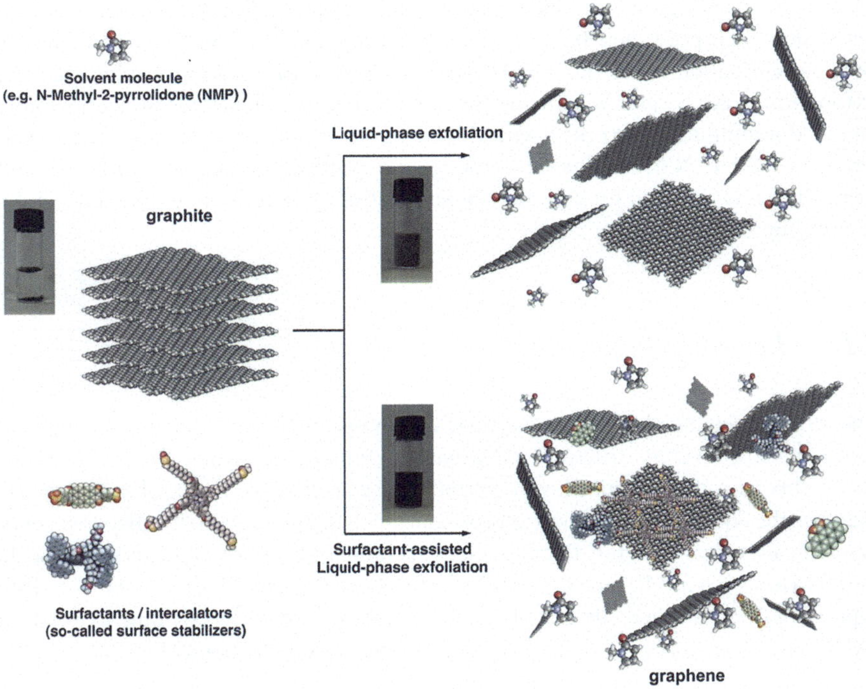

Fig. 4.5 Schematic visualization sonicated-ALE from graphite to GR. Reproduced with permission from [32]

the solvents described above contain high boiling point and are highly toxic. With this in mind, attempts were therefore made to investigate the use of volatile solvents for liquid exfoliation aided by sonication. Coleman and colleagues [35] proved how to sonicate graphite in a variety of low-boiling point solvents, with isopropanol and chloroform, to yield a reasonably huge amount of GR NS suspensions. Other solvents with a low-boiling point, including propanol and acetonitrile, were employed to aid in liquid exfoliation of graphite by sonication [36, 37]. Apart from GR, Coleman, and colleagues [38] developed their technique for exfoliating other stacked bulk crystals into 2D sheets in 2011 including MoS_2, WS_2, $MoSe_2$, $NbSe_2$, $TaSe_2$, $NiTe_2$, $MoTe_2$, h-BN, and Bi_2Te_3. An excellent match of surface tension between stacked crystals was seen in experimental and predicted in the theoretical results; moreover, the solvent choice is a critical component in ensuring effective exfoliation with the lowest possible energy consumption [38]. Further, the solvent is critical for stabilizing the exfoliated NSs and preventing them from restacking and aggregating. Zhang and colleagues [39] also demonstrated the efficacy of a solution of H_2O and C_2H_5OH in exfoliating and dispersing TMDC sheets. It is worth mentioning that because of the high surface energy of TMDCs, pure water or ethanol is ineffective at exfoliating them. However, recent research has shown that pure water may be used to exfoliate ML bulk crystals at room temperature by simply heating them to a high temperature (e.g., 60 °C) [40]. Water may now exfoliate bulk layered structures of graphite, h-BN, MoS_2, WS_2, and $MoSe_2$ into NS while also stabilizing the exfoliated NS attributed to platelet surface charges produced through edge functionalization or inherent polarity. One explanation for the edge functionalization or intrinsic polarity-induced platelet surface charge was indeed the high solubility of exfoliated sheets in water. The ability of stacked bulk crystals to exfoliate in clean water implies this process may be utilized soon to generate 2DM for various applications.

Sonication Type and Power

Probe sonication (Fig. 4.6a) allows for direct application of higher energy signals generated through tip and power to dispersive medium. These sonoprobes have been developed to produce significantly large amounts of GR dispersions in a shorter time period (2 $mgmL^{-1}$) after 6 h than ultrasonic baths (0.6 $mgmL^{-1}$) after 100 h; even so, these sonoprobe dispersions involve GR sheets that are lesser than those produced by bath sonication. Thus, higher power probably facilitates NS scission; comparable results have been later reported in MoS_2 and WS_2 exfoliation [41–44]. The acoustic power input is determined by the dispersion of the TMDCs NS required for their applications. Begimai et al. [45] synthesized highly exfoliated TMDCs NS up to monolayers on a Si^{++}/SiO_2 substrate using probe sonication, as demonstrated in AFM images of the prepared sample in Fig. 4.6b–e.

To expedite the exfoliation process, a few studies have recognized research on the refining of ultrasonic power applied using sonoprobes. Using an ultrasonic horn-probe, Zhong and colleagues [46] examined the exfoliated MoS_2 concentration and

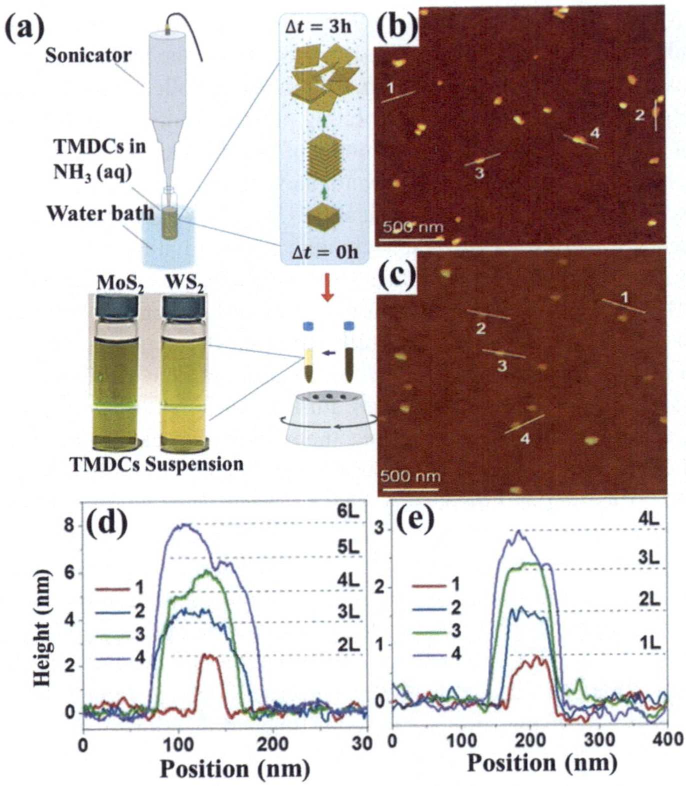

Fig. 4.6 **a** Sonication mechanism, **b, d** AFM photographs as well as line scans owing to MoS_2 nanoflakes via probe sonication, **c, e** for WS_2 nanoflakes over Si^{++}/SiO_2 substrate. Reproduced under the terms of CC BY 3.0 license [45], https://creativecommons.org/licenses/by/3.0/

morphology at various ultrasonic powers with NMP. As a result, a nonlinear correlation between the lateral dimensions (average) and concentration of NSs have been seen when an array of ultrasonic powers has been applied inside the MoS_2 dispersion. Within their system, the maximum power was measured to be between 250 and 320 W. Additionally, the impact of decreased or increased power was found. Low power results in less exfoliated energy due to the inhibition of cavitation, whereas at high power (greater than 320 W), cavitation forces are suppressed by increasing the concentration of cavitation bubbles, thereby protecting NS from them. Additionally, at lower powers, the largest flakes dimensions are comparably thick sheets were formed.

Sonication Time

Coleman and colleagues on the other hand studied reaction time effect on NSs exfoliation on various occasions [38, 47, 48]. Several solvents were used to achieve efficient exfoliation of a large collection of TMDCs using an ultrasonic horn-probe. By waiting up to 200 h for sonication, it is possible to get a dispersion concentration of MoS_2 as high as 40 $mgmL^{-1}$ in the presence of NMP [48]. It is important to note that increasing the sonication duration may result in increased scissoring of NS, decreasing their thickness to an SL, and increasing their lateral dimension of exfoliated sheets around the sub-100 nm scale. Correlations increased between size and concentration NS (average dimensions of 500–800 nm) are often evident after 48–64 h of continuous sonication. Ultrasonic baths have been used to get parallel reflections [38]. These data demonstrate that in order to control the size or thickness of exfoliated TMDC, it is better to practice prudent and discerning control overexfoliation duration.

4.1.7.2 Shear Force-Assisted Liquid Exfoliation

The sonication-ALE method is quite well-recognized for a wide range of 2DM with production rate greater compared to micromechanical cleavage. While sonication-ALE may produce 2DM with concentrations up to 1 $mgmL^{-1}$ under optimal circumstances, its production rate is insufficient for commercial processes. To increase production rates even further, shear force-ALE was created for scaling up the manufacturing of 2DM. Coleman and colleagues [49] produced excellent yield and large generation of GR flakes utilizing a high-shear rotor–stator mixer based on their extensive expertise with sonication-ALE (Fig. 4.7a–c). It was previously assumed that when the mixer rotated at a high speed, it could create significant shear strengths in liquids comprising bulk layered crystals [49, 50]. The shear forces that might be induced with high-shear strength can initiate the exfoliation of layered bulk crystals. As with sonication-ALE, selecting the appropriate polymer and solvent may lower the power consumption of the exfoliation procedure and stabilize the exfoliated sheet, therefore increasing the efficiency of this exfoliation procedure. The shear force apparatus consists of a mixing head comprised of a rotor and a stator, a straightforward and commonly accessible configuration. Graphite may be exfoliated into GR using a lateral dimension of 300–800 nm in NMP solvent with this device by rotating the mixing head at a high speed (Fig. 4.7d–h) [49]. Additionally, this technique was expanded for the exfoliation of BP bulk crystals into FL-NSs [51, 52]. Notably, the rotator diameter may be adjusted to regulate the number of GR sheets produced. To further examine the shear force-ALE process, other features, i.e., effects of rotor diameter and mixed induced shear force, were carefully investigated. Unsurprisingly, shear rate plays a crucial role in the ML- materials exfoliations. When the shear rate is $<10^4$ S^{-1}, GR has very low exfoliation efficiency and good exfoliation may be produced when the shear rate is more than 10^4 S^{-1}. Thus, GR exfoliation may be accomplished using any apparatus that generates shear rates $>10^4$ S^{-1}.

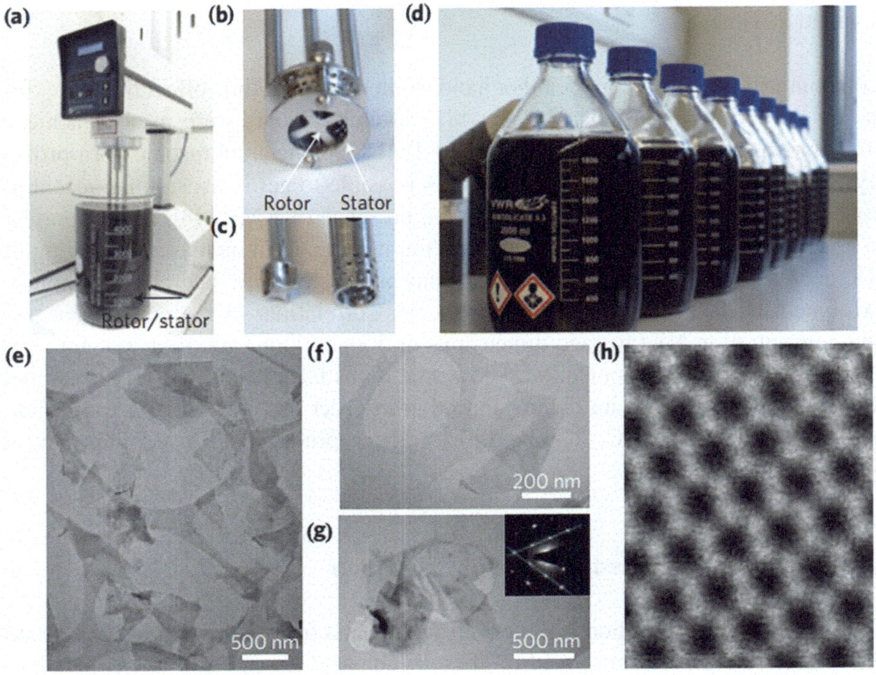

Fig. 4.7 Visulization of GR via shear mixing: **a** silver-son-model L5M high-shear-mixer along with mixing head into 5 **l** beaker for GR dispersion. Close-up view of **b** $D = 32$ mm, whereas **c** $D = 16$ mm of mixing head for left rotor isolated by **c** stator, **d** GR-NMP via shear-ALE, **e** TEM photograph for exfoliated GR-NSs, **f–h** TEM photographs for isolated NSs. Reproduced with permission from [49]

Currently, graphite exfoliation might be accomplished in aqueous volumes ranging from hundreds to thousands of liters with 1.44 gh^{-1} rate is far quicker relative to any previously documented prepared technique. To achieve high-shear strength in all areas, two separate groups reported exfoliating graphite into GR using a kitchen blender in an aqueous surfactant solution [53]. A kitchen blender is basic and easily accessible and is available for purchase rotating-blade mixer capable of creating turbulence throughout the container. Through adjusting the mixing parameters, for instance, the NS concentration, the aqueous volume, the rotor speed, and the mixing duration, the production strength can be increased to the order ofmg min^{-1}. The resulting NSs have (2–12 layers) thickness and (40–200 nm) length. Coleman and colleagues [54] have shown the ability to manufacture WS_2, h-BN, and MoS_2 NSs with a domestic blender. These findings suggested the industrial rotating-blade stirred tank reactors are a viable option to produce GR on a wide scale. In general, shear-ALE is an extremely promising process that has been commercialized for the manufacture of GR.

4.2 Ion Intercalation Exfoliation

Another prominent top-down method for fabrication 2DM is ion intercalation-ALE. Basically, this approach involves intercalating less ionic radius cation (Li^+, Na^+, K^+, or Cu^{2+}) into the bulk layered crystals interspacing to generate ion-intercalated composites, which considerably increase the interspacing and diminish the vdW interaction between neighboring layers. Indeed the ion-intercalated composites can easily be exfoliated into SL or FL-NS by moderate sonication in a fixed solvent (i.e., water) for 10 min. Intercalated ions (Li^+/Na^+) may usually interact with the solvent (H_2O) for H_2 gas production could help separate adjacent layers during sonication, boost exfoliation efficiency [55, 56]. After additional purification to remove thick flakes by centrifugation, a high yield of SL or FL-NS may be produced. The chemical Li intercalation-ALE technique is a classic example of this approach that utilizes organometallic compounds (n-BuLi) as intercalators [55, 56]. In a critical method, bulk layered structures were refluxed for many days (e.g., three days) in a hexane solution containing n-BuLi, where this species interacted with the bulk layered crystals to create the Li-intercalated composite. The Li-intercalated composite has been dissolved in a common solvent (H_2O) and then sonicated to produce the NSs suspension. It is important to remember that the Li^+ can combine with H_2O to produce H_2 gas throughout the process to support exfoliation process to succeed even more [56]. After centrifugation purification, large yields of single or FL-NS may be produced. The first effort for TMDCs layered exfoliation with chemical Li intercalation-ALE technique dates all the way back to the 1980s when Morrison and colleagues [56] employed n-BuLi as the intercalator for the exfoliation of bulk MoS_2 into SL-MoS_2. Apart from n-BuLi, several additional compounds ($LiBH_4$) have been demonstrated to be efficient intercalators for exfoliating 2DM [57]. While a high yield of SL-2DM may be achieved, the resulting NSs have a relatively small lateral dimension, less than 1 μm.

4.2.1 Intercalation Routes

Morrison and colleagues devised the original procedure [56, 58] have been widely used in studies to manufacture monolayer TMDCs NS [59, 60]. Through soaking bulk MoS_2 the intermediate [$Li_x^+MoS_2$] is prepared in an n-BuLi solution at a high temperature (80 °C+) for up to 4 days in an inert environment. A solvent-free method was developed for Li intercalation via heating TMDC and $LiBH_4$ in an argon envelope. Though solvent-free method and dispersion route were demonstrated only for MoS_2 [61, 62], $MoSe_2$ [60], WS_2 [63, 64], and ReS_2 [65, 66], the Morrison technique has been demonstrated to be useful for a many kind of layered composites [67], with hydrotic-TiS_2 [68, 69], TaS_2 [70, 71], $NbSe_2$ [72, 73] chemical and electrochemical intercalation routes are the two primary techniques for intercalation.

4.2.1.1 Chemical Intercalation

These two techniques enable the maximum exfoliation of many bulk MoS_2 crystals into NS; Li intercalation-ALE provides much higher yields than any further top-down approach, with examples commonly reporting near-quantitative bulk material conversion to NS. The NSs produced are often isolated in the range of 100s nm to sub-micron seem to be far thinner than those generated by other top-down techniques. MoS_2 dispersions might be virtually fully produced as monolayers [56]. For months, exfoliated NSs were kept at low-boiling solvent dispersions in liquid to survive. Liquid media dispersions enhance the exfoliated TMDCs surfaces activity containing organic ligands and the production of new composites in a way similar to liquid exfoliated dispersions for diverse applications [74, 75]. Therefore, there are many problems with exfoliation and intercalation routes. Numerous technical problems arise as a result of the source of Li^+: n-BuLi poses several hazards when used, and the intercalation method appears at elevated temperatures, posing problems for huge-scale processing; the lengthy lithiation times needed for its usage to also be cited as a concern. While MoS_2 dispersions are readily available, Li intercalation approach is less effective for telluride and selenide exfoliation [48, 76]. Additionally, problems of assessing and regulating the intercalation degree, meanwhile incomplete intercalation produce poor yields of monolayer products, while the intermediate Li_xTMDC with matching metal NPs and chalcogens decompose during insertion.

4.2.1.2 Electrochemical Intercalation

Electrochemical exfoliation is used to exfoliate and/or intercalate layered material into single or ML-NSs, as described in Table 4.1 [77, 78]. Electrochemical processes on an ML electrode result in intercalation and/or electrode exfoliation. Electrochemical exfoliation has several advantageous characteristics, including a short cycle duration, simplicity of action, control, and the possibility for scaling up. The quantity and applied voltage of the electrolyte have a significant effect on the consistency of exfoliated NS [15]. This is accomplished using a setup similar to that of a battery test device operated with galvanostatic discharge mode that has a constant current. The anode is metallic Li foil, the cathode is bulk TMDCs powder, and the electrolyte is $LiPF_6$ with ethylene and diethyl carbonates [79]. Li^+ is injected into the TMDC's interlayer vdW gap during the intercalation cycle and the incoming electrons from the external circuit converted them into Li atoms (Fig. 4.8a). Following the production of Li_xTMDC using the desired Li content, the remaining exfoliation process is analogous to chemical process. By monitoring galvanostatic discharge curves, the extent of Li intercalation is worthily controlled. Figure 4.8b shows that as the MoS_2 lithiation cycle development, the voltage falls monotonically and the phase shift from 2H to 1T occurs among (1.2 and 1.1 V), at which point a simple discharge plateau may be created and the intercalated Li value approaches 0.28 [80]. Electrochemical procedures are classified into two types based on the potential employed. The first kind is cathodic exfoliation that is carried out in organic solvents as dimethyl

Table 4.1 Summary of electrochemical exfoliation of various bulk TMDCs

Bulk material	Solvent	Electrolyte	Voltage	Layer thickness	Application	Ref
MoS_2	AN	THAB	−5 to −10 V	3.8 ± 0.9 nm	FET	[84]
MoS_2	EC/DMC	(5 $mgmL^{-1}$)	(1 h)	Monolayer	FET	[79]
MoS_2	H_2O	$LiPF_6$ (1 M)	0.05 mA	1–5 layers	FET	[76]
MoS_2	H_2O	Na_2SO_4 (0.5 M)	+10 V (0.5–2 h)	1–3 layers	FET	[85]
MoS_2	AN	H_2SO_4	+10 V	~2 nm	Catalysis	[86]
MoS_2	H_2O	TBAP (0.1 M)	−10 V	1–2 layers	Electrocatalysis	[87]
MoS_2	PC	LiTFSI (0.1, 1,	+5 V	6–10 layers	Biosensors	[88]
WS_2	EC/DMC	5 wt. %)	−5 V	Monolayer	FET	[79]
WSe_2	EC/DMC	TBA.H_2SO_4	0.05 mA	3–10 layers	Thermoelectric	[89]
TiS_2	EC/DMC	$LiPF_6$ (1 M)	0.025 mA	Monolayer	devices	[79]
TaS_2	EC/DMC	$LiPF_6$ (1 M)	0.05 mA	Monolayer	FET	[79]
ZrS_2	EC/DMC	$LiPF_6$ (1 M)	0.05 mA	Monolayer	FET	[79]
$NbSe_2$	EC/DMC	$LiPF_6$ (1 M)	0.05 mA	3–10 layers	FET	[89]
VSe_2	PC	$LiPF_6$ (1 M)	0.025 mA	1–5 layers	Thermoelectric	[90]
Sb_2Se_3	EC/DMC	$LiPF_6$ (1 M)	−4 V to −2 V	3–10 layers	devices	[89]
Bi_2Te_3	EC/DMC	TPA.Cl	0.025 mA	3–10 layers	Ferromagnets	[89]
		$LiPF_6$ (1 M)	0.025 mA		Thermoelectric	
		$LiPF_6$ (1 M)			devices	
					Thermoelectric	
					devices	

Acronym definitions: tetraheptylammonium bromide (THAB), acetonitrile (AN), propylene carbonate (PC), lithium bis-trifluoromethyl sulfonamide (LiTFSI), tetrapropylammonium (TPA), Tetrabutylammonium hexafluorophosphate (TBAP)

sulfoxide (DMSO) and propylene carbonate with alkyl-NH_4/Li salts as the electrolyte. The other method is anodic exfoliation that is performed using ionic liquid or H_2O mixtures or acidic liquid solutions, i.e., H_2SO_4, $HClO_4$, H_3PO_4, and $H_2C_2O_4$ [81–84].

4.2.1.3 Intercalation Chemistry of 2D Materials

A dynamic process, intercalation involves the interchange or insertion of guests into host crystals. Consequently, intermediary stages form in 2D compounds and are mobile within the host lattice. The kinetics of intercalation can be affected by macroscopic variables, for instance, deviations in crystal size displacements, pore mouth blockage, and stacking faults. At low guest concentrations, redox intercalation reaction and interlayer defects at partially occupied host lattices layer sites indicate two promising activity types in TMDCs [91, 92]. Each layer contains a uniform distribution of guest species at intermediate compositions that are randomly distributed throughout accessible interlayer sites. Up to a certain concentration of inserted content, two-phase activity is seen, implying that just a small amount of reaction energy is needed to open the vdW gap among layers. In comparison with edge intercalation, Zhang et al. [93] showed intercalation of Na^+ and Li^+ ions in

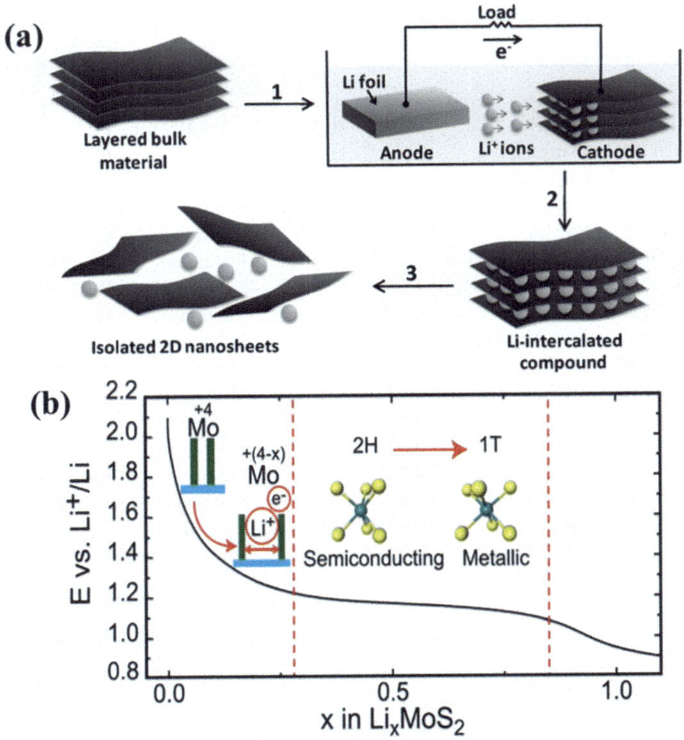

Fig. 4.8 **a** Electrochemical exfoliation format. Reproduced with permission from [79], **b** galvanostatic discharge graph electrochemically Li-intercalated. Reproduced with permission from [80]

ML-MoS_2 using a top surface method that exhibits superior controllability, stability, and reversibility. These discoveries are significant because electrochemical control is an efficient technique for exploiting the characteristics of 2D objects. Stable intercalation and reversible sequences, on the other hand, are critical for accurate in situ characterization of physiochemical changes during the intercalation procedure. As demonstrated in Fig. 4.9a, certain alternative chemical reagents with lower alkali metals activities, such as immediate sodium/potassium naphthalide and benzophenone solutions in THF [94], are used in place of Li reagents with varying reduction potentials [91]. Other reactions that occur as a result of staging offer a method for lowering energy required for layer separation. Figure 4.9b illustrates the preferred staging consequences in the production of compounds using irregular configurations of empty or occupied interlayer sites. Using this fundamental concept, the fine transitions among even and odd-numbered stages observed in graphite intercalation compounds are not understood easily [91, 92]. Thus, Daumas and Herold predicted a theory in which crystal stages are projected to take the shape of local domain [91, 95]. Due to the existence of staging, the energy associated with a homogenous cations distribution in the interlayer region is greater at low visitor concentrations than at

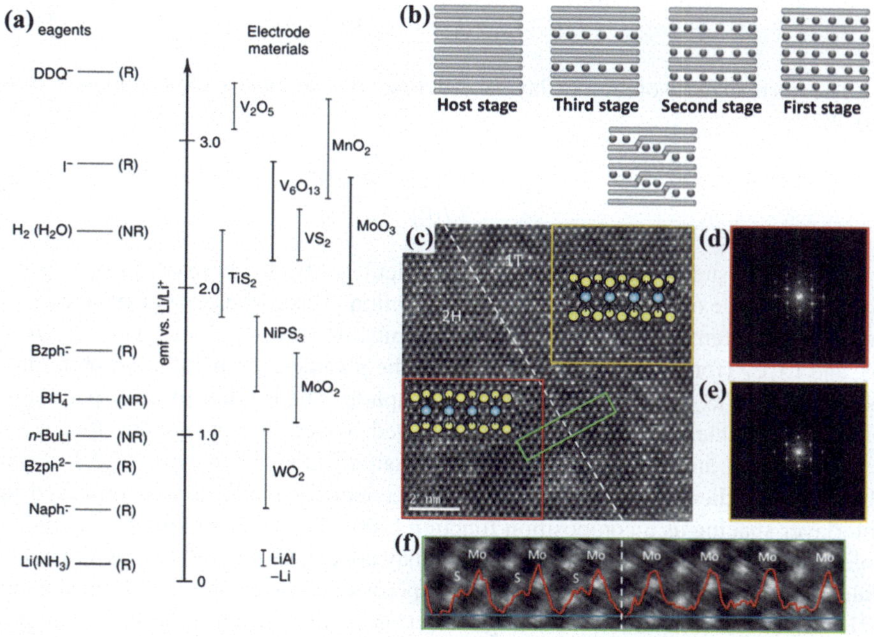

Fig. 4.9 Representation of **a** reduction trend of different host lattices toward various common Li reagents. Reproduced with permission from [91], **b** intercalation staging demonstration with below part showing isolation of island/domain model, solid lines showing intercalation of host crystal, **c** STEM view attributed to boundary area showing intercalated 2H as well as 1T phases of MoS_2, **d** Fourier transform picture for 2H, **e** whereas 1**T** phases, **f** and green rectangular zoomed area in image **c** with 3.9 nm width. Reproduced with permission from [99]

high guest concentrations when guests are grouped in a proportion of interlayer sites. When layers in staged structure come into contact, layer deformation occurs most frequently seen in the compounds formulations with flexible layers. Moreover, the uptake isotherms established through TMDCs structures for gaseous intercalation generally exhibit the sigmoid behavior of a nucleated phase. These curves can also exhibit plateaux as a result of intermediate phase's formation. During the early phases of intercalation experiments, numerous investigations have demonstrated the higher stages of participation. For instance, NH_3 intercalation in TaS_2 of a bulk polycrystalline sample was proven by in situ neutron diffraction to continue with the formation of higher stages that coexist in a powder sample. Besides, it was demonstrated that a single-crystal analysis of Ag intercalated TiS_2 includes the first stage early formation phase following electrochemical reaction [96, 97]. The rigidity of the layer plays a significant role in controlling the intercalation procedure, and Solin and colleagues [98] emphasized the importance of layer rigidity. The typical composition of intercalated composites is $G_xG_1 - x'H$, where intercalatants G and G′ have an unoccupied site; in Eq. 4.1, layer separation appears to be a mole fraction x function of the guest material.

$$d_n(x) = 1 - (1 - x)^p \tag{4.1}$$

The normalized isolation of layers is defined by the layer rigidity function p in Eq. 4.2;

$$d_n(X) = \frac{[d(x) - d(0)]}{[d(0) - d(1)]} \tag{4.2}$$

The host system's rigidity parameter is calculated by working out the composition dependence of normalized interlayer isolation. This technique is expanded to a number of systems, involving Cs_xRb_{1-x}-vermiculite (for $p = 7$), Li_xTiS_2 (for $p = 5$), and Li_xC_6 (for $p = 2$). As observed with the p values, vermiculite layer rigidity is more than TiS_2 and hence greater than graphite. The rigidity of layer paradigm assumes that intercalated ions are not distorted, which do not possible for layers of perovskites, and the layers are more rigid significantly than those of 2:1 clays; therefore, another model was produced to account for the variations observed in interlayer spacing as a composition function [100]. The layer rigidity has a crucial role in determining whether or not staging may appear. Compared to graphite intercalation compounds and TMDCs, clays and perovskites layers do not exhibit staging [91, 92]. From a metallic ion, a single electron is transferred to the lowest empty energy level in the metallic d-band correlates to the MX_2 crystal [91]. This charge move in the electronic band arrangement results in consequences that are mostly dependent on the metallic center's d-electron, and group VI undergoes particular changes. When WS_2, WSe_2, MoS_2, and $MoSe_2$ structures are intercalated with n-BuLi, TMDCs crystal is transformed into the $[Li_x^+MX_2^{\delta-}]_n$ specie. The d^3 electrons impose a vigorous penalty on metallic center reduced by shifting charge in metal's coordination geometry to an octahedral charge stabilized biasing (1T-) geometry. What is more, the successive reaction of $[Li_x^+MX_2^{\delta-}]_n$ through water eliminates the formal changing inside the metallic cations and NS. This in turn boosts the exfoliation process results in the formation of neutral MX_2 NSs with a 1T higher energy shape. When electronic modifications were identified, significant changes have been seen in the characteristics of TMDCs group VI. Intercalation and modification of the octahedral geometry remove the bandgap (pure) that confers metallic character on the 1T-part. The degree of Li intercalation provides a clear indication of the extent of the transition into 1T-structure (Fig. 4.9c–f). The 1H-1T interfaces produced inside the slightly intercalated metastable TMDCs NS of group VI were identified as a novel electronic HJ across a homogeneous chemical layer, signifying a possible way for molecular electronic systems [99, 101, 102].

4.2.1.4 Mechanism

Chemical intercalation is illustrated in Eq. 4.3, where I denotes an intercalant (molecule/cation, anion), and n denotes the intercalant numbers. During the process, intercalants move from the edge to the center of TMDCs layers [95, 103].

$$\mathrm{nI + TMDCs \rightarrow I_nTMDCs} \quad (4.3)$$

Guest intercalates inserted directly into host crystals, intercalate species inject into restacked TMDCs, guest species insertion after intercalation of solvent also guest and solvent cointercalated into the host crystal [104–106]. Thermodynamic and kinetic characteristics of processes are important to chemists because they can control them. Evans et al. [107] investigated the kinetics of cobaltocene intercalation in TiS_2, 2H-$NbSe_2$, 2H-$SnSe_2$, 2H-TaS_2, and ZrS_2, obtaining rate constants using Avrami–Erofeyev relation (Eq. 4.4) and achieved reaction order through Sharp–Hancock plot (Eq. 4.5) where, k and m denote nucleation strength and nuclei growth dimensions correspondingly, and t denotes reaction duration.

$$\alpha(t) = 1 - \exp\left(-(kt)^m\right) \quad (4.4)$$

$$\ln[-\ln(1 - \alpha(t))] = \ln(kt) \quad (4.5)$$

They concluded that rate constants are independent of intercalated quantity and temperature, solvent, and host lattice size. Mortazavi et al. [108] observed that tetrahedral and octahedral sites provided intercalation sites for Li^+/Na^+ using 2H- and 1T-TMDCs; however, the octahedral site shows more conductivity as compared to lithiation intercalation than that of tetrahedral site.

4.3 Oxidation-Assisted Exfoliation

Another frequently used technique for graphite exfoliation is the oxidation-ALE method called modified Hummers' method [109]. The main concept behind this approach is to oxidize graphite with powerful oxidizing chemicals to create graphite oxide. The graphite oxidation may result in the formation of many oxygen-containing functional groups on every GR layer surface, considerably increasing the interlayer separation of bulk graphitic structure and therefore weakening the vdW interaction between adjacent layers [109]. The enlarged graphite oxide may indeed be exfoliated into SL-GO through a further sonication treatment [109, 110]. The oxidizing agent most frequently employed is a solution of $KMnO_4$ permanganate and strong H_2SO_4 acid [110]. This approach enables the high yield and wide-range production of SL-GO in the solution stage. One of the primary drawbacks in this procedure is use of powerful oxidizing chemicals provides less safety than other techniques, such as sonication-ALE. While this method is extremely successful for GO-NS production, it is difficult to apply it to other layered materials. The abundance of O_2-containing groups, involving carboxyl, epoxy, and hydroxyl groups produced on the GR oxide surface as a result of oxidation is of special interest [111]. After reduction, these O_2-containing functional groups on GO sheet might be partially detached to provide rGO NSs. Numerous techniques exist for reducing GO-NS, including chemical reduction through reducing agents, thermal annealing, photochemical and electrochemical reduction. Even after reduction, residual functional groups on rGO NSs surface can persist. The GO and rGO NS exhibit distinct physical, chemical, and electrical characteristics when compared to GR in its natural state or GR produced by CVD [112–114]. Even after reduction to its rGO state, the electrical conductivity of rGO sheets is inferior to that of mechanically exfoliated or CVD-grown sheets. Furthermore, GO NSs are amphiphilic and readily disperse in water, whereas GR is very hydrophobic [115]. However, GO and rGO were described as GR derivatives rather than GR, even though GR has been frequently used to refer to rGO NS in several published publications. In this situation, GO and rGO sheets can be considered a distinct class of carbonaceous 2DM, exhibiting unique characteristics not found in other 2DM. Due to the abundance of functional groups on GO-NS, they can be easily changed via covalent interactions with organic or polymeric molecules [116]. Electrical conductivity of GO-NS can be readily adjusted from insulating to highly conductive by carefully controlling the quantity of O_2-containing functional groups on their surface. Functional groups on GO surfaces can operate as nucleation sites for other nanocrystals (noble metals, metal oxides, metal chalcogenides, or MOFs), making those suitable dispersible templates for solution-phase assembly of functional composites [57, 117].

4.4 Wet-Chemical Syntheses

WC-syntheses, which are bottom-up methods, are also viable options for producing 2DM in large quantities [118, 119]. All synthetic procedure that depends on chemical reactions of particular precursors in solution phase under proper experimental circumstances is referred to as WC-syntheses [120]. Due to its superior controllability, WC-syntheses were recognized as a part of efficient and repeatable techniques for the manufacture of 2DM with regulated thickness and size that are potentially available for industrial use. WC-syntheses were widely utilized to create a variety of non-layer structured 2DM that cannot be prepared by top-down techniques. Due to the ease with which the produced 2DMs may be disseminated in an aqueous or organic medium, they are well-suited for a variety of applications. WC-syntheses can be used utilized for the production of 2DM having layered structural characteristics [118]. Unlike the previous techniques, there are no general rules that govern any WC-synthesis process. One WC-synthesis technique might be as distinct from the other synthetic routes. This section will introduce these common WC methods for the production of 2D-sheet and will illustrate each protocol with representative examples.

4.4.1 *Hydro/Solvothermal Synthesis*

Aqua or an organic solvent is used as a reaction media in a vessel that is sealed at a temperature above the solvent's boiling point in the hydro/solvothermal synthesis process [121]. For a closed system reaction, a high-pressure solvent is auto-generated to drive reaction and enhance crystallinity of produced nanocrystals. Many 2DM, especially inorganic, were made using this method [119]. Using DMF as the reductant, still, another notable example is Li and coworkers [122] easy solvothermal synthesis of SL-Rh NSs having an edge length of 500–600 nm using PVP as the surfactant. Interestingly, this research is the first to demonstrate the successful synthesis of SL-metal NS. Along with metal NS, this technique has been used to manufacture a huge number of metal oxides chalcogenides with 2D-NS [123, 124]. For example, Dou and coworkers [124] exhibited a TiO_2, ZnO, Co_3O_4, WO_3, Fe_3O_4, and MnO_2 NSs with thicknesses ranging from 1.6 to 5.2 $_{nm}$, TiO_2 (~200 nm), ZnO (1–10 μm), Co_3O_4 (1–10 μm), WO_3 (1–10 μm), Fe_3O_4, and MnO_2 NSs with thicknesses ranging from 1.6 to 5.2 nm, (equivalent to 2–7 monolayers) were produced using this method. Xie and colleagues [123] produced $CeCO_3OH$ NSs utilizing the hydrophobic properties of oleate tails. Through heat treatment for 2 min in air at 400 °C, the $CeCO_3OH$ NS may be converted into CeO_2 NSs with a 3-atom-layer thickness (see Fig. 4.10). The MoS_2 NSs, as a typical member of TMDCs, may also be produced through a simple hydrothermal process. Xie and colleagues [125] employing thiourea in large quantity demonstrated the synthesis of an FL defect-rich MoS_2 NS. The identical group [126] also produced an O_2-combined MoS_2 NS

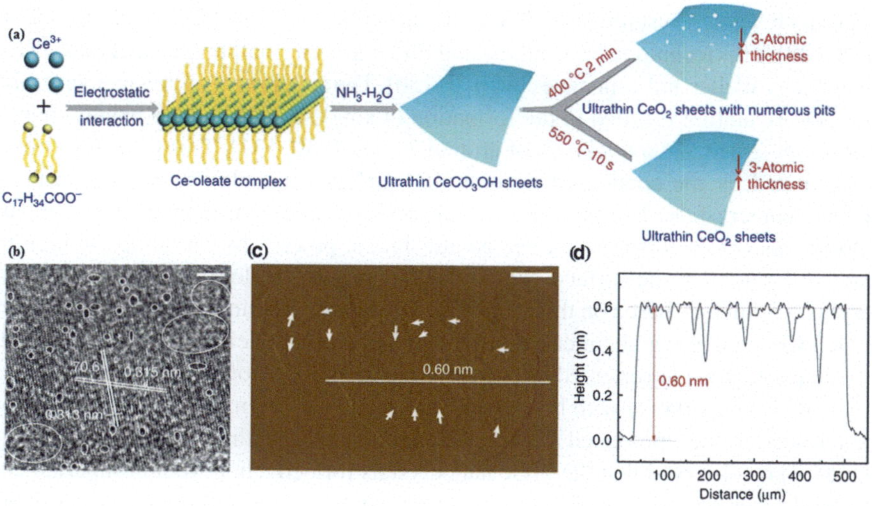

Fig. 4.10 **a** Schematically illustrated CeO_2 sheets containing various surface pits with intact CeO_2 sheets, **b** HRTEM view, **c** AFM outcome, and **d** height portray allied with white line highlighted in figure c. Reproduced with permission from [123]

in H_2O, using a related hydrothermal technique at 200 °C, When compared to the 2H-MoS_2 in pristine state, interlayer spacing of 6.15 Å, this resulted in an enhanced interlayer spacing of 9.5 Å. Later, Wei and colleagues [127] used a similar synthetic technique to create 2H-MoS_2 NSs. A second solvothermal treatment can partially convert the 2H-MoS_2 NS into the metallic 1T phase. At ambient temperature, due to the existence of metallic 1T phase, the MoS_2 NS with mixed 2H and 1T phases displayed high ferromagnetism. Hydro/solvothermal fusion is a simple and viable method for producing 2DM in high yield and at a low cost. Furthermore, understanding the growth mechanism for the hydro/solvothermal synthesis approach is difficult due to the fact that all reactions occur in an autoclave that is sealed, making it impossible to build experiments for different materials. Hydro/solvothermal process is highly dependent on experimental circumstances like precursor concentration, solvent systems, polymers or surfactants employed, and temperature makes it difficult to accurately regulate the structure between laboratories and batches. The majority of the 2D NSs produced using the hydro/solvothermal technique was ML rather than SL.

4.4.2 2D-Oriented Attachment

Another common WC-synthesis technique for 2DM is the 2D-oriented attachment [119]. The 2D attachment, in contrast to the usual technique, demonstrates a unique growth process and results in nanostructures with a well-defined shape. By having a crystallographic aspect in common, neighboring nanocrystals are bonded and fused to produce single-crystalline 2D NS, eliminating high-energy facets and interfaces [128, 129]. This technique for synthesizing PbS NS from small PbS nanocrystals was invented by Weller and colleagues in 2010 [130]. The attachment of PbS nanocrystals in a 2D-orientation resulted in the formation of single-crystalline NS having diameter of 0.8–2 μm and a thickness of around 2.2 nm [130, 131]. This growth process is facilitated by the condensed packing of $C_{18}H_{34}O_2$ ligands on the [98] surface of PbS nanocrystals. The presence of chlorine-containing substances that contain chlorine also helps to start the directed attachment process by revealing the highly reactive portions of the surface. Sun and colleagues [132] developed PbS NS with varying thicknesses based on this work by adjusting the reaction temperature and the related Cl_2-containing chemicals. Vanmaekelbergh and colleagues [133] described a two-step technique (oriented attachment and assembly) for fabricating 2D-PbSe sheets with a superlative with honeycomb structure, as shown in Fig. 4.11. To begin their research, they produced PbSe nanocrystals ended with [98, 107], and [108] planes. They observed that the PbSe nanocrystals formed a honeycomb superlattice in two dimensions with octahedral symmetry when they were connected via their [98] facets. A cation exchange process can convert the 2D-PbSe honeycomb structure to a 2D-CdSe superlattice. Currently, 2D attachment is regarded as the only method for producing 2D nanocrystals with unlayered structures. The 2D-oriented attachment approach may as well be used to fabricate various 2D NSs with layered

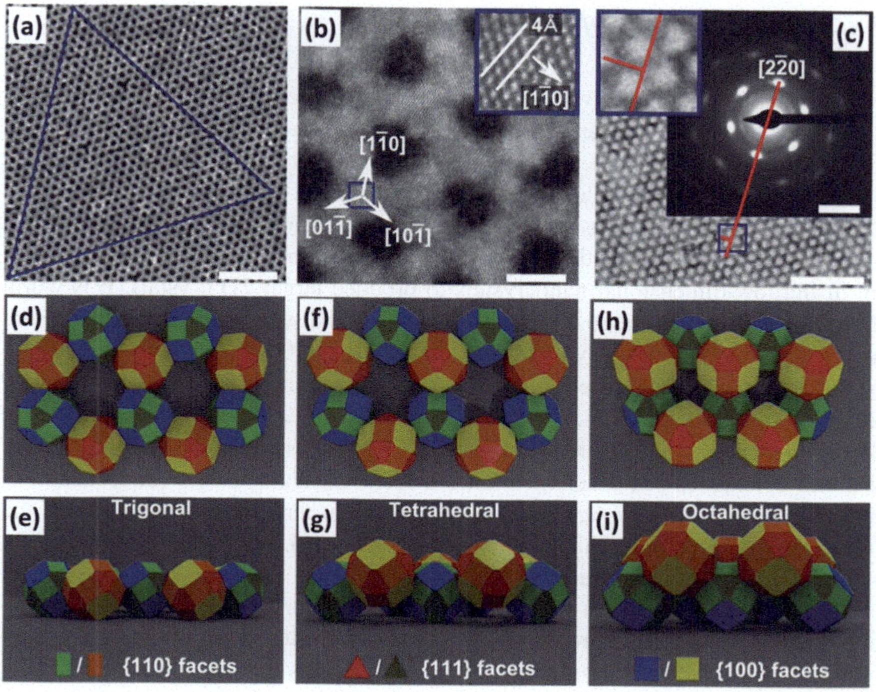

Fig. 4.11 Visulization of PbSe honeycomb crystallographic structure produced from 2D-oriented attachment: **a** HAADF-STEM owing to honeycomb structure with a triangle showing long-range order structure, **b** HR-HAADF-STEM shows ⟨111⟩ nanocrystal axes perpendicular to honeycomb plane, whereas three ⟨110⟩ axes perpendicular to nanocrystal bonds. Inset zoomed area indicating atomic-columns with blue box, **c** electron diffracted image showing improved crystallinity with inset SAED pattern, **d, e** top as well as side view owing to trigonal structure, **f, g** tetrahedral structure, **h, i** and octahedral structure. Reproduced with permission from [133]

structures. Jeong and colleagues [134] created a Bi_2Se_3 NS using an easy and rapid solution method. The directed attachment is part of the development technique of tiny nanocrystals after which epitaxial recrystallization occurs of the nanocrystals into NSs. This form of 2D-oriented attachment was pushed by the intrinsic inclination of a layered structure with negatively charged Se atoms. In their investigation, they discovered that amine-containing surfactants containing amine tend to enhance the lateral dimension and so form larger NS. Wang and coworkers [135] synthesized an SL-SnSe NSs with lateral dimensions of ~300 nm and an approximately 1 nm thickness using OLA as a solvent and surfactant.

4.4.3 Self-assembly of Nanocrystals

Nanocrystal self-assembly was already devised as an efficient method generating an ordered and steady nanoarchitecture with internal nanocrystals, where presynthesized nanocrystals instinctively arrange with one another by non-covalent interactions such as vdW interactions, electrostatic interactions, and others [136]. It was proven that assembling low-dimensional nanocrystals including NPs and nanowires is a viable approach for fabricating 2DM [119]. Nanocrystals are assembled primarily by fusing low-dimensional nanocrystals to produce larger 2D crystals. It is worth discussing separately due to its extensive research and widespread application with a variety of materials. Kotov and coworkers [137] in 2006, the self-assembly of a polycrystalline 2D-CdTe sheet with a thickness of 3.4 nm was reported for the very first time (diameter). The main principles of self-assembly were thought to be the dipole moment, small positive charge, and attraction is hydrophobically directed. Wu and colleagues self-assembled Cu NS from Cu nanoclusters (dodecanethiol-capped) [138, 139]. Although the initial Cu nanoclusters are non-luminescent, the formed Cu NS exhibits a high luminescent intensity. Later in the same year, the same group [140] discovered that the 2D-Au NS may be made by self-assembly of dodecanethiol-capped Au nanoclusters. Adjusting the Au nanocluster concentration and reaction temperature, the thickness of the produced Au NS may be varied between 1.5 and 100 nm. Similarly, Cheng and colleagues [141] self-assembled, at the liquid/air contact, a BL-Au sheet serves as an artificial membrane using polystyrene-capped Au NPs. Apart from NPs, nanowires were utilized as building blocks in the assembly approach to fabricate 2D NSs. Acharya and colleagues [142] synthesized 2D PbS NSs through PbS nanowire coalescence. To begin with, supercrystalline-assembled PbS nanowires have been synthesized using the capping ligand trioctylamine. The supercrystalline nanowires were then arranged into single-crystalline NS with a constant rectangular shape at the air/water interface using temperature and surface pressure. Yao and colleagues [143] used a similar approach to manufacture Eu_2O_3 NS by self-assembly of Eu_2O_3 nanowires. The Eu_2O_3 nanowire bundles were initially produced in their experiment as a capping and structure-directing agent, 1,5-pentanediol was used. The nanowire bundles self-assembled into porous structures similar to sheets after soaking in water, eventually forming a rectangular Eu_2O_3 NS with a thickness of 3.8 nm. By varying the time, the bundles are immersed in water, the lateral size of Eu_2O_3 sheet may be adjusted from a few hundred nanometers to tens of micrometers.

4.4.4 2D-Templated Synthesis

The term "templated synthesis" refers to an operative approach for growing nanostructures with anisotropy. It relates to the procedure of using presynthesized material or bulk substrates as templates for directing/containing the formation of certain nanostructures. Over the last several years, considerable progress has been achieved in the synthesis of 2D NSs using a 2D-template [144, 145]. For example, using the GO sheet as a template, Zhang and colleagues [146] developed an hcp Au square NS with a diameter of 200–500 nm and a thickness of ~2.4 nm (Fig. 4.12b–e) using such a WC-synthesis technique (Fig. 4.12a). Generally, the most stable phase of gold is the fcc structure may be easily produced by WC techniques. To be clear, this is the first time a solution-based approach has been employed to create a stable pure hcp Au structure in ambient circumstances. Phase transformation initiated by e-beam irradiation in a TEM, a subsequent growth step, exchange surface ligand, or coating with a thin metal layer converts hcp Au sheets to the fcc phase [147, 148]. Using a CuO nanoplate as a template, Wei and coworkers [149] showed the production of a half-unit-cell freestanding α-Fe_2O_3 sheet. After forming an iron hydroxide layered sheet on the exterior of a freestanding CuO nanoplate, iron hydroxide NS may be

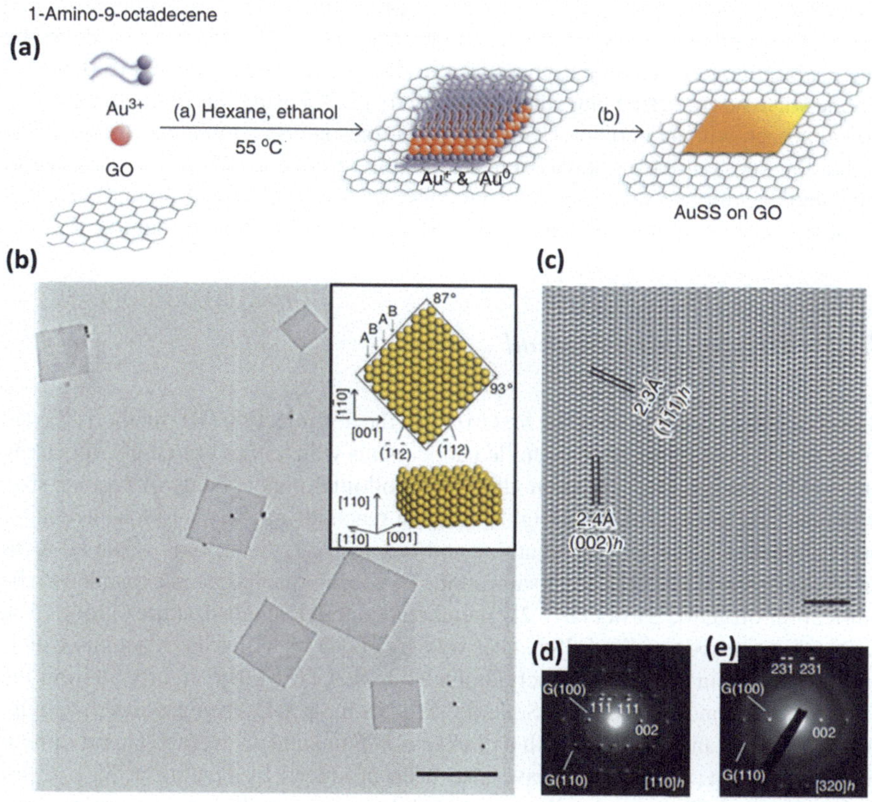

Fig. 4.12 **a** Schematic view presenting synthesis owing to hcp Au square type NSs over GO template, **b** TEM photograph of ~ 2.4 nm thick with black dots identifying NPs byproduct in spherical Au NPs of 10–25 nm, inset crystallographic images **c** HRTEM image for small region oriented toward $[107]_h$, **d** SAED pattern involving diffraction rings for GO with spots for $[107]_h$ zone-axis, **e** and SAED of $[320]_h$ zone-axis. Reproduced with permission from [146]

produced by gradually etching the CuO template. Following that, a simple thermal annealing procedure was used to convert the iron hydroxide NSs into a freestanding α-Fe_2O_3 NS. The resulting α-Fe_2O_3 NS is up to 1 μm in length and 0.55–0.59 nm in thickness. Li and coworkers produced NiO NSs with a diameter of up to a micrometer and a thickness of <2 nm through thermal annealing utilizing layered α-$Ni(OH)_2$ sheet as a template [150]. Prasad and his team investigated that metal thiolates occur in their natural state as lamellar assemblies [151]. Diluting Pd thiolates in toluene allowed for the isolation of the lamellar structure into a single NS. Additionally, Pd NS can be produced by heat treatment of a single NS of Pd thiolates deposited on a suitable substrate. Numerous 2D semiconductor sheets were also synthesized using the 2D template synthesis approach. Zhang and colleagues [152] synthesized ternary and quaternary chalcogenide containing Cu, nanoplates comprising $CuInS_2$, $CuIn_xGa_{1-x}S_2$, and Cu_2ZnSnS_4, through a process of cation exchange utilizing a prefabricated CuS nanoplate as a template. The nanoplates (ternary and quaternary) formed as a consequence of this procedure are identical in shape, thickness, and size to the template of CuS nanoplate. Cubic-phased $Cu_{2-x}Se$ NSs were produced by heating preprepared hexagonal-phased CuSe NS in the existence of CuI [153]. h-CuS NS may also be converted into cubic-phased $Cu_{1.97}S$ NS using this phase transformation approach with a template. Using a lamellar mesophase template technique, NSs with a thickness of ~1 nm have been successfully produced [154]. The presence of the lamellar-template structure was established by low-angle XRD characterizations of the mixture containing precursors of $Pb(OAc)_2{\cdot}3H_2O$ and n-octylamine.

4.4.5 Hot-Injection Method

Bawendi and colleagues were the first to create the hot-injection (HI) method [155] in order to make cadmium chalcogenide nanocrystals which is an extremely appealing process for the synthesis of monodispersed colloidal nanocrystals of regular size, structure, and purity. Injecting highly reactive reactants quickly into a heated solution containing long-chain oleyl amine and/or $C_{18}H_{34}O_2$ surfactants was a HI technique driven by high initial supersaturation. This technique has been expanded to the fabrication of 2DM, particularly 2D metal chalcogenide nanostructures [156, 157]. Using this method, an SL-CdSe sheet was created by heating the reactants $CdCl_2$ and Se powder in a mixture of octylamine and OLA [158]. By slightly altering the experimental conditions, 2D-CdSe and CdTe NS have also been produced. Single-crystalline GeS and GeSe NS with a thickness of 5 nm and an average lateral dimension of 2–4 μm × 0.5–1 μm have also been produced by heating at 320 °C for 24 h [159, 160]. Zhang and colleagues [161] demonstrated the generalization of the preparation of metal sulfide nanocrystals in 2012 using the HI technique. Specifically, a gram quantity of CuS NS with a consistent hexagonal form was produced. The resulting CuS NS has a thickness of ~3.2 nm (two unit cells) and a lateral dimension of ~453 nm, respectively. Yan and colleagues [162] produced a variety of 2D trivalent rare-earth-Se (RE-Se) sheets, namely $RESe_2$ (RE = La, Nd) and $RE_4O_4Se_3$

(RE = Nd, Sm, Gd, and Ho), which all include Se with planar Se layers with a single valence. Take note that metal chalcogenide's each NSs described above is non-layer structured material. This technique may be used to manufacture 2D NSs of ML-metal chalcogenides, such as TMDCs. Cheon and coworkers [163] demonstrated the use of the HI technique to synthesize SL-$MoSe_2$ and WSe_2 NS having lateral dimensions of 200–400 nm. They observed that capping ligands seemed critical for controlling the number of layers in NS throughout their studies. By substituting $C_{18}H_{34}O_2$ for oleyl alcohol and OLA as capping ligands, the thickness of the NS might be adjusted from SL to 2–3 and 4–8 layers. Further, the same group synthesized a variety of TMs such as sulfide and selenide NS in groups IV and V, including TiS_2, ZrS_2, HfS_2, VS_2, NbS_2, and TaS_2, as well as $TiSe_2$, $ZrSe_3$, $HfSe_3$, VSe_2, $NbSe_2$, and $TaSe_2$, CS_2 and Se powders, were used as injection chalcogen precursors [164–166]. Ozin and coworkers [167] achieved the production of 2D-WS_2 NSs by changing the injected precursors a controlled crystal phase which can be achieved. SL-metallic WS_2 NSs with a deformed octahedral (1T′) phase were generated by injecting the CS_2 and WCl_6 precursors into a heated OLA solution (320 °C). The synthesized NS has a circular morphology with a diameter of about 100 nm (Fig. 4.13). By contrast, a small quantity of HMDS has been added to the reaction solution to produce a flower-like WS_2 NS with a typical hexagonal phase. This approach may also be used to produce ternary layered nanostructures in contrast to binary NSs. Gupta and colleagues [168] synthesized a $CuSbS_2$ NS with a thickness that may be varied from one to many layers. The HI approach enables the production of colloidal 2D metal chalcogenide sheets with excellent disparity, pureness, shape, and size uniformity. But, there are still some drawbacks to this approach. First of all, the HI procedure is often carried out with oleyl amine having long chains and/or $C_{18}H_{34}O_2$ surfactants that are difficult to eliminate and unsuitable for uses including electronics, energy storage, and catalysis. Second, due to the system's use of a high boiling point solvent, the reaction temperatures for the hot-inject technique are frequently quite high (300–350 °C). Third, due to the precursor injection's limits, this technique is

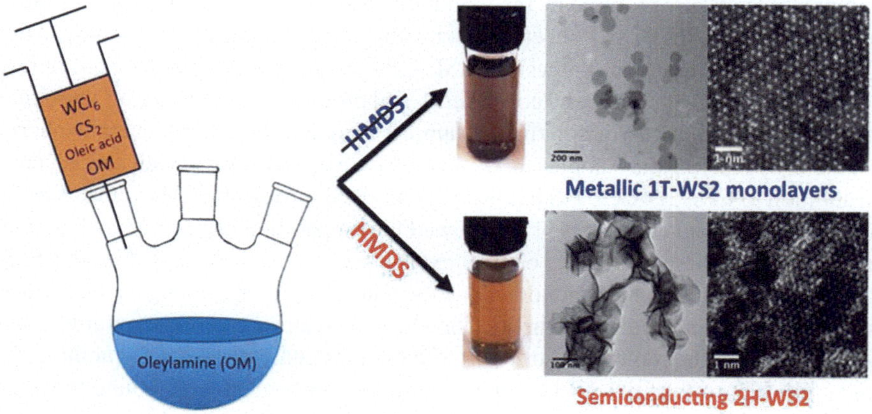

Fig. 4.13 Schematic view giving synthesis for 2H and 1T′ WS_2 NSs followed by HI route. Reproduced with permission from [167]

often unsuitable for mass manufacturing. By contrast, large-scale manufacturing may be accomplished using the one-pot heating technique that involves mixing all of the reactants at ambient temperature into a solvent having a high boiling point (OLA, $C_{18}H_{34}O_2$, or octadecene) and then heating the mixture to the desired temperature [169, 170].

4.4.6 Interface-Mediated Synthesis

Another frequent WC-synthesis strategy for 2DM, notably metal coordination polymers (CPs) and polymers, is interface-mediated (IM) synthesis [171, 172]. Schlüter and colleagues [173] demonstrated the first time in 2011 that an SL-CP NS could be produced near the water/air interaction. The fundamental concept is to restrict the ligand to the interface of water/air prior to producing CP NS by reacting with water dissolved metal salt. A common procedure involved dissolving a tpy-based organic ligand in chloroform and dropping the solution on top of a metal saltwater solution. Due to the chloroform's rapid evaporation and the ligand's limited solubility in water, the organic ligand has been restricted in 2D design at interface of water/air. Thus, the ligand present at interface interacted with water dissolved metal ions to form dense SL-CP NS, namely $[M_2^{2+}(tpy)^{4+}]_x$ $(0 < x < 3$; M = Fe, Ru, and Zn), at the water/air contact may be readily transported to different substrates. It is necessary to better describe the crystal structure of 2D-CP NS. Subsequently, later on, this approach was expanded to the production of a variety of 2D-CP sheets by the interaction of a terpyridine-based ligand with an oligo-functionalized with various TM salts (Fe^{2+}, $Co^{2+/3+}$, Zn^{2+}, and Pb^{2+}) at the interface of water/air. Furthermore, it was revealed that the as-synthesized 2D-CP sheets may be used as a basic component for fabricating ML assemblies on substrates of solid through the layer-by-layer deposition approach. Using the inter-IM approach, Nishihara and coworkers [174] synthesized an SL-Ni-bis(dithiolene) NS at the interface aqua/air. In their case, instead of the compressed pressure required by the Langmuir–Blodgett process, they used ambient pressure to create the SL-CP sheet. Further, Feng and colleagues [175] employed the IM synthesis technique to synthesize an SL-CP sheet of Ni-bis(dithiolene) complex. On the other hand, nishihara and coworkers [176] produced, at the water/air interface, 2D-CP NSs of bis(dipyrrinato)-zinc(II) laminar complex (Fig. 4.14), at atmospheric pressure, by reacting the complex three-way dipyrrin ligand directly with zinc acetate. In addition to CPs, 2D-COF NS has been synthesized through the IM synthesis technique. Schlüter and coworkers [177] showed the preparation of COF NS at aqua/air interface for the first time, based on their extensive expertise with the IM synthesis of 2D-CP NS. An amphiphilic monomer that is capable of pre-organizing on the surface of water by intermolecular interactions had been developed and manufactured. This astute design of the molecular structure of monomer guarantees that the anthracene units pile face-to-face. Under UV light irradiation, monomers were bound together through a [4+4]-cycloaddition process to create the SL-COF NS. Later, King and colleagues [177] utilized a similar approach to fabricate another SL-COF NSs, UV light irradiation of a monomer comprised of three anthracene blades and diethylene

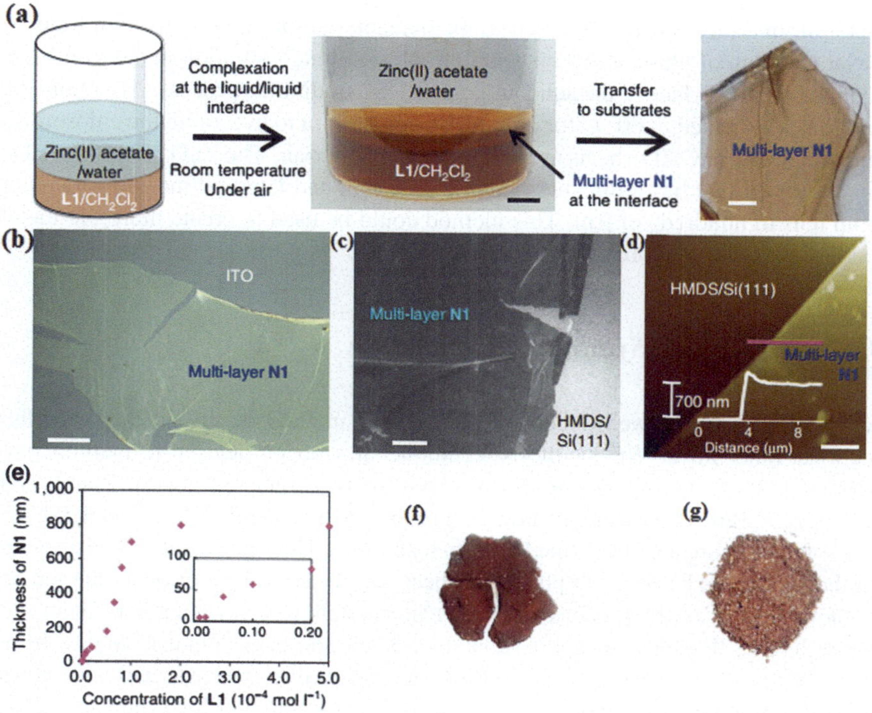

Fig. 4.14 **a** Schematic view for liquid/liquid interfacial fabrication with ML-CP NS from bis (dipyrrinato)zinc(II) laminar complex diverted toward ITO substrate, **b** optically collected image over ITO substrate showing scale bar = 50 mm, **c** FE-SEM pattern over HMDS/Si(111) with 20 mm, **d** AFM pattern over HMDS/ Si(111) involving cross-sectional analysis with magenta line-scale bar = 5 mm, **e** control of the thickness based on the concentration of **L1** in the liquid/liquid interfacial synthesis. The inset shows a close-up of the low concentration region. Reaction time, 4 days. Temperature, room temperature. Container, cylindrical glass vial with a diameter of 3.2 cm. Volume of the upper (aqueous) layer, 20 ml. Volume of the lower (dichloromethane) layer, 10 ml. Concentration of zinc(II) acetate in the upper layer, 0.05 mol l−1, **f, g** photographs of the products of single-phase reactions in dichloromethane at room temperature and in N,N-dimethylformamide at 105 °C, respectively. Reproduced under the terms of CC BY 4.0 license [176], https://creativecommons.org/licenses/by/4.0/

glycol chain functionalized with a trypticene core at the water/air contact [177]. The resulting COF 2D-sheet can have a lateral dimension of up to 1 mm^2. Not restricted to the water/air interface, Bao and colleagues [178] described the formation of a COF NS based on an imine at the DMF/air interface. But contrasting water/air interface combination, reaction solution was DMF, contained residual monomers from the COF powder synthesis, and had been stored undisturbed in a water-saturated environment. As a result, after a specific time period, a thin COF NS has been formed at the DMF/air interface. By adjusting the time of incubation, the COF NS's thickness may be simply modified from 1.8 to 29 nm. Additional to CPs and polymers, 2D inorganic sheet may be synthesized through the IM synthesis method. Wang and colleagues [179] described the synthesis of ZnO NSs in freestanding state at the water/air interface. $Zn(NO_3)_2$ and hexamethylenetetramine are dissolved in water in a vial made up of glass, and then a sodium oleyl sulfate's solution in chloroform is dropped on the surface of the aqueous solution in a traditional process. After that, the solution was heated to 60 °C for around 1 h and 40 min to evaporate the chloroform and obtain the ZnO 2D-sheets at the interface of water/air. The ZnO NS produced is single crystal that has a thickness ranging 1–2 nm, and lateral dimensions ranging from tens to hundreds of μm. This method could be used to create more 2D-metal oxide sheets in the future at the water/air interface.

4.4.7 Other WC-Synthesis Methods

Apart from the now well-documented synthesis method described in the earlier sections, other novel WC-synthesis techniques have been devised to manufacture particular 2DM that do not conform to any of the other categories. Zheng and colleagues [180] demonstrated how to produce a freestanding hexagonal Pd NS in a solution using a CO-confined growth technique. The significant CO adsorption on the Pd sheet's basal (111) planes has been suggested to be a critical component in the restricted development of 2D Pd nanocrystals to synthesize nanosheet. Pd sheets having thickness of approximately 1.8 nm and edge lengths ranging from 20 to 160 nm have been produced [180]. In another example, Xie and colleagues [181] described a facile and environmentally friendly method for synthesizing SL β-$Co(OH)_2$ NSs using $CoCl_2$ and C_2H_7NO ambient air conditions aqueous solution. Yan and colleagues [182] proposed a colloidal synthesis in a single pot technique for the production of various RE-oxide NS. Gd_2O_3, Ho_2O_3, and Y_2O_3 2D NS with size range of more than 100 nm and thickness less than 1 nm have been produced. The synthesis of synthetic 2D polymer single crystals in layered form utilizing the single-crystal-to-single-crystal photopolymerization (PTP) process and succeeding exfoliation into 2D polymer sheet was proven by two distinct groups [183, 184]. To begin the production of the hydrocarbon 2D polymer sheet containing fluorine, King and colleagues [183] produced a single-crystalline monomer based on triptycene and composed of three photoreactive tetrafluoroanthraceno "blades" linked by a [2.2.2] bicyclic core. The monomer is single crystalline that has been PTP through [4+4] cycloaddition by first irradiating it with 460 nm light and then 400 nm light. After PTP, it was possible to create a bulk crystal of polymer with a structure having layers similar to graphite. The resulting bulk crystal polymer was subsequently exfoliated into SL or FL-NSs by several days of rotation in NMP at 50 °C. PTP can be caused by UV irradiating a polyhedral monomer single crystal at 465 nm, leading to an ML polymer bulk crystal. Using a technique comparable to that used for the above-mentioned 2D polymer NSs, this bulk crystal polymer may be readily flaking into single or FL-NSs in NMP or perfluorinated carboxylic acids. UV irradiation and heating were used to reverse the transition between the monomer and the bulk crystal of the polymer. Yang and coworkers [185] were the first to report on the solution technique for the manufacture of square-shaped atomically thin single-crystalline 2D NSs of organic–inorganic hybrid perovskites.

4.5 Chemical Vapor Deposition

Chemical vapor deposition is a bottom-up strategy that offers numerous benefits, including precursors that are relatively simple and diverse, as well as a rapid rate of growth, making it a viable method for the applied synthesis of 2D atomic layers on a variety of substrates. Numerous publications have proven that CVD methods may produce high-quality 2DM and HS with scalable sizes, controlled thickness, and good physical and chemical characteristics, opening the path for the full promise of 2DMs to be realized (Fig. 4.15) [186]. Growth of CVD is utilized to produce low-defect density, larger-area 2DMs with the same quality as exfoliated materials and fine control over layer numbers, domain size, and shape as exfoliated materials. Precursors' excitation to the necessary phase having vapors, transit, in a CVD process, reactive gaseous species, as well as heterogeneous surface reactivity on substrates, all contribute to the formation of solid-phase materials [187]. Specific CVD methods, including small-pressure CVD [188] and plasma-enhanced CVD (PECVD) [189], are used to increase the homogeneity and decrease the growth temperature of 2DMs, respectively.

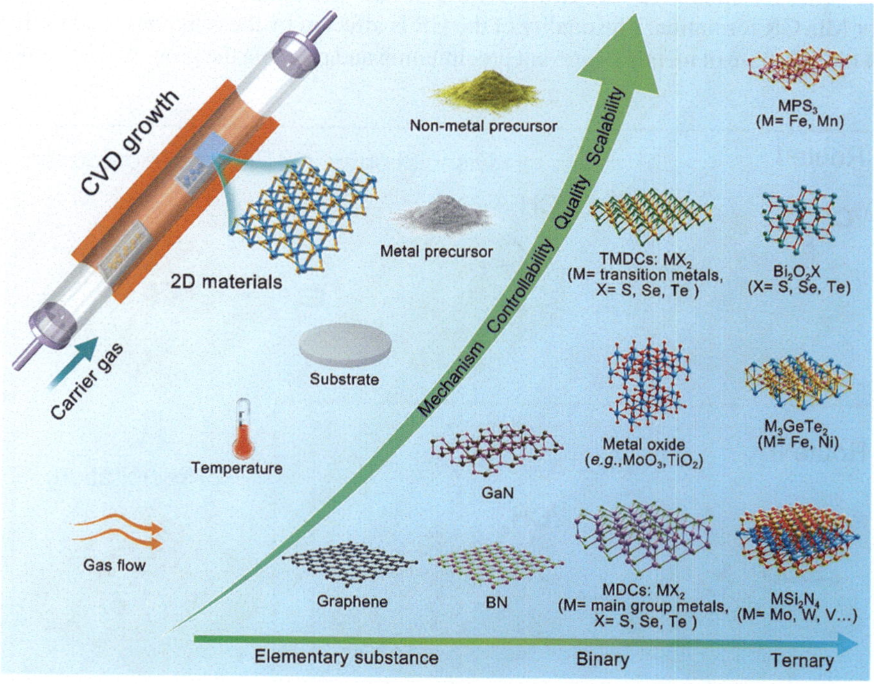

Fig. 4.15 Schematic diagram with basic parameters of CVD growth for 2DMs enveloping elementary substances comprising binary as well as ternary toward complex species. Reproduced with permission from [186]

4.5.1 Graphene and Hexagonal Boron Nitride

A classic method for CVD growth of GR or h-BN is to deposit these materials on the surfaces using specified precursors supported by or blended with unreactive gases including Ar or H. The substrate of metal, precursor, temperature growth, and pressure are all factors that influence the process of growth.

4.5.1.1 Effect of Substrate

The growth of GR and h-BN is highly dependent on the substrate. TMs have been initially studied as substrates for catalysis. Numerous TMs, including Cu, Ni, Pt, and Fe, were shown to be appropriate growth substrates [190, 191]. The growth mechanism of GR is dependent on the metal's C solubility. As seen in Fig. 4.16, under normal growing conditions, C-atoms incline to form SL-GR directly on the surface of poor C solubility metals such as Cu [190, 192]. Further, C-atoms can dissolve into metals with high C solubility, such as Ni, and subsequently precipitate on the surface of the substrate, forming a layer of graphite. Metals with a low C solubility are better for monolayer GR growth, while metals with a high C solubility are better for ML-GR formation. The quality of the GR is affected by the roughness and purity of the substrate of metal. To prevent precipitation and promote the growth of ML-GR,

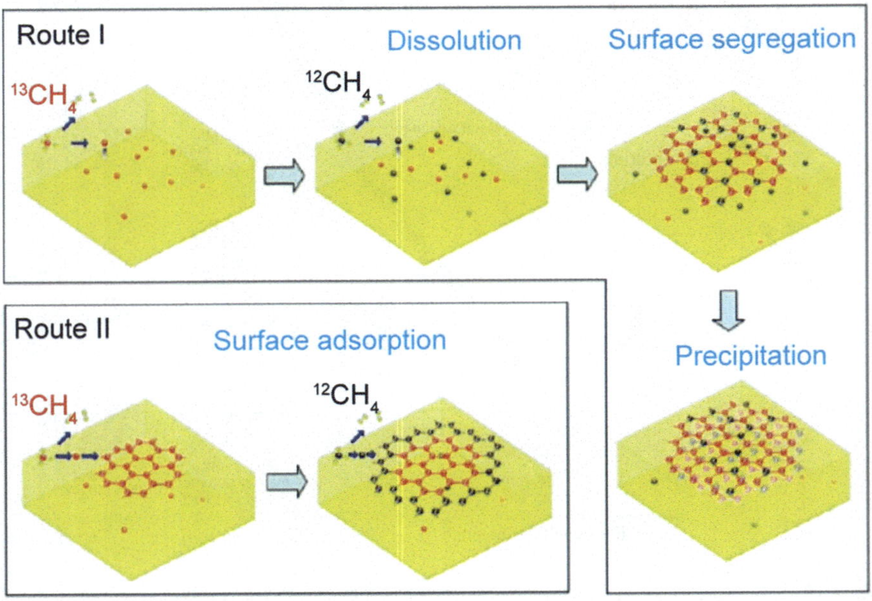

Fig. 4.16 Growth mechanism of graphene by CVD on metal substrates, schematic diagram of the two representative growth routes of graphene on metal catalysts with high and low carbon solubility. Reproduced with permission from [192]

alloy substrates with a high C solubility are utilized in addition to single elemental metals. Furthermore, the crystallinity of the substrate of metal has a major impact on the standard of quality of GR. Cu (111) has a greater than 1%; in comparison with GR, there is a lattice mismatch. While the growth process remains epitaxial, on different crystal facets, the growth behavior differs. For instance, GR sites of nucleation on Cu (100) have an arbitrary orientation [193]. For Cu (111), nucleation of GR islands occurs with the same crystal orientation and eventually combines to form a continuous film. CVD h-BN has a comparable growth process to GR, metal substrates along with Cu and Pt with low N-solubility promote, and as a result, they are good for monolayers with surface-mediated growth. h-BN growth at elevated temperatures under low pressure [194]. Conversely, a substrate including Fe foil that is very soluble in B and N generally acts as a catalyst for precipitation development, resulting in ML h-BN. Kim et al. [195] established the feasibility of growing large-area ML h-BN films with a thickness of 5–15 nm utilizing Fe foil as the substrate and borazine as the precursor, respectively. When adopting this precipitation growth procedure, it has been demonstrated that controlling the cooling rate is crucial for the thickness of as-grown h-BN films [195, 196].

4.5.1.2 Effects of Precursor and Pressure

Methane, the most commonly utilized fuel for GR growth due to its relatively high pyrolysis temperature enables the creation of higher grade material (at higher temperatures). Yet, the CVD approach's economic efficiency is constrained by a high growing temperature (1000 °C). Thus, other C sources such as benzene and toluene have been used to prepare GR at low temperatures (300–400 °C) [197, 198]. In the context of h-BN, the quality of the compound is largely dependent on precursor and rate at which it flows. Due to the 1:1 stoichiometry of B and N in solid BH_3NH_3 and liquid $B_3H_6N_3$, these are the most often utilized precursors. The CVD and precursor system for producing SL h-BN are shown schematically in Fig. 4.17a. At room temperature, BH_3NH_3 is a solid that has a vapor pressure very low. To convert BH_3NH_3 into hydrogen, $B_3H_6N_3$, and BH_2NH_2, a pyrolysis process is required [199]. As a liquid precursor, $B_3H_6N_3$ can be introduced into the growth chamber through a bubbler with carrier gas; however, it is a combustible liquid that reacts readily with water, necessitating additional handling efforts. Furthermore, the growing pressure has a major influence on h-BN's thickness because it influences the reactant concentration and rate of mass transport. Song et al. [200] demonstrated that even at atmospheric pressure, h-BN films with large area with a thickness of 2–5 layers may be produced. Moreover, Jang et al. [201] demonstrated that by adjusting the precursor's partial pressure and growth period, h-BN films of varied thicknesses may be produced. Usually, a low $B_3H_6N_3$ partial pressure results in monolayer h-BN, high partial pressure of $B_3H_6N_3$ results in ML h-BN with a thickness ranging to 100 nm. The h-BN growth on Cu at a wafer scale with thicknesses ranging from 1 to 20 layers, high structural, and electrical homogeneity has been described utilizing a

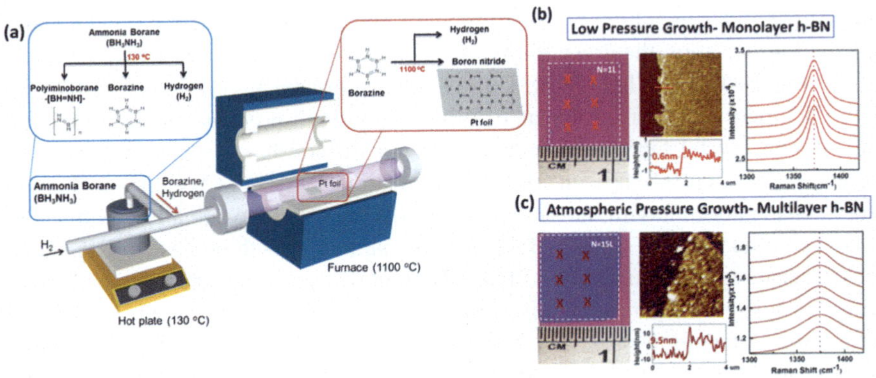

Fig. 4.17 Schematic view for **a** CVD system with h-BN growth involving ammonia borane precursor. Reproduced with permission from [203], **b** photograph of h-BN film SL over 285 nm SiO_2/Si substrate, whereas AFM and Raman spectra photographs, **c** photograph of ML h-BN film over 285 nmsubstrate of SiO_2/Si, whereas consistent AFM images and Raman spectrum. Reproduced with permission from [202]

hybrid low pressure and atmospheric pressure CVD method, as seen in Fig. 4.17b, c [202].

4.5.1.3 Wafer-Scale Growth

Wafer-scale growth is crucial for upcoming 2DMs. Cai et al. [204] revealed an industrial Cu foil to Cu (111) foil roll-to-roll technique for producing meter-scale single-crystal GR. For large h-BN area growth, polycrystalline Cu foil is favored technique. They demonstrated ML h-BN growth on polycrystalline Cu foil with diameters higher than 7-inch with low pressure, roll the Cu foil substrate into a cylindrical shape. A massive ML h-BN film with a thickness of roughly 25-inch has been generated by shaping a Cu foil substrate into a mainspring shape and supporting it with a multiprong quartz fork. A dense grain boundary (GB) defect population is seen in polycrystalline h-BN produced on polycrystalline Cu foils. To achieve high effectiveness 2D electrical and optoelectronic devices, the CVD synthesis of high-area and single-crystal h-BN is critical. To increase the size of domain of h-BN grains, single-crystalline substrates can be used to promote epitaxial CVD development of ML h-BN by lowering the GB density or even smoothly linking multiple domains [205]. The large-area epitaxial growth of monolayer h-BN on a Cu (111) thin film placed on c-plane sapphire resulted in highly aligned triangular h-BN grains that finally bonded together to form a continuous coating that covered the entire Cu surface. A 2-inch sapphire substrate has also been used to illustrate the epitaxial CVD development of ML h-BN films. The tiny thermal expansion coefficient difference between sapphire and h-BN leads to wrinkle-free growth [206].

4.5.2 *Transition Metal Dichalcogenides*

Numerous methods for large-scale CVD development of different TMDCs have been proven during the last decade [207–210]. Precursor kinds and concentrations, carrier gas flow rate, seeding and growth stimulants, and substrate nanoscale morphologies all have a substantial impact on the morphology, domain size, homogeneity, and electrical performance of the emerging TMDCs. Various crucial parameters are discussed in this section [187].

4.5.2.1 Effects of Precursor and Seed

Liu et al. [211] used an $(NH_4)_2MoS_4)$ solution to produce high-area MoS_2 films on SiO_2/Si and sapphire substrates in 2012. The thermolysis of $(NH_4)_2MoS_4$ was triggered using a high-temperature (500–1000 °C) annealing method yielding BL and trilayer MoS_2 sheets. During the conversion process, oxygen has a major effect on as-grown MoS_2 quality. More crystalline quality (stronger Raman and PL intensities) was observed in the MoS_2 sheet compared to SiO_2/Si sheet produced on the chemically inert sapphire substrate. According to the concerned scientists, oxygen from SiO_2 might impact the thermal decomposition of $(NH_4)_2MoS_4$ during the high-temperature procedure, therefore degrading the quality of as-grown MoS_2. Furthermore, it was found that in the presence of sulfur vapor, the following annealing, the as-grown MoS_2 films' electron mobility was enhanced by magnitude of two orders (from 0.02 to 4.7 $cm^2 \cdot V^{-1} \cdot s^{-1}$). Sulfur treatment of exfoliated samples resulted in a comparable increase in MoS_2 electron mobility [212]. Another two-step strategy for large-area TMDC formation, in addition to the solution-based precursor, is to deposit a solid metal's thin film or oxides of metals on the target substrates, followed by a sulfurization/selenization/tellurization procedure to create the necessary TMDC thin films. According to Zhou et al. [209, 213], the phases of as-grown $MoTe_2$ are dependent on the chemical conformation of the Mo sources. When Mo was employed as the source, tellurization of the MoO_3 precursor caused a semiconducting 2H-$MoTe_2$ with field-effect hole mobility and p-type conduction of 1 $cm^2 \cdot V^{-1} \cdot s^{-1}$, but when Mo has been utilized as the source, a metallic 1T' phase was produced. While the two-step technique described above is simple and accessible, the MoS_2 thin films produced using this method typically have poor mobility of electron (0.004–0.8 $cm^2 \cdot V^{-1} \cdot s^{-1}$) [214]. The inability to control the thicknesses and maintain uniformity across a broad region further restricts the quality of this synthesis process. However, vaporous-phase reactions have been frequently utilized to create extremely crystalline TMDC single crystals and regulate the forms or edges of amorphous metal oxide powders (MoO_3 or WO_3), chalcogen elements (S or Se), and sources of metal–organic and organic compounds [$Mo(CO)_6$ and $C_4H_{10}S$]. Using MoS_2 as an example, it is suggested that the reaction may be accomplished in two straightforward steps (Eqs. 4.6 and 4.7) [215].

$$MoO_3 + \frac{x}{2S} \rightarrow MoO_{3-x} + \frac{x}{2SO_x} \quad (4.6)$$

$$MoO_{3-x} + \frac{7-x}{2S} \rightarrow \frac{3-x}{2MoS_2} \quad (4.7)$$

Variations in the concentrations of local precursors along the substrate's growth have a significant effect on the morphologies and terminations of edge of as-grown TMDC grains. The rates of growth of Mo-zigzag and S-zigzag terminations (Mo edge's rate is three times that of the S) are said to favor the formation of 3-point-star MoS_2 at low temperatures or in S-rich environments. To avoid this, the 3-point-star domain's edges should be terminated with S-zigzags. In contrast, Mo-rich or heated environments produce truncated triangles with Mo- and S-zigzag edges terminations. With the use of a large carrier gas flow rate in a non-equilibrium growth state, dendritic-like grains can be detected, indicating a more efficient mass transfer system and a faster rate of expansion $MoCl_5$ was also employed as a precursor for the large-area growth of ML and FL-MoS_2 films [216, 217]. The chemical interaction between $MoCl_5$ and S was formerly thought to be a one-step process without the formation of intermediate byproducts including such MoO_2 or $MoOS_2$. As-grown MoS_2 films demonstrated excellent continuity and homogeneity. Furthermore, electron mobility is low as 0.003 to 0.03 $cm^2 \cdot V^{-1} \cdot s^{-1}$ [216].

4.5.2.2 Substrate Engineering

Substrates are critical in the development of TMDCs. Epitaxy of TMDC atomic layers is largely influenced by nanoscale form, termination of atoms, lattice constant, and symmetry of growth substrates. The morphologies of MoS_2 flakes have been shown to be significantly dependent on the substrate's lattice symmetry. MoS_2 single crystals having sharp edges are often found on SiO_2/Si or Si_3N_4 substrates [218]; however, a quartic symmetry $SrTiO_3$ substrate can generate dendritic MoS_2 flakes with three-fold symmetry [219] due to anisotropic diffusion of precursors on substrates. Sixfold symmetric MoS_2 crystals were also produced on Si-carbide (0001) substrates with a naturally sixfold symmetric lattice [220]. To achieve high-quality vdW epitaxy of TMDCs, the surface must be atomically flat, chemically inert, and hexagonal in plan. Various substrates have been utilized for huge-area vdW epitaxy of TMDCs, including mica [221], sapphire [222], h-BN [223], and GR [217]. Ji et al. [221] demonstrated the ability to grow cm-scale uniformly aligned MoS_2 ML flakes with comparable domain sizes on mica. According to data, a greater average domain size correlates with a lower density of surface nucleation, implying that the size of MoS_2 flakes is determined by the nucleation density, not the spreading capacity of MoS species on mica. Dumcenco et al. [224] demonstrated that up to the single crystals 91.5% of MoS_2, the c-plane sapphire domain is well-aligned, with 0° and 60° edge orientations prevailing greatly decreasing GB density in film. During early development, because of the poor vdW contact between MoS_2 and sapphire, the MoS_2

nuclei can be rotated and translated relative to the sapphire surface. The mobility of field-effect of the as-grown MoS_2 film was 25 $cm^2 \cdot V^{-1} \cdot s^{-1}$ and not dependent on channel length (4–80 μm). Apart from stiff protecting substrates, Shi et al. [225] demonstrated that ML-MoS_2 growth is dependent on the Au foils crystallography. Single-crystal MoS_2 grains with a large domain (115 μm) are preferred to form on Au (100) and Au (110) facets at high temperatures (111). The substrate facets and MoS_2 binding energies impact this growth behavior. The substrate-facet effect and vdW epitaxy displayed excellent domain size and orientation control, paving the door for large-domain, low grain-boundary density TMDC films to be developed on a massive scale. TMDC crystal orientations and GB resolution have also been determined using second-order nonlinear optical interactions [226]. Direct CVD development of 2DMs on flexible polymer substrates is necessary for scalable production of 2D flexible devices. No polymer substrate, on the other hand, would be thermally stable at normal CVD growth temperatures (600–900 °C). Recent advances [189, 227] have been made in overcoming this obstacle through the use of PECVD or by prudently picking a precursor with a low temperature of decomposition. PECVD reduces the reaction energy barrier by introducing accelerated energetic electrons, excited molecules and atoms, and other highly active species, lowering the temperature required to generate the target materials. Ahn et al. [189] show that utilizing a Mo-deposition/sulfurization PECVD method, large-area ML-MoS_2 (3–4 nm in thickness) thin films can be produced on substrates of polyimide at temperatures as low as 150–300 °C (mechanism shown in Fig. 4.18). This is owing to the greater

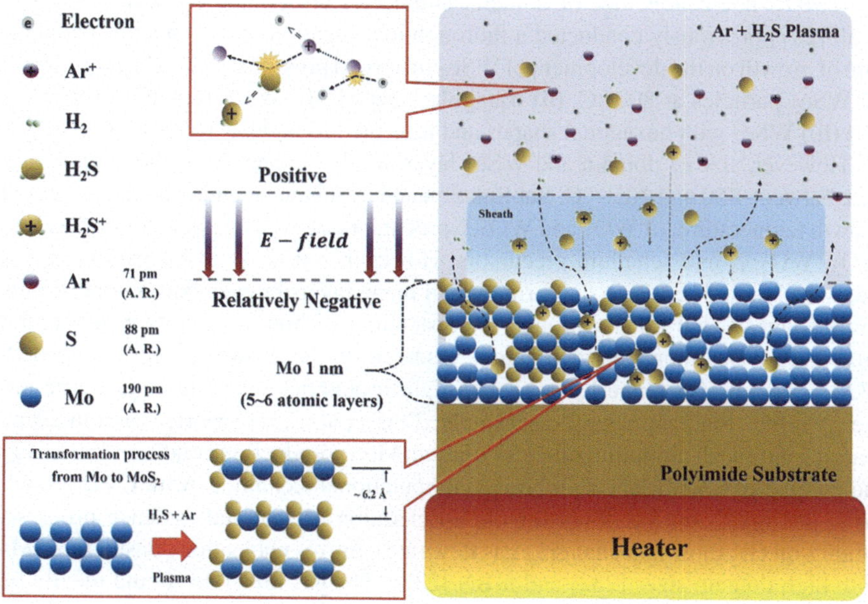

Fig. 4.18 Synthesis mechanism for investigating MoS_2 thin-layer film via PECVD. Reproduced with permission from [189]

crystalline quality and lack of transfer processes in MoS_2 thin films created using traditional CVD methods with a two-step approach. MoS_2 thin films produced using conventional CVD methods and following a similar two-step strategy had average mobilities of 2–3.71 $cm^2\ V^{-1}\ s^{-1}$. Single crystals of monolayer MoS_2 (>100 μm) have also been produced on polyimide substrates at 450 °C (small temperature) [227]. Along with their impacts on form and morphology, recent studies show that underlying substrates have the potential to cause unintended strain and doping into as-grown TMDCs. Additionally, theoretical research indicates that the energy required to generate sulfur vacancies may be highly dependent on the type (tensile or compressive) and of the strain's strength applied on the TMDCs [228, 229]. For instance, tensile strain applied to MoS_2 lattice, sulfur vacancies become less energetically advantageous. This discovery covers the way for the fabrication of high-performance TMDCs with a low-defect density. Furthermore, strain-engineered growth has been used to modulate the band structure of and brighten dark excitons in WSe_2 [230].

4.5.2.3 Effect of Temperature and Gas

Apart from the nonmetal precursor, and engineering of substrate described earlier, there are a few additional factors that affect the CVD development of 2D compound materials, like temperature of growth and flow of gas. These effects are discussed here to provide favorable circumstances for the development of 2D compound material. During thermodynamically regulated development, growth temperature influences the TMDC nucleation, size of domain, number of layers, and crystallization. Liu et al. [207] previously conducted a thorough investigation of the effect of temperature of growth on the development of WSe_2 in normal pressure CVD. They developed **(i)** WSe_2 particles at 800 °C, **(ii)** triangular-shaped ML-WSe_2 domains at 950 °C, and **(iii)** WSe_2 with hexagonal shapes and truncated triangle at 1050 °C (Fig. 4.19a, b). However, size of domain and WSe_2 layer number enhanced as the temperature rose between 850 and 1050 °C. Sublimation and diffusion are aided by the increased growth temperature, of WO_3 and WO_{3-x} precursors, as well as interactions between WO_3, WO_{3-x}, and selenium precursors. However, a process of nucleation that is excessively rapid results in the formation of many layers or even bulk samples. This study suggests that for a thorough understanding of how temperature affects the growth of WSe_2 more thermodynamic research on the growth process is needed. We can also use machine learning and high-throughput calculations to figure out the right reaction parameter. For example, Tang et al. [231] reported on a machine learning approach for optimizing CVD-grown MoS_2 production conditions, reducing the number of CVD trials. Combining computational techniques with CVD growth processes opens up the doors for the acceleration of materials science progress. Studies on GR show that changing gas flow rate controls nucleation density in 2DMs with high domain sizes [232] and h-BN [233]. The gas flow rate should ideally be adjusted to create viscous laminar flow to maintain a constant environment during vapor deposition. However, the growth chamber has a velocity gradient of gases, which lowers to zero at the surface of substrate, causing a still layer to form on top

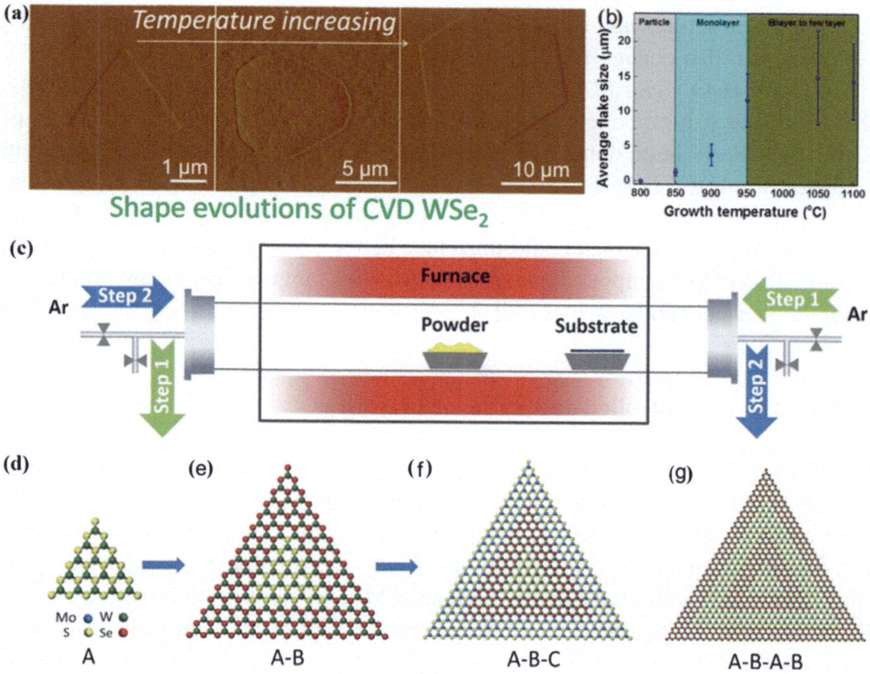

Fig. 4.19 Representation of temperature and gas flow over 2D compound materials generation: **a, b** temperature effect showing on domain size, layer number as well as morphology owing to CVD-fabricated WSe_2. Reproduced with permission from [207], **c** a modified CVD format utilizing gas flow reversely toward epitaxial development of various 2D-HS with growth outcomes for **d** an SL seed A, **e** A and B HS, **f** from A to C multi-HS, whereas **g** depiction of A-B-A-B superlattices. Reproduced with permission from [236]

of the substrate [234]. Since the precursor concentration on the substrate surface influences the nucleation density, its control is important for modulating the nucleation of the 2D material. An alternate strategy for developing 2DMs with a great domain size and outstanding quality is provided by creating a microreactor with a confined area and a constant development environment. In this context, Tang et al. [235] synthesized 2D In_2Se_3 in a limited microreactor comprised of two slanted mica substrates placed face to face. As a result of the tightly regulated concentration of precursor, the as-prepared In_2Se_3 grew to a diameter of more than 200 μm. Furthermore, developing growth reactors that control gas flow is a viable method for growing additional 2D molecules. Zhang et al. [236] described a general method for the development of various HS, multi-HS, and superlattices using a reverse gas flow design to accurately manage the precursor supply (Fig. 4.19c–g). By quickly blowing away surplus precursor, the reverse flow minimizes uncontrolled nucleation during subsequent vapor deposition growths, as well as thermal degradation of the reversed gas due to exhaust heat. The manufacture of various 2D-HS and junctions is improved by an epitaxial growth process that can be used to construct novel

2D complex structures and superlattices, such as perovskite junctions and hybrids. Additives can also change the microenvironment for CVD growth like O_2 [237], F [238], and H_2O vapor [239] into the processes to alter the chemistry of 2DMs. These methods exhibit benefits for achieving required structure and morphology of 2DM, particularly huge domain sizes and HS. Moreover, external fields including the usage of a PACVD system to construct the Janus structure of TMDCs, plasma, laser, microwave, magnetic, and electric fields might be employed to regulate the CVD growth process, which could introduce better CVD reactor designs enabling the formation of 2D compound materials with a wide range of shapes and properties, including thermal conductivity [240].

4.5.2.4 Layer-Controlled and Patterned Growth

Since TMDCs exhibit a wealth of layer-dependent characteristics, large-area production of ML-TMDC crystals/films with controlled thicknesses was extensively explored [209, 213, 241]. It was demonstrated that by treating SiO_2/Si substrates with oxygen plasma for varying durations of time, it is possible to manufacture scalable MoS_2 films with controllable thickness (ML, BL, and triple-layer). For ML, BL, and trilayer MoS_2 transistors, the as-grown ML films had homogeneous thicknesses and field-effect motion values of 3.6, 8.2, and 15.6 $cm^2 \cdot V^{-1}$ s^{-1}, respectively. Plasma treatment lowers the surface energy of the SiO_2 substrate by forming dangling Si-O or Si.OH bonds on the surface, making it hydrophilic and boosting reactivity of surface that could facilitate MoS_2's diverse nucleation and layer growth rates as hypothesized by researchers. Indeed, the oxygen plasma method may be used to achieve patterned growth of TMDCs and associated HS [242, 243]. This technique would permit the direct development of TMDCs on high-k dielectric materials having controlled thicknesses for electronic uses.

References

1. Whitener KE, Sheehan PE (2014) Diam Relat Mater 46:25
2. Torres L, Armas LG, Seabra AC (2014) Graphene 3(1):42074
3. Abdelkader AM, Kinloch IA (2016) ACS Sustain Chem Eng 4:4465
4. Frindt RF (1966) J Appl Phys 37:1928
5. Valota AT, Toth PS, Kim Y-J, Hong BH, Kinloch IA, Novoselov KS, Hill EW, Dryfe RAW (2013) Electrochim Acta 110:9
6. Novoselov KS, Jiang D, Schedin F, Booth TJ, Khotkevich VV, Morozov SV, Geim AK (2005) Proc Natl Acad Sci USA 102:10451
7. Ataca C, Şahin H, Ciraci S (2012) J Phys Chem C 116:8983
8. Najmaei S, Liu Z, Ajayan PM, Lou J (2012) Appl Phys Lett 100:013106
9. Niu Y et al (2018) Nanomaterials 8
10. Wang F, Kinloch IA, Wolverson D, Tenne R, Zak A, O'Connell E, Bangert U, Young RJ (2016) 2D Mater 4:015007
11. Liu W, Kang J, Sarkar D, Khatami Y, Jena D, Banerjee K (2013) Nano Lett 13:1983

12. Castellanos-Gomez A et al (2013) Nano Res 6:191
13. Horzum S, Çakır D, Suh J, Tongay S, Huang YS, Ho CH, Wu J, Sahin H, Peeters FM (2014) Phys Rev B 89:155433
14. Novoselov KS, Castro Neto AH (2012) Physica Scripta T146:014006
15. Parvez K (2019) In: Nurunnabi M, McCarthy JR (eds) Biomedical applications of graphene and 2D nanomaterials. Elsevier, p 1
16. Hu Z, Liu Z-B, Tian J-G (2020) Chin J Chem 38:981
17. Castellanos-Gomez A, Buscema M, Molenaar R, Singh V, Janssen L, van der Zant HSJ, Steele GA (2014) 2D Mater 1:011002
18. Yu Y, Jiang S, Zhang G, Zhou W, Miao X, Zeng Y, Wang J, He J, Zhang L (2012) Appl Phys Lett 101:073113
19. Yu Y, Jiang S, Zhou W, Miao X, Zeng Y, Zhang G, Liu S (2013) Sci Rep 3:2697
20. Ibrahem MA, Lan T-W, Huang JK, Chen Y-Y, Wei K-H, Li L-J, Chu CW (2013) RSC Adv 3:13193
21. Liang X, Chang ASP, Zhang Y, Harteneck BD, Choo H, Olynick DL, Cabrini S (2009) Nano Lett 9:467
22. Gacem K, Boukhicha M, Chen Z, Shukla A (2012) Nanotechnology 23:505709
23. Brent JR, Savjani N, O'Brien P (2017) Prog Mater Sci 89:411
24. Radisavljevic B, Radenovic A, Brivio J, Giacometti V, Kis A (2011) Nat Nanotechnol 6:147
25. Lee HS, Min S-W, Chang Y-G, Park MK, Nam T, Kim H, Kim JH, Ryu S, Im S (2012) Nano Lett 12:3695
26. Kim S et al (2012) Nat Commun 3:1011
27. Del Rio Castillo AE et al (2018) Mater Horizons 5:890
28. Ibrahem MA et al (2014) J Mater Chem A 2:11382
29. Liu G, Komatsu N (2016) ChemNanoMat 2:500
30. Nicolosi V, Chhowalla M, Kanatzidis MG, Strano MS, Coleman JN (2013) Science 340:1226419
31. Niu L, Coleman JN, Zhang H, Shin H, Chhowalla M, Zheng Z (2016) Small 12:272
32. Ciesielski A, Samorì P (2014) Chem Soc Rev 43:381
33. Hernandez Y et al (2008) Nat Nanotechnol 3:563
34. Hamilton CE, Lomeda JR, Sun Z, Tour JM, Barron AR (2009) Nano Lett 9:3460
35. O'Neill A, Khan U, Nirmalraj PN, Boland J, Coleman JN (2011) J Phys Chem C 115:5422
36. Choi E-Y, Choi WS, Lee YB, Noh Y-Y (2011) Nanotechnology 22:365601
37. Qian W, Hao R, Hou Y, Tian Y, Shen C, Gao H, Liang X (2009) Nano Res 2:706
38. Coleman JN et al (2011) Science 331:568
39. Zhou K-G, Mao N-N, Wang H-X, Peng Y, Zhang H-L (2011) Angew Chem Int Ed 50:10839
40. Kim J et al (2015) Nat Commun 6:8294
41. Peng B, Zhang H, Shao H, Xu Y, Zhang X, Zhu H (2016) RSC Adv 6:5767
42. Nguyen TP et al (2015) Sci Adv Mater 7:700
43. Štengl V, Tolasz J, Popelková D (2015) RSC Adv 5:89612
44. Jha RK, Guha PK (2016) Nanotechnology 27:475503
45. Adilbekova B et al (2020) J Mater Chem C 8:5259
46. Qiao W, Yan S, He X, Song X, Li Z, Zhang X, Zhong W, Du Y (2014) RSC Adv 4:50981
47. Cunningham G, Lotya M, Cucinotta CS, Sanvito S, Bergin SD, Menzel R, Shaffer MSP, Coleman JN (2012) ACS Nano 6:3468
48. O'Neill A, Khan U, Coleman JN (2012) Chem Mater 24:2414
49. Paton KR et al (2014) Nat Mater 13:624
50. Yi M, Shen Z (2015) J Mater Chem A 3:11700
51. Woomer AH, Farnsworth TW, Hu J, Wells RA, Donley CL, Warren SC (2015) ACS Nano 9:8869
52. Xu F et al (2016) 2D Mater 3:025005
53. Yi M, Shen Z (2014) Carbon 78:622
54. Varrla E, Backes C, Paton KR, Harvey A, Gholamvand Z, McCauley J, Coleman JN (2015) Chem Mater 27:1129

55. Dines MB (1975) Mater Res Bull 10:287
56. Joensen P, Frindt RF, Morrison SR (1986) Mater Res Bull 21:457
57. Tan C et al (2017) Chem Rev 117:6225
58. Divigalpitiya WMR, Morrison SR, Frindt RF (1990) Thin Solid Films 186:177
59. Miremadi BK, Morrison SR (1988) J Appl Phys 63:4970
60. Gordon RA, Yang D, Crozier ED, Jiang DT, Frindt RF (2002) Phys Rev B 65:125407
61. Voiry D, Salehi M, Silva R, Fujita T, Chen M, Asefa T, Shenoy VB, Eda G, Chhowalla M (2013) Nano Lett 13:6222
62. Ramakrishna Matte HSS, Gomathi A, Manna AK, Late DJ, Datta R, Pati SK, Rao CNR (2010) Angew Chem Int Ed 49:4059
63. Ghorai A, Midya A, Maiti R, Ray SK (2016) Dalton Trans 45:14979
64. Rohaizad N, Mayorga-Martinez CC, Sofer Z, Pumera M (2017) ACS Appl Mater Interfaces 9:40697
65. Fujita T, Ito Y, Tan Y, Yamaguchi H, Hojo D, Hirata A, Voiry D, Chhowalla M, Chen M (2014) Nanoscale 6:12458
66. Sun Z, Lv H, Zhuo Z, Jalil A, Zhang W, Wu X, Yang J (2018) J Mater Chem C 6:1248
67. Whittingham MS, Gamble FR (1975) Mater Res Bull 10:363
68. Lin C et al (2013) J Am Chem Soc 135:5144
69. Dahn J, Haering RR (1979) Mater Res Bull 14:1259
70. Ayari A, Cobas E, Ogundadegbe O, Fuhrer MS (2007) J Appl Phys 101:014507
71. Yu Y et al (2015) Nat Nanotechnol 10:270
72. Reynolds KJ, Frey GL, Friend RH (2003) Appl Phys Lett 82:1123
73. Hitz E et al (2016) ACS Appl Mater Interfaces 8:11390
74. Wu Y-C, Liu W-R (2020) J Alloys Compd 813:152074
75. Seo J-WT, Zhu J, Sangwan VK, Secor EB, Wallace SG, Hersam MC (2019) ACS Appl Mater Interfaces 11:5675
76. Liu N, Kim P, Kim JH, Ye JH, Kim S, Lee CJ (2014) ACS Nano 8:6902
77. Chen D, Wang F, Li Y, Wang W-W, Huang T-X, Li J-F, Novoselov KS, Tian Z-Q, Zhan D (2019) Chem Commun 55:3379
78. Kovalska E, Luxa J, Hartman T, Antonatos N, Shaban P, Oparin E, Zhukova M, Sofer Z (2020) Nanoscale 12:2638
79. Zeng Z, Yin Z, Huang X, Li H, He Q, Lu G, Boey F, Zhang H (2011) Angew Chem Int Ed 50:11093
80. Wang H et al (2013) Proc Natl Acad Sci 110:19701
81. Wu W, Zhang C, Zhou L, Hou S, Zhang L (2019) J Colloid Interface Sci 542:263
82. Parvez K, Wu Z-S, Li R, Liu X, Graf R, Feng X, Müllen K (2014) J Am Chem Soc 136:6083
83. Bianca G et al (2020) ACS Appl Mater Interfaces 12:48598
84. Lin Z et al (2018) Nature 562:254
85. You X, Liu N, Lee CJ, Pak JJ (2014) Mater Lett 121:31
86. Li F, Xue M, Zhang X, Chen L, Knowles GP, MacFarlane DR, Zhang J (2018) Adv Energ Mater 8:1702794
87. Gopalakrishnan D, Damien D, Li B, Gullappalli H, Pillai VK, Ajayan PM, Shaijumon MM (2015) Chem Commun 51:6293
88. Zhang P et al (2019) Small 15:1901265
89. Zeng Z et al (2012) Angew Chem Int Ed 51:9052
90. Yu W et al (2019) Adv Mater 31:1903779
91. Murphy DW, Christian PA (1979) Science 205:651
92. Jacobson AJ, Nazar LF (2006) In: King RB, Crabtree RH, Lukehart CM, Atwood DA, Scott RA (eds) Encyclopedia of inorganic chemistry, vol 1. https://onlinelibrary.wiley.com/doi/abs/10.1002/0470862106.ia098
93. Zhang J et al (2018) Nat Commun 9:5289
94. Zheng J, Zhang H, Dong S, Liu Y, Tai Nai C, Suk Shin H, Young Jeong H, Liu B, Ping Loh K (2014) Nat Commun 5:2995
95. Li Y, Lu Y, Adelhelm P, Titirici M-M, Hu Y-S (2019) Chem Soc Rev 48:4655

96. Titov AN, Yagafarova ZA, Bikkulova NN (2003) Phys Solid State 45:2067
97. Titova SG, Titov AN (2007) Phys Solid State 49:63
98. Solin SA (1997) Ann Rev Mater Sci 27:89
99. Eda G, Fujita T, Yamaguchi H, Voiry D, Chen M, Chhowalla M (2012) ACS Nano 6:7311
100. Solin SA, Hines D, Jacobson AJ, Mahanti SD (1996) Solid State Commun 100:143
101. Cheng Z, Yu Y, Singh S, Price K, Noyce SG, Lin Y-C, Cao L, Franklin AD (2019) Nano Lett 19:5077
102. Andrews K, Bowman A, Rijal U, Chen P-Y, Zhou Z (2020) ACS Nano 14:6232
103. Zhang Q, Mei L, Cao X, Tang Y, Zeng Z (2020) J Mater Chem A
104. Liang Y, Yoo HD, Li Y, Shuai J, Calderon HA, Robles Hernandez FC, Grabow LC, Yao Y (2015) Nano Lett 15:2194
105. Xu J, Zhang J, Zhang W, Lee C-S (2017) Adv Energ Mater 7:1700571
106. Leng K, Chen Z, Zhao X, Tang W, Tian B, Nai CT, Zhou W, Loh KP (2016) ACS Nano 10:9208
107. Evans JSO, Price SJ, Wong H-V, O'Hare D (1998) J Am Chem Soc 120:10837
108. Mortazavi M, Wang C, Deng J, Shenoy VB, Medhekar NV (2014) J Power Sour 268:279
109. Stankovich S, Dikin DA, Piner RD, Kohlhaas KA, Kleinhammes A, Jia Y, Wu Y, Nguyen ST, Ruoff RS (2007) Carbon 45:1558
110. Zhu Y, Murali S, Cai W, Li X, Suk JW, Potts JR, Ruoff RS (2010) Adv Mater 22:3906
111. Dreyer DR, Park S, Bielawski CW, Ruoff RS (2010) Chem Soc Rev 39:228
112. Jung I, Dikin DA, Piner RD, Ruoff RS (2008) Nano Lett 8:4283
113. Pei S, Zhao J, Du J, Ren W, Cheng H-M (2010) Carbon 48:4466
114. Compton OC, Nguyen ST (2010) Small 6:711
115. Bi H et al (2012) Adv Func Mater 22:4421
116. Georgakilas V, Otyepka M, Bourlinos AB, Chandra V, Kim N, Kemp KC, Hobza P, Zboril R, Kim KS (2012) Chem Rev 112:6156
117. Huang X, Yin Z, Wu S, Qi X, He Q, Zhang Q, Yan Q, Boey F, Zhang H (2011) Small 7:1876
118. Han JH, Lee S, Cheon J (2013) Chem Soc Rev 42:2581
119. Tan C, Zhang H (2015) Nat Commun 6:7873
120. Zhuang Z, Peng Q, Li Y (2011) Chem Soc Rev 40:5492
121. Shi W, Song S, Zhang H (2013) Chem Soc Rev 42:5714
122. Duan H et al (2014) Nat Commun 5:3093
123. Sun Y et al (2013) Nat Commun 4:2899
124. Sun Z, Liao T, Dou Y, Hwang SM, Park M-S, Jiang L, Kim JH, Dou SX (2014) Nat Commun 5:3813
125. Xie J, Zhang H, Li S, Wang R, Sun X, Zhou M, Zhou J, Lou XW, Xie Y (2013) Adv Mater 25:5807
126. Xie J et al (2013) J Am Chem Soc 135:17881
127. Cai L et al (2015) J Am Chem Soc 137:2622
128. Zhang X, Zhang J, Zhao J, Pan B, Kong M, Chen J, Xie Y (2012) J Am Chem Soc 134:11908
129. Evers WH, Goris B, Bals S, Casavola M, de Graaf J, van Roij R, Dijkstra M, Vanmaekelbergh D (2013) Nano Lett 13:2317
130. Schliehe C et al (2010) Science 329:550
131. Wang Z, Schliehe C, Wang T, Nagaoka Y, Cao YC, Bassett WA, Wu H, Fan H, Weller H (2011) J Am Chem Soc 133:14484
132. Bhandari GB, Subedi K, He Y, Jiang Z, Leopold M, Reilly N, Lu HP, Zayak AT, Sun L (2014) Chem Mater 26:5433
133. Boneschanscher MP et al (2014) Science 344:1377
134. Min Y, Moon GD, Kim BS, Lim B, Kim J-S, Kang CY, Jeong U (2012) J Am Chem Soc 134:2872
135. Li L, Chen Z, Hu Y, Wang X, Zhang T, Chen W, Wang Q (2013) J Am Chem Soc 135:1213
136. Simon P et al (2014) Adv Mater 26:3042
137. Tang Z, Zhang Z, Wang Y, Glotzer SC, Kotov NA (2006) Science 314:274
138. Wu Z, Li Y, Liu J, Lu Z, Zhang H, Yang B (2014) Angew Chem Int Ed 53:12196

139. Wu Z et al (2015) J Am Chem Soc 137:12906
140. Wu Z, Liu J, Li Y, Cheng Z, Li T, Zhang H, Lu Z, Yang B (2015) ACS Nano 9:6315
141. Rao S, Si KJ, Yap LW, Xiang Y, Cheng W (2015) ACS Nano 9:11218
142. Acharya S, Das B, Thupakula U, Ariga K, Sarma DD, Israelachvili J, Golan Y (2013) Nano Lett 13:409
143. Zhong Y, Yang Y, Ma Y, Yao J (2013) Chem Commun 49:10355
144. Seo J-W, Jun Y-W, Park S-W, Nah H, Moon T, Park B, Kim J-G, Kim YJ, Cheon J (2007) Angew Chem Int Ed 46:8828
145. Bi W, Zhou M, Ma Z, Zhang H, Yu J, Xie Y (2012) Chem Commun 48:9162
146. Huang X et al (2011) Nat Commun 2:292
147. Huang X, Li H, Li S, Wu S, Boey F, Ma J, Zhang H (2011) Angew Chem Int Ed 50:12245
148. Fan Z et al (2015) Nat Commun 6:6571
149. Cheng W et al (2014) J Am Chem Soc 136:10393
150. Zhu Y, Cao C, Tao S, Chu W, Wu Z, Li Y (2014) Sci Rep 4:5787
151. Busupalli B, Kummara S, Kumaraswamy G, Prasad BLV (2014) Chem Mater 26:3436
152. Wu X-J, Huang X, Qi X, Li H, Li B, Zhang H (2014) Angew Chem Int Ed 53:8929
153. Wu X-J, Huang X, Liu J, Li H, Yang J, Li B, Huang W, Zhang H (2014) Angew Chem Int Ed 53:5083
154. Morrison PJ, Loomis RA, Buhro WE (2014) Chem Mater 26:5012
155. Murray CB, Norris DJ, Bawendi MG (1993) J Am Chem Soc 115:8706
156. Deng Z, Cao D, He J, Lin S, Lindsay SM, Liu Y (2012) ACS Nano 6:6197
157. Chatterjee A, Biswas K (2015) Angew Chem Int Ed 54:5623
158. Son JS et al (2009) Angew Chem Int Ed 48:6861
159. Xue D-J, Tan J, Hu J-S, Hu W, Guo Y-G, Wan L-J (2012) Adv Mater 24:4528
160. Vaughn DD, Patel RJ, Hickner MA, Schaak RE (2010) J Am Chem Soc 132:15170
161. Du Y, Yin Z, Zhu J, Huang X, Wu X-J, Zeng Z, Yan Q, Zhang H (2012) Nat Commun 3:1177
162. Gu J, Zhao Z-Q, Ding Y, Chen H-L, Zhang Y-W, Yan C-H (2013) J Am Chem Soc 135:8363
163. Jung W, Lee S, Yoo D, Jeong S, Miró P, Kuc A, Heine T, Cheon J (2015) J Am Chem Soc 137:7266
164. Jeong S, Han JH, Jang J-T, Seo J-W, Kim J-G, Cheon J (2011) J Am Chem Soc 133:14500
165. Jeong S, Yoo D, Jang J-T, Kim M, Cheon J (2012) J Am Chem Soc 134:18233
166. Yoo D, Kim M, Jeong S, Han J, Cheon J (2014) J Am Chem Soc 136:14670
167. Mahler B, Hoepfner V, Liao K, Ozin GA (2014) J Am Chem Soc 136:14121
168. Ramasamy K, Sims H, Butler WH, Gupta A (2014) J Am Chem Soc 136:1587
169. Wu Y, Yuan B, Li M, Zhang W-H, Liu Y, Li C (2015) Chem Sci 6:1873
170. Altavilla C, Sarno M, Ciambelli P (2011) Chem Mater 23:3879
171. Boott CE, Nazemi A, Manners I (2015) Angew Chem Int Ed 54:13876
172. Rodríguez-San-Miguel D, Amo-Ochoa P, Zamora F (2016) Chem Commun 52:4113
173. Bauer T, Zheng Z, Renn A, Enning R, Stemmer A, Sakamoto J, Schlüter AD (2011) Angew Chem Int Ed 50:7879
174. Kambe T et al (2013) J Am Chem Soc 135:2462
175. Dong R, Pfeffermann M, Liang H, Zheng Z, Zhu X, Zhang J, Feng X (2015) Angew Chem Int Ed 54:12058
176. Sakamoto R et al (2015) Nat Commun 6:6713
177. Murray DJ, Patterson DD, Payamyar P, Bhola R, Song W, Lackinger M, Schlüter AD, King BT (2015) J Am Chem Soc 137:3450
178. Feldblyum JI, McCreery CH, Andrews SC, Kurosawa T, Santos EJG, Duong V, Fang L, Ayzner AL, Bao Z (2015) Chem Commun 51:13894
179. Wang F et al (2016) Nat Commun 7:10444
180. Huang X, Tang S, Mu X, Dai Y, Chen G, Zhou Z, Ruan F, Yang Z, Zheng N (2011) Nat Nanotechnol 6:28
181. Gao S, Sun Y, Lei F, Liang L, Liu J, Bi W, Pan B, Xie Y (2014) Angew Chem Int Ed 53:12789
182. Zhang Q, Yan B (2012) Chem Eur J 18:5150
183. Kissel P, Murray DJ, Wulftange WJ, Catalano VJ, King BT (2014) Nat Chem 6:774

184. Kory MJ, Wörle M, Weber T, Payamyar P, van de Poll SW, Dshemuchadse J, Trapp N, Schlüter AD (2014) Nat Chem 6:779
185. Dou L et al (2015) Science 349:1518
186. Tang L, Tan J, Nong H, Liu B, Cheng H-M (2021) Acc Mater Res 2:36
187. Shen P, Lin Y, Wang H, Park J, Leong WS, Lu A, Palacios T, Kong J (2018) IEEE Trans Electron Dev 65:4040
188. Khan S, Ali J, Harsh, Husain M, Zulfequar M (2017) In: Jain VK, Rattan S, Verma A (eds) Recent trends in materials and devices. Springer International Publishing, Cham, pp 119
189. Ahn C et al (2015) Adv Mater 27:5223
190. Li X et al (2009) Science 324:1312
191. Gao L et al (2012) Nat Commun 3:699
192. Li X, Cai W, Colombo L, Ruoff RS (2009) Nano Lett 9:4268
193. Rasool HI, Song EB, Mecklenburg M, Regan BC, Wang KL, Weiller BH, Gimzewski JK (2011) J Am Chem Soc 133:12536
194. Kim KK et al (2012) Nano Lett 12:161
195. Kim SM et al (2015) Nat Commun 6:8662
196. Park J-H et al (2014) ACS Nano 8:8520
197. Zhang B, Lee WH, Piner R, Kholmanov I, Wu Y, Li H, Ji H, Ruoff RS (2012) ACS Nano 6:2471
198. Li Z et al (2011) ACS Nano 5:3385
199. Han J, Lee J-Y, Kwon H, Yeo J-S (2014) Nanotechnology 25:145604
200. Song L et al (2010) Nano Lett 10:3209
201. Jang SK, Youn J, Song YJ, Lee S (2016) Sci Rep 6:30449
202. Lin W-H, Brar VW, Jariwala D, Sherrott MC, Tseng W-S, Wu C-I, Yeh N-C, Atwater HA (2017) Chem Mater 29:4700
203. Kim G, Jang AR, Jeong HY, Lee Z, Kang DJ, Shin HS (2013) Nano Lett 13:1834
204. Wu C, Soomro AM, Sun F, Wang H, Liu C, Yang X, Kang J, Cai D (2016) physica status solidi (b) 253:829
205. Uchida Y, Iwaizako T, Mizuno S, Tsuji M, Ago H (2017) Phys Chem Chem Phys 19:8230
206. Jang AR et al (2016) Nano Lett 16:3360
207. Liu B, Fathi M, Chen L, Abbas A, Ma Y, Zhou C (2015) ACS Nano 9:6119
208. Zhao Y, Lee H, Choi W, Fei W, Lee CJ (2017) RSC Adv 7:27969
209. Zhou L et al (2015) J Am Chem Soc 137:11892
210. McCreary KM, Hanbicki AT, Jernigan GG, Culbertson JC, Jonker BT (2016) Sci Rep 6:19159
211. Liu K-K et al (2012) Nano Lett 12:1538
212. Leong WS, Li Y, Luo X, Nai CT, Quek SY, Thong JTL (2015) Nanoscale 7:10823
213. Zhou L et al (2016) Adv Mater 28:9526
214. Lin Y-C, Zhang W, Huang J-K, Liu K-K, Lee Y-H, Liang C-T, Chu C-W, Li L-J (2012) Nanoscale 4:6637
215. Li B, Yang S, Huo N, Li Y, Yang J, Li R, Fan C, Lu F (2014) RSC Adv 4:26407
216. Yu Y, Li C, Liu Y, Su L, Zhang Y, Cao L (2013) Sci Rep 3:1866
217. McCreary KM, Hanbicki AT, Robinson JT, Cobas E, Culbertson JC, Friedman AL, Jernigan GG, Jonker BT (2014) Adv Func Mater 24:6449
218. Sanne A, Ghosh R, Rai A, Movva HCP, Sharma A, Rao R, Mathew L, Banerjee SK (2015) Appl Phys Lett 106:062101
219. Zhang Y et al (2014) ACS Nano 8:8617
220. Wu D, Min T, Zhou J, Li C, Ma G, Lu G, Xia M, Gu Z (2017) Sci Rep 7:15166
221. Ji Q et al (2013) Nano Lett 13:3870
222. Zhan Y, Liu Z, Najmaei S, Ajayan PM, Lou J (2012) Small 8:966
223. Wang S, Wang X, Warner JH (2015) ACS Nano 9:5246
224. Dumcenco D et al (2015) ACS Nano 9:4611
225. Shi J et al (2015) ACS Nano 9:4017
226. Gao Y, Goodman AJ, Shen P-C, Kong J, Tisdale WA (2018) Nano Lett 18:5001
227. Gong Y et al (2017) 2D Mater 4:021028

228. Choi M (2018) Phys Rev Appl 9:024009
229. Sensoy MG, Vinichenko D, Chen W, Friend CM, Kaxiras E (2017) Phys Rev B 95:014106
230. Ahn GH et al (2017) Nat Commun 8:608
231. Tang B et al (2020) Mater Today 41:72
232. Li X, Magnuson CW, Venugopal A, Tromp RM, Hannon JB, Vogel EM, Colombo L, Ruoff RS (2011) J Am Chem Soc 133:2816
233. Liu J, Yu L, Cai X, Khan U, Cai Z, Xi J, Liu B, Kang F (2019) ACS Nano 13:2094
234. Bhaviripudi S, Jia X, Dresselhaus MS, Kong J (2010) Nano Lett 10:4128
235. Tang L, Teng C, Luo Y, Khan U, Pan H, Cai Z, Zhao Y, Liu B, Cheng H-M (2019) Research 2019:2763704
236. Zhang Z, Chen P, Duan X, Zang K, Luo J, Duan X (2017) Science 357:788
237. Chen W et al (2015) J Am Chem Soc 137:15632
238. Liu C et al (2019) Nat Chem 11:730
239. Sahoo PK, Memaran S, Xin Y, Balicas L, Gutiérrez HR (2018) Nature 553:63
240. Lu A-Y et al (2017) Nat Nanotechnol 12:744
241. Xu X et al (2017) Science Bulletin 62:1074
242. Chen X, Park YJ, Das T, Jang H, Lee J-B, Ahn J-H (2016) Nanoscale 8:15181
243. Kim H-J, Kim H, Yang S, Kwon J-Y (2017) Small 13:1702256

Chapter 5
2D-Heterostructures

The development of 2D layered materials has recently experienced a remarkable trajectory arising from their potential uses. In contrast to bulk materials, atomically thin 2DMs feature a diverse set of fundamental building components with distinct optical, electrical, and thermal characteristics. The interaction between the vdW layers permits the exfoliation and reassembly of diverse 2DM in arbitrarily layered HSs. Recently, the organized growing vapor phase of 2DMs also paves the door for vertical HJs (V-HJs) and lateral HJs (L-HJs) to be manufactured. This chapter offers a concise introduction to preparatory techniques for 2D-HSs and their potential applications.

5.1 Advances in 2D-Van Der Waals Heterostructures

The 2DM development facilitates HSs generation with a broad variety of applications. Since the first synthesis of GR, a semimetal with no overlap, single-atom-thick crystals have evolved into a large family of compounds with a diverse characteristics range. Among these 2D crystals, families are metals ($NbSe_2$), semiconductors (MoS_2), and insulators (h-BN). We have concluded some processes since not everyone is comfortable with these materials at room temperature. Amazingly, the characteristics of these 2DMs prominently vary compare to their 3D counterparts [1, 2]. When many 2D crystals are V-stacked, a wealth of options becomes apparent. Such HSs are held together by vdW interactions allow boosted combinations than any conventional method of development, various types of HSs are shown in Fig. 5.1 [3]. However, the HSs that are formed with atomic precision are often complicated as the family of 2D crystals grows daily [3, 4]. The synergetic effects become very powerful when stacking different crystals together. The charge redistribution can occur in the first-order approximation among the neighboring (and farther away) crystals in the stack. Structural changes can also occur in nearby crystals. Further modifications

M. Ikram et al., *2D-Materials for Energy Harvesting and Storage Applications*, Nanostructure Science and Technology,
https://doi.org/10.1007/978-3-030-96021-6_5

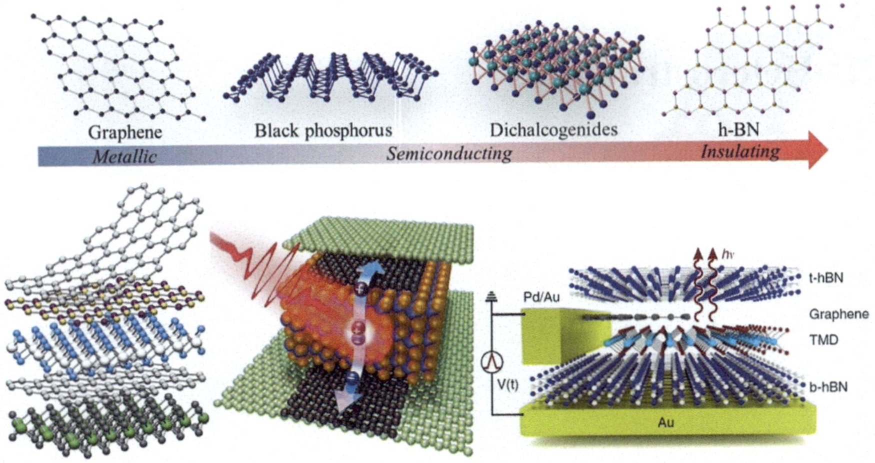

Fig. 5.1 Representation of 2D-HSs with electronic properties and applications (Reproduced with permission from [3])

could be made by altering the relative orientation of individual elements. These kinds of HSs have contributed to observation of many fascinating physical processes that are important in a variety of applications. The application of semiconducting monolayers produces optically active HSs, for example, the reconstruction of GR's spectrum using h-BN enabled multiple organizations to research topological currents and the Hofstadter butterfly effect. The crystals in this HS are relatively near (but regulated) to one another, allowing tunneling and drag phenomena to be investigated [2, 5, 6]. The application of semiconducting monolayers produces optically active HSs. Similarly, a wide range of HSs features produces a variety of applications. This capturing GR with h-BN is now known for the highest mobility GR transistors. Similarly, photovoltaic and light-emitting devices were developed using optically active semiconducting layers with GR as transparent electrodes [6].

5.2 Properties

5.2.1 Band Tuning

The fundamental property of 2D-HSs is band alignment. The band structures, especially the band alignment, are responsible for the majority of charge transport behavior and luminance properties. Band energy may be calculated theoretically and evaluated via μ-XPS [7]. Figure 5.2a depicts a monolayer semiconducting TMDCs band alignments and a monolayer SnS_2 computed using PBE-Perdew–Burke–Ernzerhof with SOC-spin–orbit coupling, and Fig. 5.2b shows the structure of TMDC. The results demonstrate that conduction band minimum (CBM) and valence band

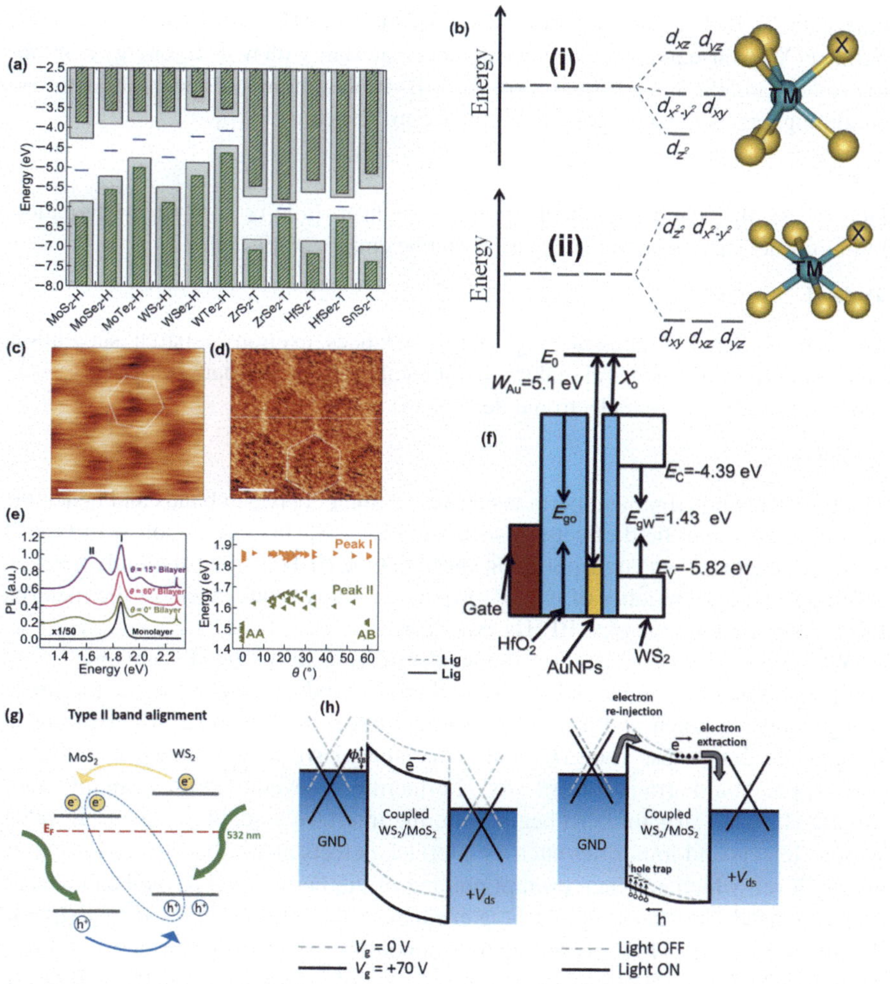

Fig. 5.2 Alignment of bands and engineering techniques: **a** theoretical positions determined by PBE-SOC for VIB and IVB-TMDCs, **b** atomic configurations of trigonal prismatic coordination (i) and octahedral coordination (ii), and the corresponding d-orbital splittings in trigonal prismatic and octahedral crystal fields. "TM" and "X" in atomic structures represent transition metal atoms and chalcogen atoms, (Reproduced with permission from [7]). **c** STM measurements of h-BN/GR-HS layered well with varying rotation angles of **c** 5° **d** and 0° (Reproduced with permission from [14]), **e** MoS_2 monolayer and BL-PL spectra with various twist angles, Peak I corresponds to the direct transition at K valley, whereas peak II might be the indirect E_g transition (Reproduced with permission from [16]), **f** floating-gate device band structure based on SL-WS_2 (Reproduced with permission from [20]), **g** V-stacked WS_2/MoS_2-HS band schematic, **h** engineering principle underlying photogate band alignment (Reproduced with permission from [19])

maximum (VBM) energy for monolayer TMDCs in MX_2 form rises with atomic number (X). Consequently, when two distinct materials with wide E_g energy combine to produce an HS, three distinct forms of I–III band alignments occur, alluded to as straddling gap, staggering gap, and broken gap, respectively [8].

Type I

Due to fact that band alignment promotes electron–hole ($e^{-}-h^{+}$) recombination, luminous devices including such LEDs can be employed [9, 10].

Type II

The type II band alignment (Fig. 5.2g) enhances excitons' spatial separation, increasing their lifespan and making them an attractive candidate for use in $e^{-}-h^{+}$ separators and other optoelectronic devices [8].

Type III

The type III band alignment permits carriers to tunnel between bands and hence the operation of tunnel field-effect transistors (TFET) [8]. In comparison with type II band alignment, $e^{-}-h^{+}$ transportation speed in type III-HSs is substantially quicker, resulting in a high number of $e^{-}-h^{+}$ dispersing into various layers of the material this activity making the type III HJs excellent.

When taken into consideration, the band arrangements of the HSs are much more complex. When two semiconductors combine to produce a type I band alignment p–n junction, a notch and a peak are formed when the V-band and C-band bend in opposite directions. The ΔE_c and ΔE_v values have no symmetry, resulting in various potential barriers for $es^{-}-hs^{+}$. Furthermore, several features can influence the 2D-HS band structure. To begin with, the interface's quality is acute. In other words, defects and impurities such as interface defects and impurities can promote energy level, which can then prompt pinning of the Fermi level as well as intrinsic metal-induced gap states. Similarly, gap states at the interface caused by extrinsic disorder can do this [11]. According to Bampoulis et al. [12], subsurface metal-like defects in MoS_2/metal-HSs lead to higher levels of Fermi level pinning and hence reduction in the Schottky barrier height (SBH) [13]. There was a roughly 120 mV shift in the spectra of scanning tunneling spectroscopy (STS) performed on flakes on SL-GR and BL GR substrates. The second consideration is the materials' lattice constants and stake orientation. When two 2DMs with comparable lattice constants come into contact, including GR and h-BN and SnS_2/MoS_2, HSs such as Moiré patterns are created [14, 15]. STM observations on GR /h-BN-HSs with rotations less than and greater than 1 degree are shown in Fig. 5.2c, d. Like the displays above, although with a much broader Moiré pattern as a result of GR's, high Young's modulus vdW interpenetration triggers an extra periodic capacity for GR to adjust its band structure. With the stacking orientation, the indirect band distance changes considerably [16]. The highest redshift is seen for stacked BLs AA (twist angle 0°) and AB (twist angle 60°), whereas all other twist angles exhibit a miniature but continuous redshift (Fig. 5.2e). Changing the band orientation using the structural elements mentioned above, on the other hand, is challenging and inflexible. An HJ's band configuration can be regulated using an E_o, B_o, and light field [17]. Voltage-controlled band engineering induces 2D-HS changes from type I-III band alignments. GR is special, in that it possesses a distinctive linear band dispersive relationship using a finite DOS. As a result, the work function and Fermi level of GR may be gated, allowing it to be utilized as a contact between electrodes and channel more often of a 2DM to minimize R_c. In addition, when illuminated, the band structure might be optically manipulated to operate as a photogate, a phenomenon that seems to be a V_g addition to the device [18]. In Fig. 5.2h, the band diagram of a photodetector is shown as a WS_2/MoS_2 hetero-BL (absorption layer) and GR (electrode contact) [19]. Enlightened accumulation of holes is a positive gate on source GR electrode, decreasing Fermi level and boosting SBH [20]. Figure 5.2f shows a conventional floating-gate device band structure with an Au floating-gate layer producing energy well to collect charges.

5.2.2 *Charge Transportation in 2D-Heterostructures*

Different band alignments impart a charge transport variety properties that are then used by the energy transfer process. At the same time, new instruments based on single-particle transport, particularly interlayer tunneling effects, were developed, and as a result of these new technologies, the multibody transport and $e^- - h^+$ separation of 2D-HSs were dedicated to 2D-heterostructuring [8, 21]; this opens the door to a large range of applications in the optoelectronic sector [22–27].

5.2.2.1 Monoparticle Transports

To avoid band-edge sharpness depletion due to chemical doping, it has been observed that ultranarrow channel lengths and atomically sharp interfaces would enable electrostatic gating to achieve band-to-band tunneling [28]. WSe_2's CBM is greater than $SnSe_2$'s VBM when a small V_g is applied as seen in Fig. 5.3a–c [29]. Although electrons in $SnSe_2$ do not tunnel through WSe_2, charge diffusion happens at the interface resulting in high resistance and channel current states whenever a $\pm$ bias voltage is applied. This means the device has off state. As the V_g is increased, when the WSe_2 (CBM) is placed below the $SnSe_2$ (VBM) as illustrated in Fig. 5.3c, a tunneling window opens, allowing interlayer tunneling from $SnSe_2$ to WSe_2. In alignment of type I bands, the broad E_g layer induces spatial confinement of es^- and hs^+ in the narrow E_g "well" layer [30]; as a result, one trapping layer has a strong PL while the other has a quenched PL. (Fig. 5.3d). Based on band bending, interfacial charge trapping may be demonstrated at the Type II band alignment PN junction. In p-MSB/WSe_2, the heterostructural interfacial energy, and band bending is between p-MSB· LUMO—lowest unoccupied molecular orbit and WSe_2' CBM (see Fig. 5.3e). Cai et al. [31] have produced an electric-gating photodetector that seems to be switchable by 2D vdW epitaxial growth and functions as a light absorber. The

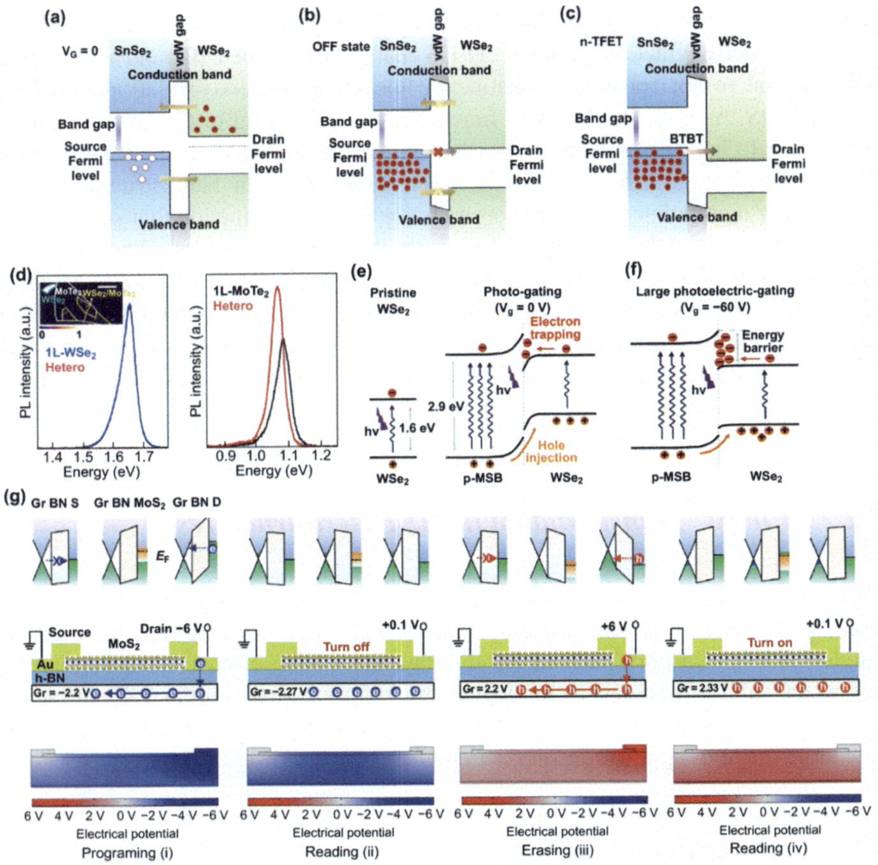

Fig. 5.3 Single-particle transport produced by distinct band structures, **a–c** TFET electrical modulation based on 2D $SnSe_2/WSe_2$-HS (Reproduced with permission from [29]), **d** PL analysis of a stacked $WSe_2/MoTe_2$-HS with appropriate annealing (Reproduced with permission from [30]), **e, f** charge transport diagram in a p-MSB/WSe_2-HS under light (Reproduced under the terms of CC BY 4.0 license [31], https://creativecommons.org/licenses/by/4.0/). **g** band alignment, charge transfer, and potential distribution inside the device (Reproduced under the terms of CC BY 4.0 license [32], https://creativecommons.org/licenses/by/4.0/)

interface energy barrier is the p-MSB HSs. The Fermi level of WSe_2 shifts downward when a $-V_g$ is applied (Fig. 5.3f), thereby increasing interface energy barrier which strengthens interfacial charge trapping mechanism. Figure 5.4d–j shows epitaxial growth of p-MSB on WSe_2, CVD schematics with substantial findings.

In plasma treatments, the interface quality is believed to be the key to avoiding a reduction in the current. A floating layer is typically placed among blocking dielectric layers and tunneling act as a charge and trapping layer between the channel and the control gate [20, 33]. Charge transfer in the floating gate structure is demonstrated in Fig. 5.3g using GR as the trapping layer and ML-MoS_2 as the channel [32]. When

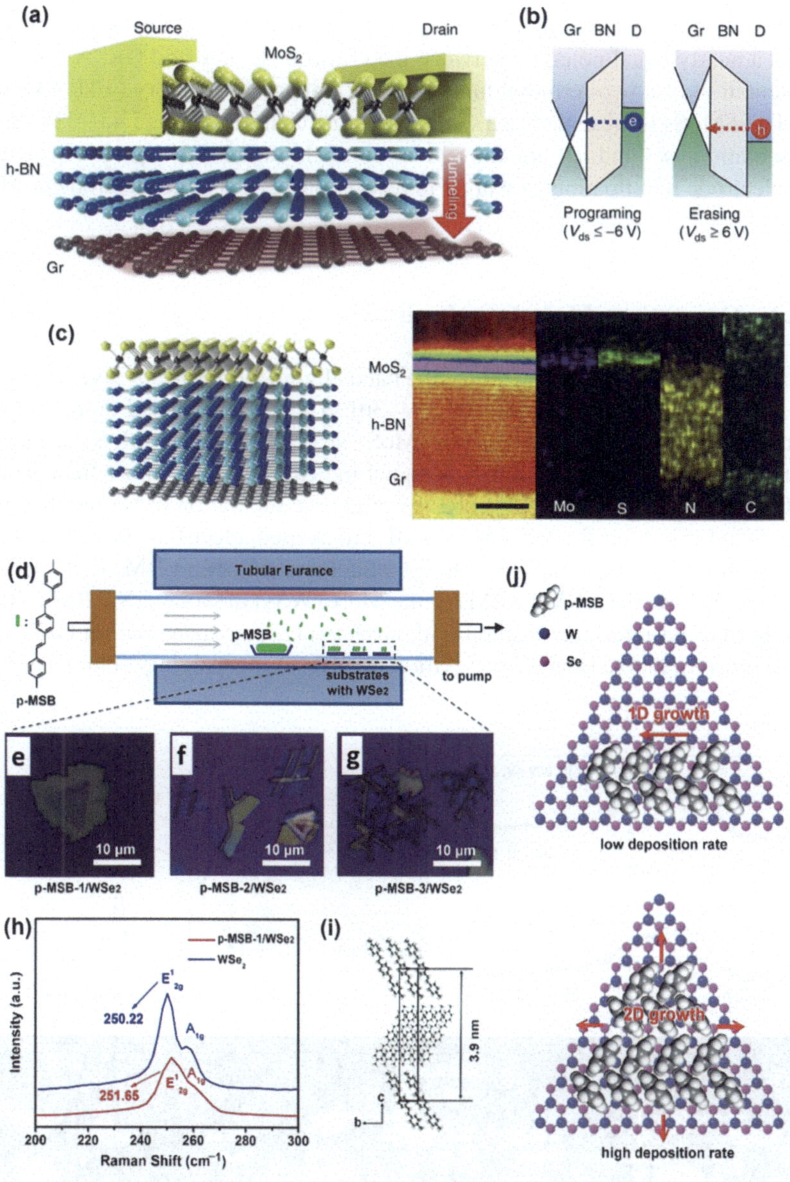

Fig. 5.4 **a** Two-TTRAM is shown schematically using monolayer MoS_2 as the semiconducting channel, h-BN as the tunneling insulator, and monolayer GR as the floating gate. **b** Drain (D)/h-BN/GR band diagrams illustrate the tunneling direction of the $es^{-}-hs^{+}$. **c** Cross-sectional bright-field STEM micrograph and EDS elemental mapping of two-TTRAM-HS with monolayer MoS_2/ML-h-BN/monolayer GR (left) (Reproduced under the terms of CC BY 4.0 license [32], https://creativecommons.org/licenses/by/4.0/), **d** p-MSB Epitaxial growth on WSe_2, CVD system schematics, **e–g** optical microscope images of the p-MSB/WSe_2-HSs grown, **h** WSe_2 and the p-MSB-1/WSe_2 Raman spectra, **i** p-MSB single-crystal unit cell structure with bc plane images, **j** On WSe_2, p-MSB crystals develop in 1D epitaxial and 2D epitaxial growth modes (Reproduced under the terms of CC BY 4.0 license [31], https://creativecommons.org/licenses/by/4.0/)

a drain bias of ±6 V is used, a large V_o between the drain and GR is generated, as well as a nearly equal potential between the electrode source and GR. A schematic representation of a two-terminal tunneling random access memory (TTRAM) with monolayer MoS_2 is presented as a semiconducting channel in Fig. 5.4a–c. From this representation, es^- and hs^+ are allowed to tunnel through h-BN layer on the side of drain electrode into floating gate layer to restrict tunnel to source electrode or MoS_2 channel [32].

5.2.2.2 Generation of Interlayer Excitons

Type II comprises CBM and VBM for the band alignment in various layers resulting in spatial separation of es^- and hs^+ [34–36]. The photo-induced signal appears instantaneously in heterogeneous WS_2/MoS_2 structure after SL-MoS_2 is pumped directly. The experimentally observed signal indicates a rise of more than 50 fs in the HS (Fig. 5.5a). The detailed processes of charge separation in HS are displayed in Fig. 5.5c [35]. After the $WS_2(MoS_2)$ ML are excited, electrons (holes) are transported within 50 fs from the $WS_2(MoS_2)$ monolayer's CBM (VBM) to the CB/VB just above (below) the CBM (VBM) of the $MoS_2(WS_2)$ monolayer's CBM (VBM), resulting in a hot interlayer exciton with a relatively long span. Within 800 fs, hot excitons are reorganized and release extra energy, resulting in tightly bound interlayer

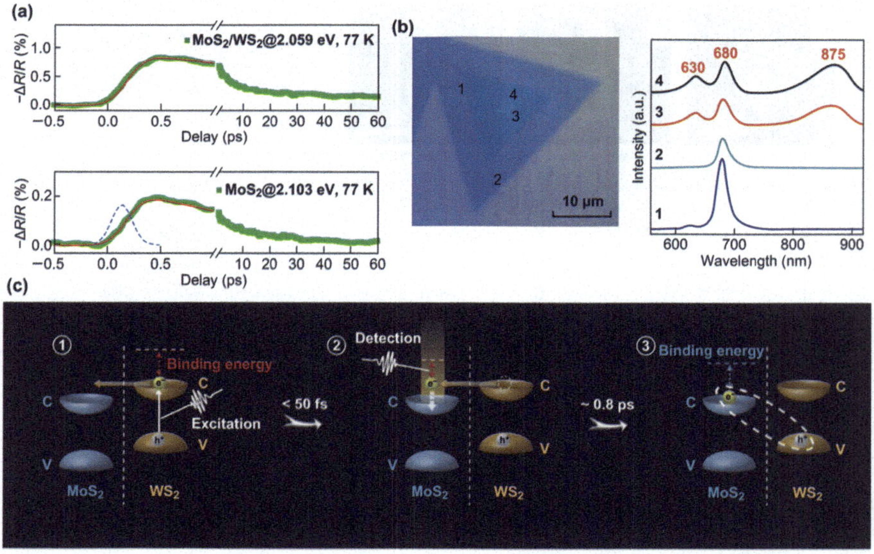

Fig. 5.5 Interlayer excitons production **a** transient absorption (Reproduced with permission from [34]). **b** PL spectra (right) acquired from regions 1–4 of a V-stacked WS_2/MoS_2-HS (left) (Reproduced with permission from [38]), **c** two procedures of interlayer excitons production (Reproduced under the terms of CC BY 4.0 license [35], https://creativecommons.org/licenses/by/4.0/)

excitons. The presence of ultimate tightly bound interlayer excitons in PL spectra can be firmly demonstrated [8, 35, 37] as depicted in Fig. 5.5b [38]. PL signal has been obtained from regions 1 and 2 of MoS_2 monolayer that correspond to A-exciton resonances. The signals from the HS regions on the other hand showed $es^{-}-hs^{+}$ resonances in MoS_2 and WSe_2, resulting in extra interlayer $es^{-}-hs^{+}$ peak. Because of charge transfer process, exciton PL signals of MoS_2 and WSe_2 have been quenched in HS [8].

5.2.3 Magnetism in 2D-Heterostructures

The layer count and grain scale are strongly related to the magnetic characteristics of low-dimensional magnetic substances, such as saturation magnetization, coercivity, and T_c (Curie temperature), among others. As a result, the magnetic properties of 2DMs frequently differ from those of their bulk form. In spintronics, valleytronics, and electromagnetics, the use of 2D magnetic materials in HSs is significant. Figure 5.6 depicts 2D magnetic crystals, and Table 5.1 lists the maximum 2D magnetic materials [8, 39]. The magnetic characteristics of 2D-HSs are required for

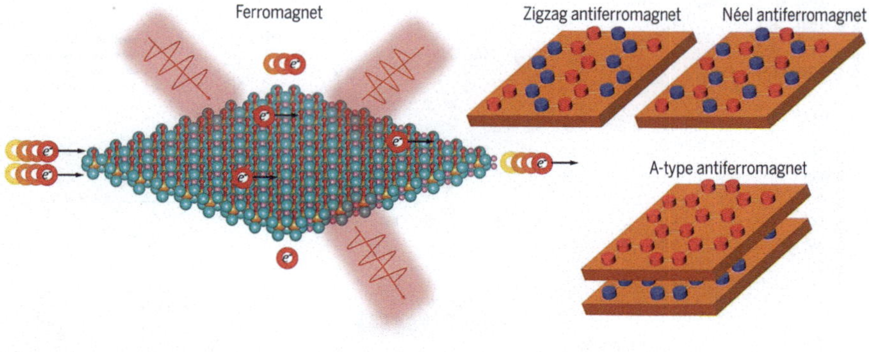

Fig. 5.6 2D-magnetic crystals: atomically thin crystalline magneto-optic and magneto-electric phenomena. 2D magnetic crystals, such as 2D-FMs (left) and 2D anti-FMs (right), the atomic Lego's red and blue protrusions represent the polar opposite local spins in magnetic layers (Reproduced with permission from [46])

Table 5.1 The vdW magnet library (Reproduced with permission from [46])

Chalcogenides	$Cr_2Ge_2Te_6$, $Cr_2Si_2Te_6$, Fe_3GeTe_2, VSe_2*, $MnSe_x$*	$Fe_2P_2S_6$, $Fe_2P_2Se_6$, $Mn_2P_2S_6$, $Mn_2P_2Se_6$, $Ni_2P_2S_6$, $Ni_2P_2Se_6$, $CuCrP_2Se_6$*, $AgVP_2S_6$, $AgCrP_2S_6$, $CrSe_2$, $CrTe_3$, $Ni_3Cr_2P_2S_9$, $MnBi_2Te_4$*, $MnBi_2Se_4$*	$CuCrP_2S_6$
Halides	CrI_3*, $CrBr_3$, GdI_2	$CrCl_3$, $FeCl_2$, $FeBr_2$, FeI_2, $MnBr_2$, $CoCl_2$, $CoBr_2$, $NiCl_2$, VCl_2, VBr_2, VI_2, $FeCl_3$, $FeBr_3$, CrOCl, CrOBr, CrSBr, $MnCl_2$*, VCl_3*, VBr_3*	$CuCl_2$, $CuBr_2$, $NiBr_2$, NiI_2, CoI_2, MnI_2 α-$RuCl_3$
Others	VS_2, InP_3, GaSe, GaS	MnX_3 (X = F, Cl, Br, I), FeX_2 (X = Cl, Br, I), MnSSe, $TiCl_3$, VCl_3	SnO, GeS, GeSe, SnS, SnSe, GaTeCl, CrN, CrB_2

external control (gate) of spin and valley (polarized) current propagation at room temperature. Vast theoretical simulations and experimental findings have shown that combining 2DMs (GR or TMDCs) with ferromagnetic (FM) materials to form vdW-HSs [40]. At the interface, a large B_o can be produced, and thus, 2DMs can be controlled to regulate the spin and valley pseudospin [41, 42]. The magnetic exchange field presence will amplify the influence of an external B_o created by an HJ's proximity effects [43]. In Fig. 5.7a, the sub-lattices of GR on EuO are shown in a variety of colors and letters. While the EuO substrate is present, the two sub-lattices of isolated GR are divided into six directories (Fig. 5.7b). The proximity effect can raise the

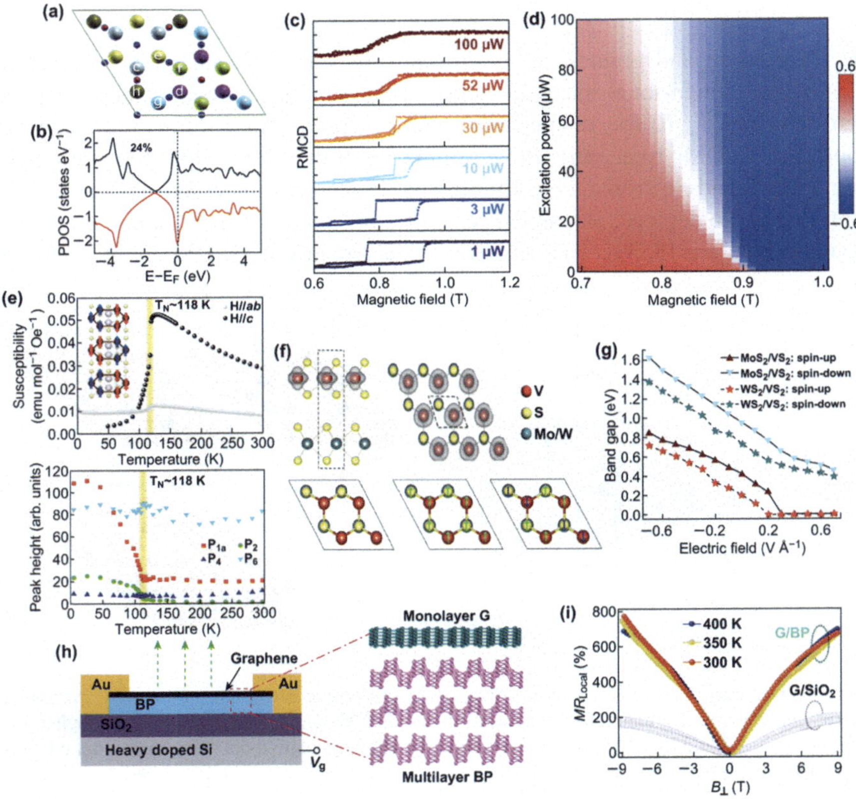

Fig. 5.7 Magnetic properties of 2D-HSs: **a** GR Broken on the EuO lattice with six sub-lattices **b** evaluated GR layer's spin polarization (Reproduced with permission from [43]), **c** The magnetic circular dichroism and **d** polarization ρ of WSe_2/CrI_3-HS under varying excitation power/voltage and sweeping external B_o (Reproduced with permission from [44]), **e** several PL peaks transition and magnetic susceptibilities (Reproduced with permission from [47]). Possible magnetic ground states (non-magnetic, FM, and anti-FM) and spin density of **f** XS_2/VS_2-HSs and **g** image depict the E_g dependency on E_o, as well as the estimated magnetic moment and relative energy (Reproduced with permission from [48]). Schematic illustration of the device based on **h** GR/ML BP-HS, **i** compared with GR on SiO_2 substrate (Reproduced with permission from [49])

surface Eu atoms surface magnetic moment, resulting in varied spin polarization on GR sub-lattices with an average observed spin polarization of around 24%, and altering GR's band structure. The current study indicates that magnetic exchange field may be regulated across a 20 T range by making small adjustments to the laser excitation strength [44]. Using reflection magnetic circular dichroism at different excitation light power, magnetization of CrI_3 in an SL-WSe_2/ML-CrI_3 HS is investigated (Fig. 5.7c). The magnetic circular dichroism's power-dependent hysteresis loop action seems to be quite identical. As seen in the PL spectra in Fig. 5.7d, an onto magnetic effect allows for power switchable valley features, in which the polarization (defined as $(I_+ - I_-)/(I_+ - I_-)$, using $I\pm$ is the intensity of peak of PL produced by a polarized laser) changes sign as the excitation strength rises. Developing acceptable 2D magnetic substances is indeed a task at the time. The above-described CrI_3 is a layered magnetic material, and the layer number has a strong correlation with the magnetic characteristics [45].

Furthermore, a new kind of magnetic HS material known as transition metal phosphorus trichalcogenides (TMPST), with TM = V, Mn, Fe, Co, Ni, or Zn and a magnetic ground state that is strongly dependent on the TM, may be easily exfoliated. From bulk to monolayer, $FePS_3$ was revealed to have an Ising-type anti-FMs ordering. Raman peaks (Fig. 5.7e) display a transition at 118 K [47] corresponding to T_N (Neel temperature). In the aforementioned HSs, FM-materials, T_c seems to be too low to add at room temperature. Based on existing theoretical equations, XS_2/VS_2-HSs are FM and have ultrahigh T_c [48]. Figure 5.7f shows non-magnetic demonstrate optimum designs, local magnetic arrangements, magnetic moments, and energy, FM, and anti-FM states. Clearly, anti-FM condition is the stronger of both HSs. The XS_2/VS_2-HSs have FM ground states and may be found using the equation below (Eq. 5.1).

$$\frac{\gamma k_B T_C \theta}{2} = E_{\text{anti-FM}} - E_{\text{FM}} \tag{5.1}$$

In the above equation, γ is the XS_2/VS_2 system dimension, k_B corresponds to Boltzmann constant, $E_{\text{anti-FM}}$ and E_{FM} are unit cell systems energies, respectively, using anti-FM and FM coupling. High T_c values obtained are 485 K for HSs MoS_2/VS_2 as well as 487 K for HSs -WS_2/VS_2. Of course, under control of the E_{ext} (Fig. 5.7g), a semiconductor/TM occurred demonstrating the ability to produce purely spin-polarized currents. Several additional 2D magnetic materials have also been tested, the majority of are MX (M = Cr, Co, V, M, Mn, and so on) and MAX (M = Cr or Fe; A = Ge; Si, Al, and Sn, and so on) and X = Te; others are semiconductors. In addition, owing to the interface state of the HSs, a magnetoresistance (MR) effect was observed in certain GR-HSs. The giants MR (defined as $\frac{[R(B)-R(0)]}{R(0)} \times 100\%$) of 775% (Fig. 5.7i), and non-local MR over 10,000% of the GR/BP gadget exhibits enhanced flavor caused by the BP channel in the Hall effect at room temperature. Liu et al. [49] have fabricated an extremely stable monolayer GR/BP device (Fig. 5.7h). These test findings provide useful knowledge to analyze magnetizing dynamics of magnetoresistive devices, like random access memories.

5.3 Fabrication

5.3.1 Mechanical Transfer Methods

V-HSs were produced via a continuous mechanical transition of arbitrary 2D layers to a substrate, e.g., with changing specific NS one by one from the top down. However, the vdW-HSs from this approach faced drawbacks, for example, in form of transport characteristics. Fortunately, a range of approaches has been devised to mitigate the transporting challenge by making large-area GR with high accuracy from its growing substrate/postexfoliation support to random substrates. In principle, these transfer techniques may also be employed for bottom-up fabrication of supreme quality vdW-HSs with several layers [50]. The 2DM is coated by a PMMA film or thermal release tape [51] and moved to target substrate with carrier foil process. The material is then liberated through dissolving the PMMA or heating the sheet Fig. 5.8 [51]. This subject has multiple versions, utilizing more layers of sacrificial and various kinds or transfer polymers thicknesses. The ideal way to exfoliate the 2D surface is to use a PMMA film over a-Si/SiO_2 substrate, or on alternative sacrificial, water-soluble polymer layer, for instance, polycarbonate (vinyl alcohol). The water-soluble layer of the three-layer structure dissolves in a bath, revealing the two-layer/PMMA film that floats on the boundary of air–water. [52]. Capillary forces between a hydrophobic polymer (e.g., cellulose acetate butyrate) and a hydrophilic substrate can be used to facilitate the release of a 2D layer/polymer film from the substrate when a water

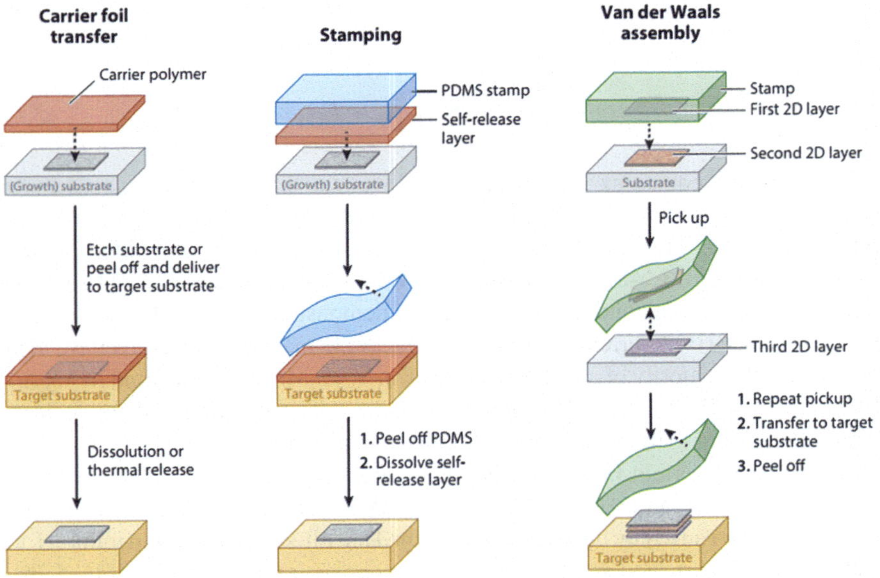

Fig. 5.8 Different vdW transfer techniques based on different types of polymer films and elastomers (Reproduced with permission from [50])

drop is introduced at the hydrophilic/hydrophobic interface, resulting in a complex liftoff of the 2D layer/polymer film (so-called wedging method) [53]. Owing to its simplicity, one major downside of this ML-assembly strategy seems to be the inability to correctly match the moving layer with the target layer due to thin polymer film mechanical instability. However, by attaching the polymer's back to a strong, transparent substrate, this limitation can be somewhat overcome. However, the use of solvents (some of which are harsh, such as KOH) in removing transition polymer and annealing phases at temperatures of approximately 100 °C/both limits the usefulness of this transfer process for fragile 2D structures. Another limitation of 2D layers is interfacial quality occurs when various types of polymers and solvents come into contact during the process and generally results in poor extraction of sacrificial polymer layers.

The second general approach for moving 2DMs from one substrate to another is PDMS microcontact stamping or other elastomer stamps. This technique is simple and does not need complicated chemical dissolving or thermal detachment operations, making it perfect for dry sequential transfer of many 2D layers over a substrate. This technique also has an alignment capability that makes use of stamp's transparency and mechanical endurance, allowing the use of preprinted stamps and, as a result, the manufacture of patterned HSs [50]. Although the technique needs a careful balance of the surface and adhesion characteristics among the stamp on one side and the 2D layer on the other because this technology does not employ capillary forces for detachment, freely hanging objects that would otherwise fall are conceivable. The stamp's quality and completeness have largely defined the interfacial quality level. Several dry stamping variations have been created to imprint 2D layers on arbitrary surfaces with little contamination from exposed interfaces, particularly soft and holey substrates. In one strategy, sacrificial polymer layers behave as a shield for the moving 2D layer while enabling the transition phase to the acceptor substrate by varying the adhesive pressures at the various interfaces.

The transfer procedures between translated 2DM and conventional PDMS stamps rely on a thin layer of polymer-poly-L lactic acid, polystyrene/other hydrophobic aromatic hydrocarbons/fluorocarbon polymers (Fig. 5.8) were devised by Li [55] and Song et al. [56]. Here, a water droplet passing among the hydrophilic Si/SiO_2 substrate and the hydrophobic polymer interface generates capillary forces that separate the PDMS/polymer/2D-layer sandwich from the Si/SiO_2 substrate [55]. The PDMS stamp is detached using a 50 °C bake and dichloromethane removal after placing peeled-off film in conformal contact with the accepter surface. Complete removal of any stamp/sacrificial polymer coating may leave tiny flaws on 2DMs exposed surface; thus, a stamped and polymer-free transfer technique looks to be the next important milestone. Dean and coworkers [54] developed a beneficial vdW assembly technique that eliminates the requirement with a stack of sacrificial layers that react with 2D components. This approach includes a vdW crystal that will be deposited on top of carrier polymer (for instance, a 1-m polypropylene carbonate layer (PPC) on Si-wafer), and afterward, PPC/2D layer will be deposited over the top of a clear PDMS stamp. The sandwiched arrangement is then put on a Si/SiO_2 wafer react with another exfoliated 2D crystal when heated to 40 °C, as seen in

Fig. 5.9. This method may be repeated as long as vdW adhesive force applied to 2D stack formed on PPC layer exceeds adhesion to substrate. Chloroform dissolves the PPC, leaving behind the sterile ML pile. This method may be successful if it avoids critical stack interfaces from coming into touch with a polymer during the whole transfer procedure, decreasing the risk of corrosion and defect entrapment at interfaces.

The mechanical transfer method usually produces turbostrometric stacks with no L-layer registry unless a micromanipulator meticulously controls the layer orientation [14]. Determining the misalignment angle of a layer, even for stacks of over two layers, is very easy by the analysis of Moire patterns in STM by unraveling hetero-ML electron diffraction patterns in STM, conductive AFM/HRTEM pictures, conductive AFM/HRTEM images (Fig. 5.10) [57]. Additionally, 2D layers mechanical transfer for different layer combinations and substrates has been investigated but typically

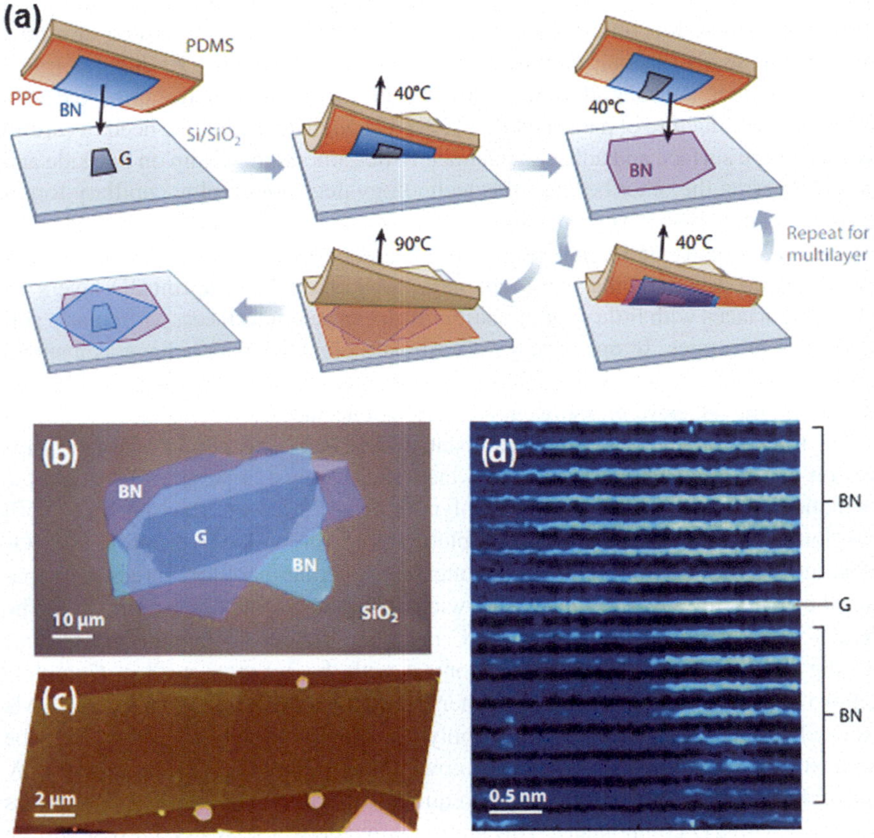

Fig. 5.9 Schematic representation of **a** polymer-free vdW assembly method **b** h-BN/G/h-BN-HS optical picture created with panel a, **c** AFM **d** HR-cross-sectional ADF-STEM, ML structure pictures revealing wrinkle-free and atomically thin clean interfaces (Reproduced with permission from [54])

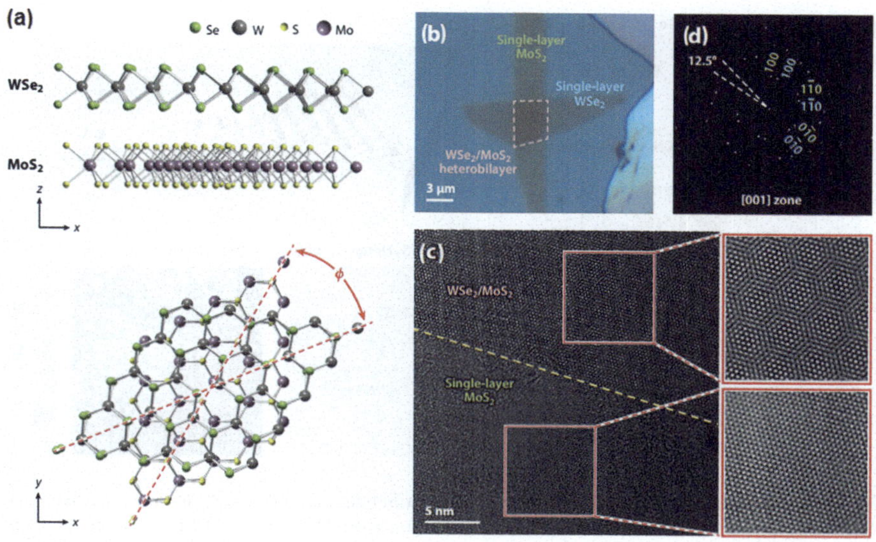

Fig. 5.10 **a** Visualization of WSe_2/MoS_2 BL structure, **b** optical microscope micrograph, **c** HRTEM micrograph, **d** SAED pattern of HS along [001] (Reproduced with permission from [57])

recorded HSs are restricted to two and four layers, with few significant exceptions (see Fig. 5.11) [58], which is testament to the mechanical transfer's complex multi-step processes. In comparison, interfacial efficiency is also extremely high and results in ML structures that are important for many device applications that have atomically clean interfaces [56]. Gorbachev and colleagues [58] used cross-sectional TEM and local EELS resolution to investigate the interfacial structure of ML GR/h-BN superlattices. According to these observations, abundant contaminations of hydrocarbons are adsorbed by 2D crystals in manufacturing processes that appear to spread over a distance of micrometers inside the stack. Resultantly, separating them into isolated pockets and making their 2D interfaces very atomically clean. Especially, interlayer

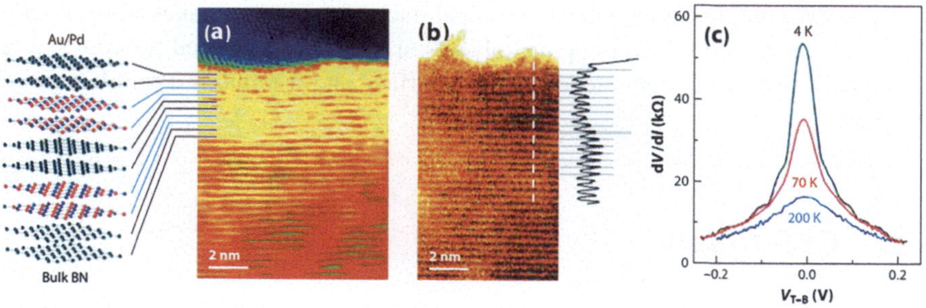

Fig. 5.11 **a** Representation of layer arrangement and accompanying bright-field stack STEM, **b** HAADF-STEM image of the same HS. The intensity line profile is displayed on the right, **c** V-transport (Reproduced with permission from [58])

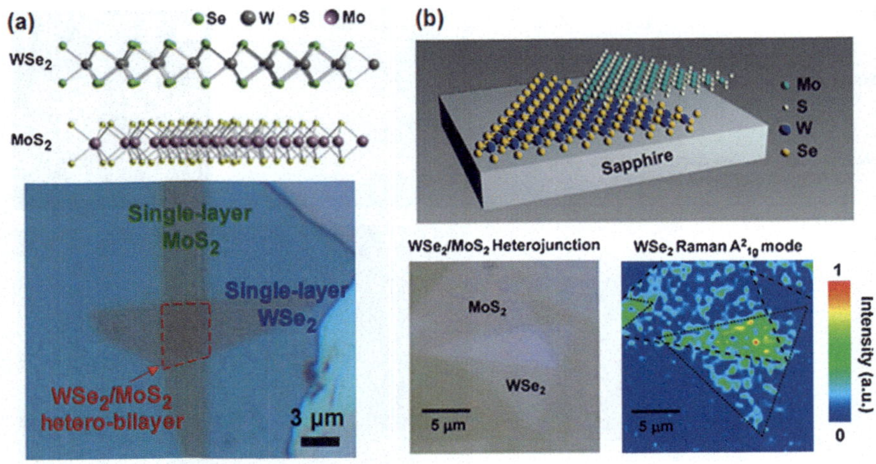

Fig. 5.12 Mechanical stacking process for V-HSs: **a** Schematic and optical microscope image of stacked WSe_2/MoS_2 HJ (Reproduced with permission from [57]), **b** CVD MoS_2/WSe_2 stacking HJ schematic diagram and optical microscope image (Reproduced with permission from [60])

isolation from basal plan distance in planes of h-BN and GR is indistinguishable, confirming that major interface contamination is impossible.

Chang and his coworkers [59] conducted a second harmonic generation (SHG) measurement to show twisting angle effect on TMDC BL stacking. The SHG from twisted two-layer system is a consistent overlay of the second harmonic fields from each stacking layer and angle only determines phase difference. Fang and colleagues [57] examined the optical transformation properties on MoS_2/WSe_2 stacking HJ for the interlayer coupling (Fig. 5.12a), The band alignment and interlayer coupling at the interface cause a 100 meV shift in the stokes, with direct absorption and indirect emission seen at HJ. The V-MoS_2/WSe_2 HJs are manufactured via stacking CVD-grown MoS_2 and WSe_2 showed that thermal treatment could be used to increase the interlayer coupling (Fig. 5.12b) [60]. Interlayer coupling with optic transition and belt alignment with STS and XPS may be used to determine HJ's band alignment and $es^{-}-hs^{+}$ binding energy. The mechanical stacking approach makes arbitrary 2D-HSs possible. Nevertheless, trapping solvents or the chemicals used for conversion can quickly impair the consistency of the interface and it remains a problem to be resolved.

5.3.2 CVD Growth

CVD synthesis is a bottom-up approach used in the production of V-2D HSs that may also be utilized to HS planar multijunction. To generate an atomically sharp HJ and a clean interface in V-stacked HSs, a variety of CVD growth methods have been studied, along with adaptable and controlled manufacturing in V- and L-HSs [1, 8]. We briefly discuss the various CVD growth approaches in the subsections below.

5.3.2.1 One-Step CVD Method

The procedures of this method are appropriate for manufacturing HSs with component materials including the same components and are shown schematically in Fig. 5.13a. Gong et al. [38] manufactured V- and L-MoS_2/WS_2 HSs to using this method, and it was discovered that IP epitaxial growth at a lower temperature of

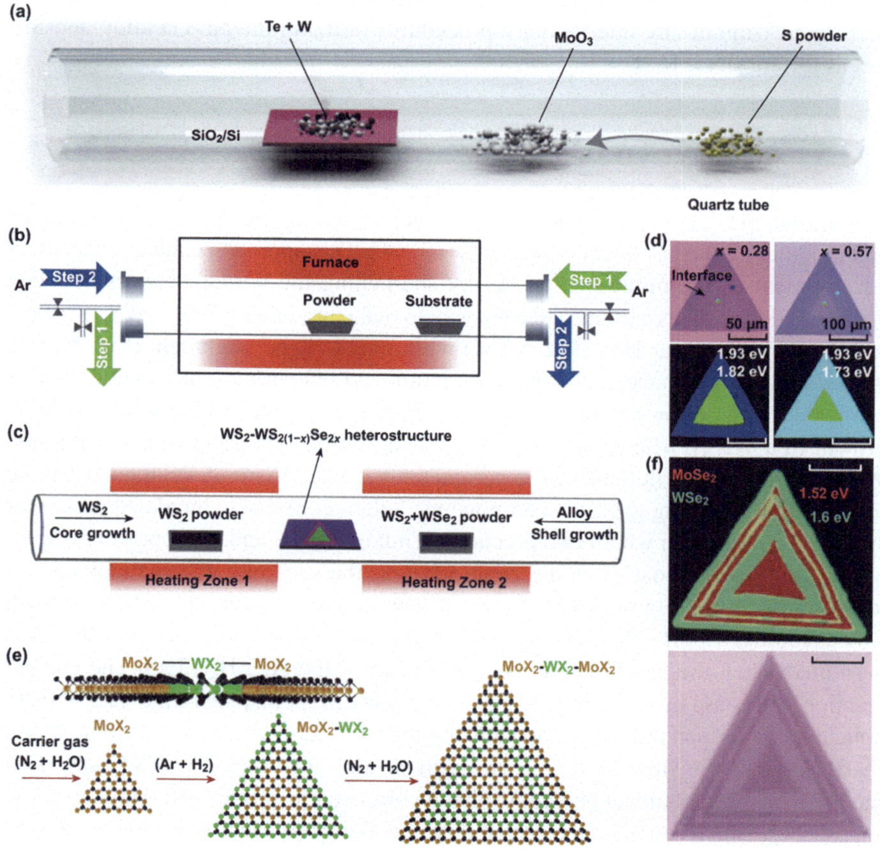

Fig. 5.13 **a** Diagrammatic illustration of the one-step CVD method for the synthesis of MoS_2/WS_2-HSs (Reproduced with permission from [38]), **b** multi-step CVD growth (Reproduced with permission from [65]), **c** modulable growth of monolayer, **d** optical picture and PL intensity mappings of HSs with different x (Reproduced with permission from [63]), **e** one-pot synthesis strategy principle, and **f** optical picture and PL intensity mappings (Reproduced with permission from [64])

650 °C was insufficient for nucleation of WS_2 on the MoS_2 surface, but that at a higher temperature of 850 °C, it could form a V-stacked structure as a vdW- HSs.

5.3.2.2 Two-Step CVD Method

This methodology, using as-grown layered crystals as a second layer substrate, is more extensively employed. Such processes have been also altered for V- and L-stacked HSs, principally impacted by gas flow rate and synthesis time. Li et al. [61] employed this two-step CVD technique to HS-2D GaSe/$MoSe_2$. Many MX/MX_2 V-HSs feature incommensurate superstructures due to the significant lattice misfit between two layers. Moreover, a stacked TMDC/h-BN HS fabrication has been performed without any intermediary procedures utilizing the Ni–Ga alloy and Mo foil as the substrate [62].

5.3.2.3 Multistep CVD Method

The multistep CVD system was developed via switching the direction (Fig. 5.13b) and modulating the gas flow components [63, 64]. This approach enables successive development of the multijunction HS by sharpening the border between HJ. Pan et al. [63] defined it as a quick and easy way to make modulable $WS_2{-}WS_{2(1-x)}Se_{2x}$ $(0 < x \leq 1)$ SL-planar HSs (Fig. 5.13c) with tunable band alignment (Fig. 5.13d). With dual heater furnace, a change in direction and temperature of chamber can be controlled. Ar-gas flow can be organized by flipping the synthesis from the pure WS_2 growth to a $WS_{2(1-x)}Se_{2x}$ $(0 < x \leq 1)$ alloy formation. The ratio of the combined WS_2/WSe_2 powders influences the parameter x. Furthermore, development can be regulated by adjusting gas flow components. Sahoo et al. [64] elucidated a one-pot synthesis approach in which two precursors (mixed MoX_2 and WX_2 powders) were placed in the same boat's heating zone, whereas the substrate was maintained at a lower temperature. Figure 5.13e shows clearly the mechanism, only MoX_2 growth was encouraged in the $N_2 + H_2O_{(g)}$ gas stream. Here, only the WX_2 growth has is permitted when the gas flow had been switched to the Ar + H_2 (5%). The current reaction will cease by reversing the gas flow. Figure 5.13f displays a 240 and 250 cm optical and a Raman intensity map showing the well-defined spatial distribution and sharp $MoSe_2$/WSe_2 interface for a HS with several junctions. The CVD approach seems to be more compact and controllable compared to single- and two-step CVD approaches, which enhances the probability of making optoelectronic devices chosen spatially by separating es^- from hs^+ into various materials.

5.3.2.4 Vertically Stacked 2D-Heterojunctions

With CVD techniques, a 2DM can be grown on another V-stacked HJ. In general, direct growth results in clean HJ interfaces. As the vdW interaction between the

2D layers is the only way the structure could be resolved via vdW epitaxy, the early experiment begins with GR and h-BN, which share a common grid. Yang and colleagues described a plasma-assisted deposition technique for growing single-domain GR on h-BN support [66] (Fig. 5.14a). Further, GR has a favored orientation on the h-BN lattice, and its size is only constrained by the field of the underlying h-BN. Shi and colleagues [67] have developed a V-layered MoS_2/GR configuration by thermally decomposing $H_8N_2MoS_4$ precursors on GR surfaces (Fig. 5.14b). Even

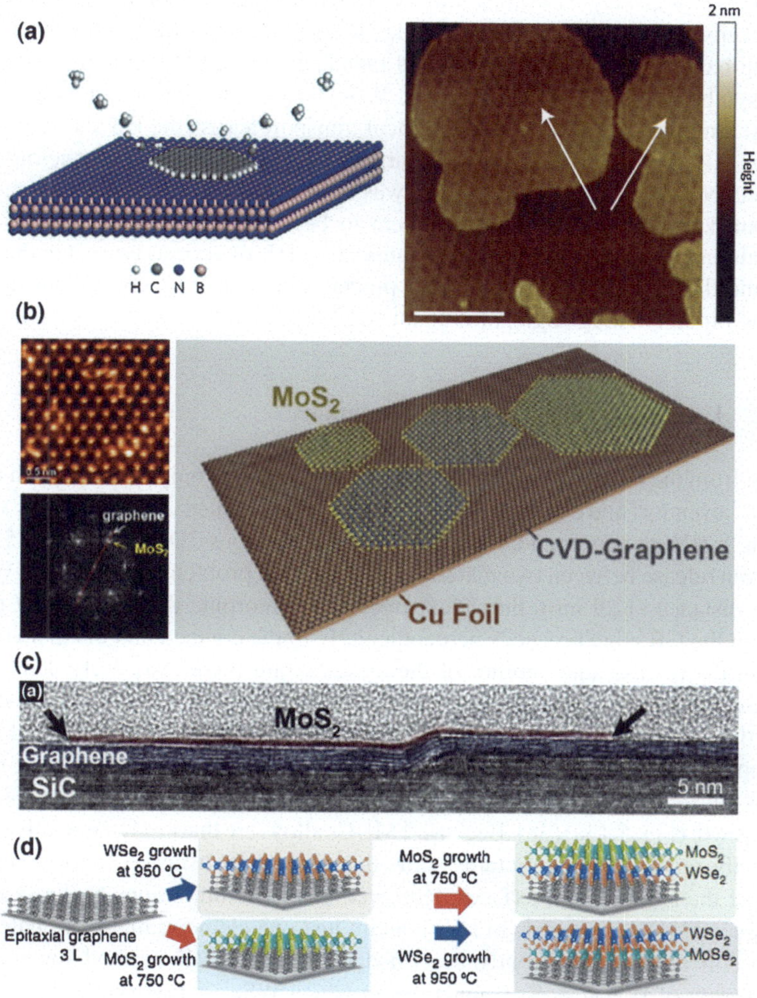

Fig. 5.14 V-HSs growth: **a** a schematic representation of the plasma-assisted deposition process (Reproduced with permission from [66]), **(b)** MoS_2 grown on GR (Reproduced with permission from [67]), **c** cross-sectional HRTEM (Reproduced with permission from [68]), **d** V-HSs of MoS_2/WSe_2/GR and WSe_2/MoS_2/GR (Reproduced under the terms of CC BY 4.0 license [69], https://creativecommons.org/licenses/by/4.0/)

though MoS_2's lattice constant is around 28% higher than GR's, GR remains a strong growth platform for MoS_2. Additionally, V-epitaxial growth is expected to withstand the crystal orientation discrepancy, whereas MoS_2 growth on GR may need strain to accommodate the lattice mismatch. Lin et al. [68] used CVD methods to show the MoS_2 direct growth, WSe_2, and h-BN on GR, where the underlying GR was grown epitaxially on SiC (Fig. 5.14c). The growth and characteristics of top HSs provide the nucleation centers for the formation of the upper layer material with strain, wrinkling, and flaws on the GR surface, which are highly influenced by the morphology of GR layers. Another area of study is the growth of TMDCs HJs. WS_2 may develop on top of MoS_2 thanks to the 2H stacking structure. On the other hand, utilizing oxide powder vaporization and metal_organic-CVD techniques show the direct synthesis of MoS_2/WSe_2/GR and WSe_2/MoS_2/GR-HSs in situ (Fig. 5.14d) [69]. At ambient temperature, the electrical transport across the HS shows a strong negative differential resistance due to charge carrier special resonant tunneling at the interface. Nonetheless, this observation was not made in junctions that are mechanically stacked. The direct syntheses seem to be theoretically applicable to other materials and open up new avenues for fabricating HJs of various types. On the other hand, the thermodynamically regulated process may constrain controllability over stacking order, twisting angle, or domain size [70, 71].

5.3.2.5 Laterally Stacked 2D-Heterojunctions

Arising from the covalent bonding of 2DMs, the L-HJ will only be made with direct growth. Levendorf and coworkers [72] revealed the L-stitched GR/h-BN HJ by CVD-growing h-BN on patterned GR, resulting in a continuous 2D HJ film (Fig. 5.15a). The strain release between two materials is the general problem of L-HJ [73]. Within a short distance (1.29 nm), the defect-free, pseudomorphic development of the h-BN from the GR edge proceeds perpendicularly to the contact, and 0D dislocations result in the release and stability of the elastic strain force. The early demonstration was accomplished with a one-pot synthesis for the L-TMDCs HJ. For instance, HJ on the L-$MoS_2/MoSe_2$ and WS_2/WSe_2 have been supplied by a physical transport vapor technique with sequential source switching, $MoSe_2/MoSe_2$ using mixed sources physical vapor transport [74], and WS_2/MoS_2 using single point multisource CVD growth (Fig. 5.13a) [38]. Because TMDCs alloys are thermodynamically stable, atomically resolved TEM pictures show that atom interdiffusion (alloy structure) occurs at the L-junction interfaces. However, given the limitations of the synthetic technique, TMDCs materials in HSs contain just one element: metal or chalcogen. Li and colleagues [75] recently described a two-step CVD method for epitaxial development of WSe_2/MoS_2 L-junctions in which WSe_2 is first synthesized on the substrate through vdW epitaxy and then epitaxially grown along the W growth front (Fig. 5.15b). The TEM image demonstrates that the WSe_2-MoS_2 interface is atomically sharp. Similarly, by a two-step CVD process, Gong and colleagues [76] found a $WSe_2/MoSe_2$ HJ. Depending on the synthesis time, the WSe_2 BL will emerge first from the SL-$MoSe_2$ edge, forming an L-junction, and then cover the $MoSe_2$

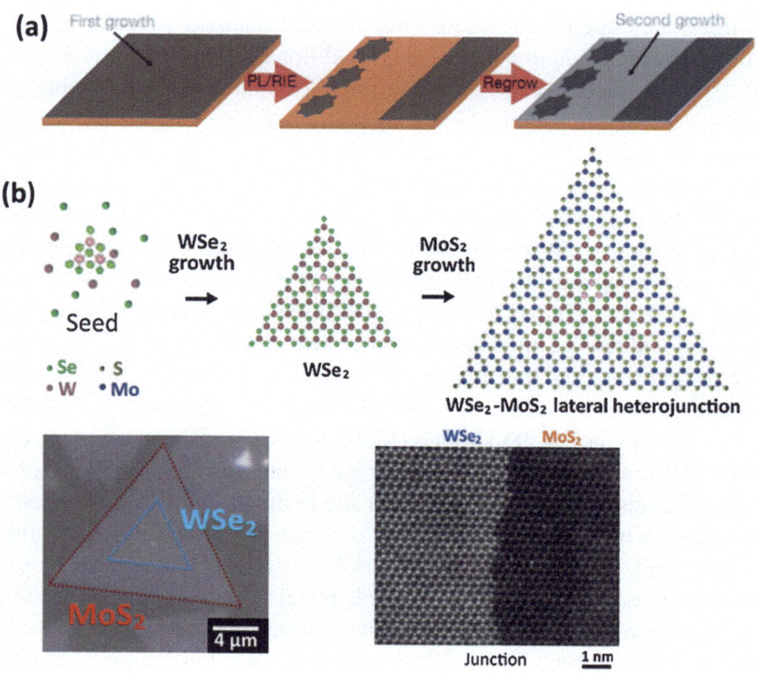

Fig. 5.15 IP epitaxial development of L-2D-HSs: **a** schematics of GR/h-BN L-HJs by utilizing CVD (Reproduced with permission from [72]), **b** depiction of two-step formation of L-HJ and SL-WSe_2/MoS_2 with optical micrograph of junction, with high-resolution STEM micrographs taken at interface displayed below right (Reproduced with permission from [75])

surface. For scientific research and application, the two-step growth method allows the creation of HJs with a variety of compositions and patterns. Besides the peculiarity of the form, this L-HJ also shows inherent p–n junction characteristics, i.e., characteristics rectification and photovoltaic effects, which are promising attributes for SL electronics [77, 78].

5.3.3 Doping and Chemical Functionalization

Controllable doping techniques are important for the flexible usage of 2DMs in next-generation optoelectronics and electronics, taking into consideration the necessity for Schottky barrier modulation at TMDCs/metal junctions, as well as the fabrication of appropriate junctions. Traditional doping procedures used in conventional bulk semiconductors, including doping diffusion and ion implantation, owing to crystal degradation, are not suited for atomic, thin 2D layers [79]. This is why the chemistries that are usually performed on organic molecules and surfaces have been employed to identify rapid and effective pathways for optical and electrical modulation as a foundation for chemically functional materials. Additional incentives for chemical alteration of 2D layers include charge storage, chemical and biological sensing, and the production of new composite materials, in addition to changing the doping amount. GR-like chemical techniques are used in a variety of ways [80–82] have been used to regularly tune the optoelectronic properties of TMDCs by modulating the charge carrier form and concentration. The MoS_2, $MoSe_2$, WS_2 or WSe_2 single and ML through the molecules physisorption [83, 84], surface functionality of metallic NPs [85, 86], solution-dependent functionalization [86, 87], and plasma-assisted doping are addressed [88, 89]. The molecular physisorption mechanism can regulate charge density over a wide range, modulating optical properties for instance PL processes. Tongay et al. [84] revealed that charge transfer occurs when electronegative gaseous molecules are physically adsorbed such as O_2 and H_2O significantly modulates the light emission efficiency of 2D TMDCs (MoS_2, $MoSe_2$, and WSe_2), with modulation of over 100-fold, whereas inert gases have a marginal impact. The physisorbed O_2 and H_2O molecules resulted in a significant rise in the PL of electro-depleted n-type materials along with MoS_2 and $MoSe_2$. In contrast, the p-type WSe_2 exhibited the reverse trend [83]. The impact of doping on MoS_2 is also investigated by utilizing electronic beam (e-beam) deposition of noble metals (Au, Ag, Pd, and Pt) NPs (i.e., metallic NPs), in which it was discovered that p-type doping results in an increasing pattern of metalwork functions as NP dosage increases. Furthermore, Mouri et al. [86] used F_4TCNQ and TCNQ as p-type chemical dopants and NADH as n-type dopants to efficiently adjust monolayer MoS_2 (PL); p-type dopants increased intensity, while n-type dopants decreased it. The p-type dopants detach electrons from MoS_2 so that the excitons recombination can become a prevailing method rather than a negative trion, whereas n-type dopant molecules inhibit this mechanism shift. Through the surface transfer method, organic molecules like benzyl viologen (BV) might form an electron transport complex with n-doped MoS_2 [90]. The BV-doped MoS_2 shows good stability in ambient air (with just a little increase in electrical characteristics after 9 days) and an electron sheet density of ~1.2×10^{13} cm^{-2} corresponds to the highest MoS_2 degenerate doping density [90]. The neutral BV molecule passes es^- to MoS_2, resulting in a two-electron transfer reaction between the two products (Fig. 5.16a). The doping process is reversible and can be used not only to selectively remove the molecule for pattern doping but rather to modulate the charge carrier density in such a way that the molecules of BV on the MoS_2 surface could be isolated by immersing it in toluene [90]. Raman spectroscopy and electrical measurements show that ML-MoS_2 is n-doped by p-TSA that was commonly used for tailoring conducting polymers' charge density [91]. TMDCs flakes may also be controlled by modifying SiO_2 substrates utilizing functionally self-assembled monolayers (SAMs); consequently, the redistribution of charge among flakes and substrates has changed [87]. The intrinsic n-type characteristic of MoS_2 flakes (p-type doping) is reduced by fluoroalkyl Trichlorosilane-SAMs with a high positive dipole moment, while 3-(trimethoxysilyl)-1-propanamine SAMs increase it (n-type doping). Due to their small dipole moment, CH_3-terminated SAMs have minimal influence on charge carrier density, while NH_2-/SH-SAMs with lone pair electrons and CF_3-SAMs with

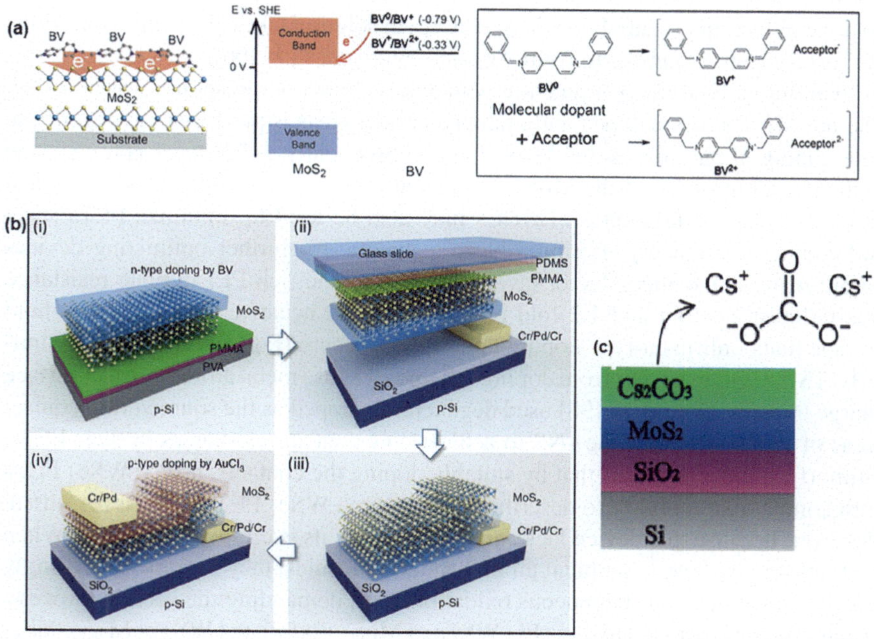

Fig. 5.16 **a** Schematic representation of energy band structure and BV redox states of MoS_2 (Reproduced with permission from [90]), **b** A FL-MoS_2 flake with a chemically doped V-p-n homogeneous junction (i) flake was moved to a PMMA/p-TSA/Si substrate, doped with BV, and then annealed. **(ii)** PMMA film containing a MoS_2 flake transferred to a PDMS/glass substrate, (iii) MoS_2 flake stamped into the SiO_2/Si substrate, (iv) MoS_2 flake with V-p–n junction after $AuCl_3$ doping and annealing (Reproduced under the terms of CC BY 4.0 license [103], https://creativecommons.org/licenses/by/4.0/), **c** schematics of MoS_2 FET configuration with a Cs_2CO_3 film on top (Reproduced with permission from [85])

high electronegativity have e^- and h^+ donating properties, respectively [87]. A large range of stable and adjustable doping capability, notably over the natural inclination of TMDCs (n-type and p-type for MoS_2 and WSe_2, respectively), is essential to generate functional junctions that are needed as essential building blocks of electrical and optoelectronic devices. Since TMDCs have such a strong predilection for a given doping form, substantial work was being put toward establishing PN connections based on V-HSs by stacking several TMDCs of distinct kinds [92, 93]. Suh et al. [94] established V-p–n junction using such a single 2DM by stacking a hole-doped MoS_2 layer (p-type) on top of an undoped MoS_2 layer (n-type). This results in a homogenous p–n junction with gate controllable current rectification. The p-doped MoS_2 is synthesized by substituting Nb for Mo atoms through a chemical vapor transport technique that employs covalent bonding in the crystal lattice to achieve stable doping. A vdW p–n junction has been formed using the surface charge transfer technique. MoS_2 was doped with BV and $AuCl_3$ to make it an n- and p-type semiconductor material, respectively (Fig. 5.16b) [79]. The holes were doped by spin-coating an $AuCl_3$ solution, which is frequently employed in GR doping because of its high positive reduction potential, in which $AuCl_4$_in solution takes electrons from MoS_2 via surface charge transfer to form Au nano-aggregates [95, 96].

In addition, L-MoS_2 *p–n* junction with an excellent photoresponse was revealed. The interface between doped and undoped crystal areas is produced via degenerate hole doping of partially veiled MoS_2 [94]. Subsequently, h-BN layer is transferred to the partial masking of the MoS_2, to prevent doping in the MoS_2 area. Chemical and molecular modulation of TMDCs may also be used to minimize FETs sheet and contact resistances, offering a basic technique for further optimizing devices output using 2D semiconductor layers. An ML-amine-rich PEI in plate resistance was reduced 2.6-fold and 1.2-fold reduced with the negative change of threshold voltage that confirms n-type doping upon PEI exposure (Fig. 5.16c) [85, 97]. Similarly, TMDCs may benefit from doping schemes that use metal atoms through surface charge transfer. Fang et al. [98] used degenerate n-doped as the source/drain contact areas of FL-MoS_2 and WSe_2-NS to achieve low contact resistances in their FETs. Similarly, the work reveals that by suitably doping the contacts, n-type WSe_2 FETs with comparable mobility to describe earlier p-type WSe_2 FETs may be permitted. However, because this device lacks air stabilization, its usefulness is limited. When K atoms are exposed to ambient air and water molecules, they become exceedingly reactive, resulting in instantaneous oxidation of K, demanding the vacuum process. In addition, Cl is utilized to dope FL-WS_2 and MoS_2. Exfoliated WS_2 or MoS_2 flakes were immersed in undiluted DCE for more than 12 h at room temperature, giving high doping densities of 6.0×10^{11} and 9.2×10^{12} cm^{-2}, respectively [99]. According to a theoretical process, the doping was performed by substituting a Cl atom for the S vacancy. Along with the advantages gained by earlier examples of degenerate chemical doping of TMDCs, the ability to design TMDC-based optoelectronic and electronic devices with wide-range charge carrier density modulation requires the development of robust and adjustable non-degenerate doping schemes that do not cause crystal damage. Lin et al. [85] non-degenerate n-doping of monolayer MoS_2 by surface functionalization with Cs_2CO_3 has previously been utilized as an electron injection material in organic-LEDs and may cause n-type doping in organic semiconductors (Fig. 5.16d) [100, 101]. By varying the concentration of ODTS, a controlled non-degenerate p-type doping approach was shown on FL and bulk types of WSe_2 [102]. The positive dipole of the methyl ($-CH_3$) functional groups in ODTS reduces the electron density in a WSe_2 crystal by concentrating charge carriers at the ODTS-WSe_2 interface. The doping quantity obtained with different ODTS concentrations varied from 1.6×10^{10} to 3.0×10^{11} cm^{-2}, indicating that doping concentration was in the non-degenerate regime.

5.3.4 Electrostatically Assembled Heterostructures

Since the 1990s, the bottom-up assembly of charged NS into artificial HSs has been intensively addressed using well-established colloidal chemistry and Chimie Douce ideas [50, 104, 105]. A key benefit was the logical creation of complex materials by deconstructing solids into their core structural building components and then reconstructing them into HSs, a process that is remarkably similar to the widely utilized idea of complete synthesis in organic synthetic chemistry [106]. Mallouk and colleagues [104] used the term molecular beaker epitaxy in the 1990s to refer to the creation of artificial and metastable solids at low temperatures that are not possible to achieve via thermodynamically driven, high-temperature solid-state synthesis. By mixing oxidic anionic and cationic NS through electrostatic self-assembly, ionically bound layered materials are less vulnerable to mechanical cleavage than vdW solids and are often exfoliated by solution processing, especially by ion-exchange intercalation and LPE. A well-known technique for effectively delaminate a range of ionic layered oxides, as well as perovskites and titanates, into particular NS involves protonation of the alkali-containing precursors followed by cation exchange with the bulky tetrabutylammonium hydroxide. The neutralization of protons by the tetrabutylammonium counterion OH is the major driving factor. Exfoliating protonated ML-oxides using primary and tertiary amines such as ethylamine and 2-(dimethylamino)-ethanol was also successful. As a result, layer charge density, swelling agent size and polarity, and the type of interlayer gallery ion all play a role in osmotic swelling and exfoliation. It was also possible to exfoliate protonated ML-oxides utilizing primary and tertiary amines like ethylamine and 2-(dimethylamino)-ethanol. As a result, osmotic swelling and exfoliation are influenced by layer charge density, swelling agent size and polarity, and the kind of interlayer gallery ion [107]. In comparison with uncharged NS, charged 2D NS can be thought of as colloids, gigantic polyions, or inorganic polyelectrolytes and are therefore susceptible to the same solution and colloidal processing processes as organic polyelectrolytes and colloids. Three distinct methods, namely flocculation (Fig. 5.17a), electrostatic layer-by-layer assembly (Fig. 5.17b), and Langmuir–Blodgett (LB) transfer (Fig. 5.17c), may be utilized to controllably assemble 2D polyelectrolytes into lamellar HSs.

5.3.4.1 Flocculation

The easiest method for heterolayering various types of NSs is to destabilize the colloidal suspension of one component by adding or introducing the second component. Overall, flocculation can be accomplished by mixing cationically and anionically charged NS suspensions, or by destabilizing a mixed solution of different anionic (or cationic) NS with an electrolyte, such as NaCl (Fig. 5.17a). While flocculation is quick and easy and produces large quantities of HSs [108], the layer structure, layer registry, and interfacial quality are all within limited control. As a result, flocculation generally results in turbostratically disordered lamellar materials that are infamous for their difficulty of analysis. The Hamaker constant, the type, and the electrolyte's ionic strength, and the surface potential of the colloid all play important roles in colloidal system stability [50]. In addition, NS dispersions are thermodynamically not stable, and NS formation is a fast, irreversible process whose kinetics is determined by the colloid contact frequency and the probability of cohesiveness during impact [50, 109]. The binary solids' degree of inner satisfaction and stacking

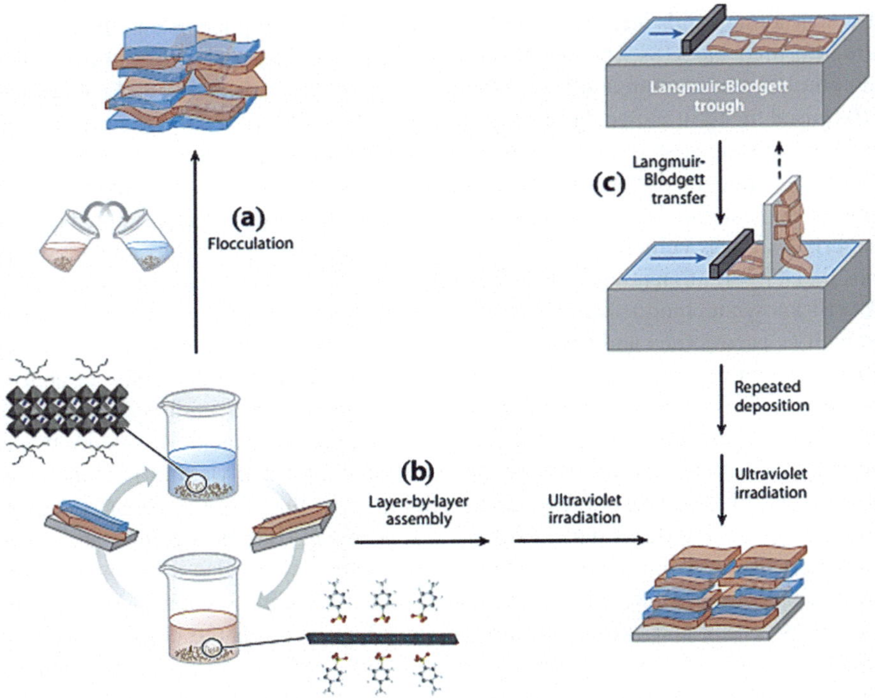

Fig. 5.17 Three WC-assembly strategies for HSs consisting of ionic and/or nonionic components are depicted schematically: **a** flocculation, **b** electrostatic layer-by-layer assembly, and **c** LB-transfer (Reproduced with permission from [50])

disorder, of coflocculated NS is rarely investigated in-depth, despite the fact that it is highly dependent on the kind of NS, their respective surface and interaction energies and their initial state of dispersion Sasaki and coworkers [110] demonstrated that restacked binary colloidal suspensions of negatively charged NS flocculated with KCl can form a variety of complexes, ranging from fully interstratified solid solutions to costacked systems with miscibility gaps to completely phase-separated systems, as simulating the diffuse scattering intensity of the resulting XRD patterns. Flocculation was used to make a number of heterolayered composites.

5.3.4.2 Langmuir–Blodgett Assembly

The LB method is a versatile tool to assemble, compress, and transfer molecularly thin films spread at the air/water interface to arbitrary substrates with precision at the (sub) nanoscale. The LB technique has recently been modified to allow the transfer of colloidal particles that have or may gain amphiphilicity, such as organoclays or MoS_2, as well as monolayers built up of amphiphilic molecules like surfactants [111, 112]. While the amphiphile is usually encapsulated in chloroform (volatile organic solvent and distributed in an LB trough on the water subphase, monolayer lepidocrocite-type titanate films can be made entirely without them. [113]. Because of the apparent hydrophobicity of $Ti_{0.91}O_2$ NS, it segregates spontaneously at the air/water interface is assisted by tetrabutylammonium cation adsorption to the negatively charged Ns. Once the surface pressure has been equilibrated by either horizontal or V-lifting of a substrate immersed in the LB trough, the NS monolayer may be transferred at a constant pace (Fig. 5.17c). The method's strength and adaptability relate to the fact that it permits the production of densely packed NS monolayers whose packing density could be accurately changed using isotherms of A (A = molecular area) and colloidal concentration adjustment. Additionally, several layers of neutral and ionic NS can be transferred with the same surface charge. This property is especially advantageous because the great majority of NS have a negative charge on their surface. Exfoliated NS suspensions, on the other hand, include cationic counterions such as tetrabutylammonium, which help with delamination. The adsorption of these positive ions onto the NS floating at the air/water interface appears to impede ML production during subsequent layer transfers to a substrate. Akatsuka et al. [114] validated the use of UV irradiation to eliminate photocatalytically organocations after each layer transfer, allowing for the clean interfaces necessary for layer development. Despite the fact that this method has been effectively used to perovskite superlattice films composed of $(LaNb_2O_7/Ca_2Nb_3O_{10})_n$ MLs with strong interfacial electronic coupling and therefore ferroelectricity [115], the approach's universality and application to non-catalytic NS have yet to be explained.

5.4 Advance Applications of 2D-Heterostructures

5.4.1 Tunneling Devices

Tunnel junctions may be created by combining GR with semiconductors and insulating 2D crystals (Fig. 5.19a) [116]. The h-BN used as a tunneling barrier is especially appealing because of its wide bandgap ($E_g = 6$ eV), the barrier has a low number of impurity states and a high breakdown field. Since the Fermi energy and density of state (DOS) in GR are being changed through an external gate, the tunneling current can also be modified, allowing these devices to be utilized as field-effect tunneling transistors (FETTs) [117]. Tunneling spectroscopy's design allows it to explore the DOS in GR as well as to see tunneling facilitated by impurities and phonons [118]. Peaks in the dI/dV_b (I, current; V_b, bias voltage) ratio are produced by elastic tunneling via impurities; the positions of the peaks are dependent on gate and bias voltages (Fig. 5.19c). Conversely, inelastic phonon-assisted tunneling is defined by a series of plateaus in dI/dV_b that are independent of V_g (which are more visible in d^2I/dV_b^2 (Fig. 5.19b)) [118]. By raising the probability of transmission and therefore tunnel conductance, an extra channel for electron tunneling opens when the bias voltage is sufficient to only emit phonons ($eV_\mathrm{b} = \hbar\omega ph$, where e is the electron charge, $\hbar = h/2\pi$ and ωph is emitted photon frequency). Tunneling via impurities and accompanying phonon emission is particularly evident when the two GR electrodes' crystalline lattices are significantly mismatched compared to one another, preventing direct electron tunneling due to the inability to meet momentum conservation criteria. By aligning the crystalline lattices of the two GR electrodes, tunneling electrons' momentum conservation may be accomplished without impurity or phonon dispersion. The two GR Brillouin zones spin in reciprocal space due to the misalignment of the two GR crystals. If the misalignment is small enough (less than 2°), the electronic states' momentum difference in the top and bottom GR layers may be altered, by bias and V_g electrostatically, resulting in the detection of negative differential resistance and resonant tunneling (Fig. 5.19d) [119]. A strong negative differential resistance characteristic enables the construction of an adjustable radiofrequency oscillator capable of reaching sub-terahertz frequencies. The highest FETTs' on–off ratio is reached when the fluctuations in the Fermi energy of GR are equal to the tunneling barrier's gap, which is achieved when h-BN is substituted with WS_2 (on–off ratio of 10^6) [120] or MoS_2 (on–off ratio of 10^3–10^4, because of the existence of impurity bands) [117]. Tunneling in vdW-HSs has been used for memory devices [121] with a floating gate, logic circuits [122], radio-frequency oscillators [119], and resonant tunneling diodes [69].

5.4.2 *Interaction with Light*

GR and other 2DM-based optoelectronic devices have extensively been investigated [123, 124]. Moreover, GR photodetectors have been established to get poor responsivity as a result of their low absorption coefficient. When alternative 2DMs are utilized for such applications, these problems are mitigated. Thus, in photodiode or photoconductor regimes, TMDCs [124], GaS [125], InSe [126], BP [127], and other materials [128] are employed as photodetectors [129]. The benefits of employing these materials are their high DOS (which ensures high optical adsorption), their flexibility, and the ability to perform local gating that allows the formation of p–n junctions [130]. Besides, such material's E_g (see Fig. 5.18) is frequently based on layers quantity [131] enabling the spectral response of such devices to be controlled.

5.4.2.1 Photovoltaic Applications

GR (as a channel material) and TMDCs are combined (considered as a light-sensitive material) to enable the fabrication of facile and effective phototransistors [133]. Combining materials with various work functions might result in indirect excitons that might result in photoexcited e^- and h^+ accumulating in distinct layers [as found for the pairings MoS_2/WSe_2 [57] and MoS_2/WSe_2 [134] (Fig. 5.19e, f)]. These excitons have a long lifetime, and the distance between the semiconductor layers may be changed to alter their binding energies. The combination of p- and n-doped

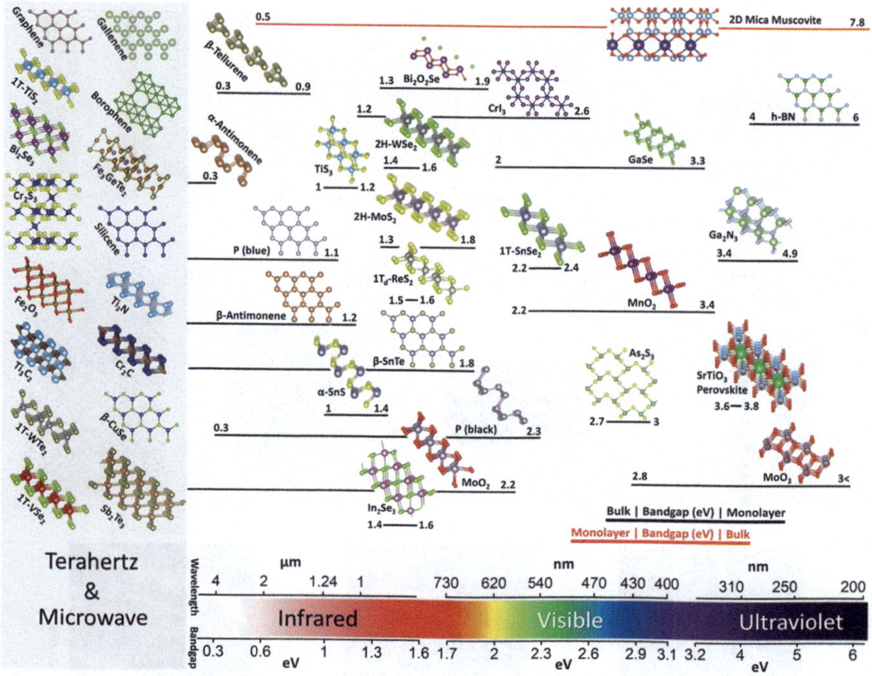

Fig. 5.18 Summary of E_g of various 2DMs and each bandgap range's related to optical spectra (Reproduced under the terms of CC BY 4.0 license [132], https://creativecommons.org/licenses/by/4.0/)

materials in these devices can result in atomically sharp $p\!-\!n$ junctions [93, 135]. Due to their high carrier separation efficiency, these devices exhibit high QE (quantum efficiency) (e.g., GaTe/MoS_2 devices have external QEs of greater than 60%) [135]. Furthermore, as previously demonstrated, the performance of BP/MoS_2 HSs may be modified externally by V_g [136]. Incorporating thin layers of TMDCs [137] or metal chalcogenides [138] with GR enables the creation of even more effective solar devices. With the photosensitive material, by sandwiching, between the GR electrodes, it is possible to extract very effective photocarriers from the device (form good ohmic contacts with the TMDCs and employed as a transparent electrode). Due to the symmetry of these structures (Fig. 5.19g), an electric field must be generated within the TMDC to achieve efficient carrier separation. This can be done via bias, external gating (because, despite its low DOS, GR does not completely screen the electric field), or by differentially doping the two GR layers.

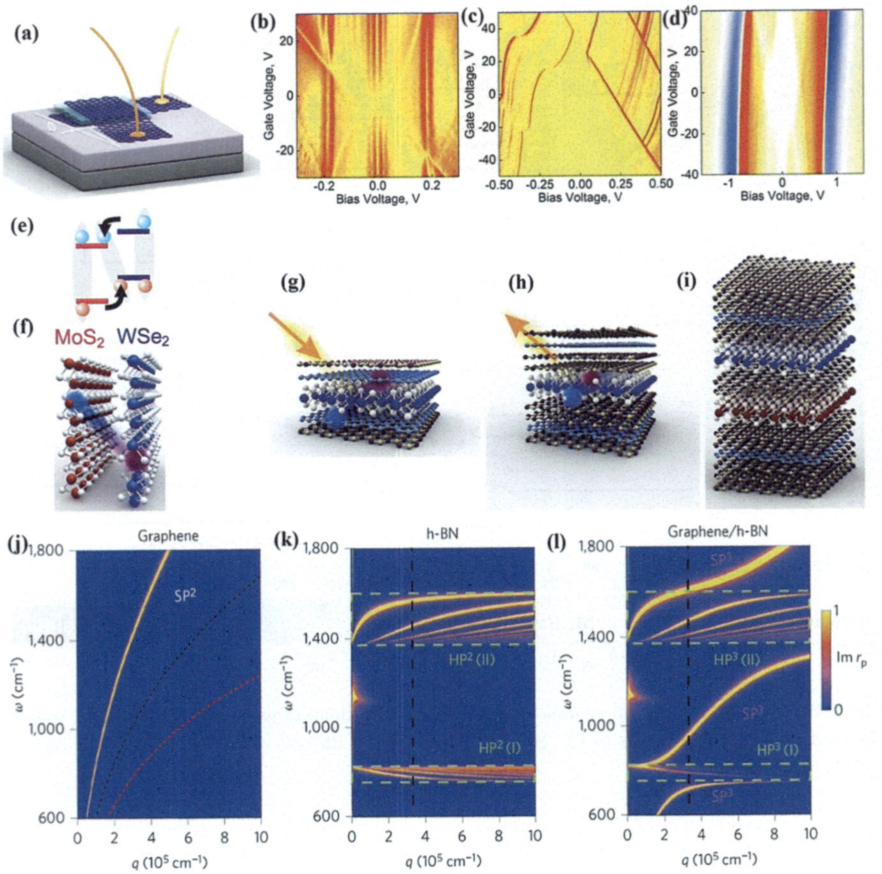

Fig. 5.19 Electronic and optoelectronic applications of vdW-HSs, tunneling in GR/h-BN/GR tunnel transistors: **a** a GR tunneling device diagram light blue h-BN tunneling barrier, dark purple GR electrodes, **b** phonon-assisted tunneling $d^2I/dV_b{}^2$ map, **c** Due to the presence of impurities in the h-BN tunnel layer, a dI/dV_b map of resonant tunneling, **d** The dI/dV_b map of resonant tunneling with momentum conservation is obtained via crystallographic alignment of two GR electrodes. The negative differential conductivity range is shown by the blue color. **e** and **f** indirect excitons in a MoS_2/ WSe_2 HS. MoS_2 accumulates photoexcited es_- from WSe_2. MoS_2 photoexcited holes accumulate in WSe_2, **g** Sandwiching TMDCs (big blue and white spheres) between GR electrodes (small light blue spheres), **h** and **i** V-LED HSs (Reproduced with permission from [1]), **j–l** polaritonic dispersions of GR, h-BN, and a GR/h-BN HS. q = polariton momentum; ω = polariton frequency; r_p = imaginary part of reflectivity (Reproduced with permission from [139])

5.4.2.2 Light-Emitting Diodes

Aforementioned p-n junctions may be operated in the charge carrier injection domain, resulting in $e^- - h^+$ recombination and light emission [93]. As a result, this structure is limited by the requirements for manufacturing p and n-type materials, which have yet to be established for all 2D crystals. In addition, the junction's resistance is analogous to electrodes resistances, making current distribution difficult to control. A simpler method is to inject carriers directly into the 2DM in a V-structure through highly conductive transparent electrodes. Nevertheless, because photoemission is a slow process compared to the usual time required to penetrate the GR-semiconductor junction, such a technique necessitates careful control of the stay duration of the injected es^- and hs^+ in the semiconductor crystal. The dwell time can be reduced by incorporating more tunnel barriers (Fig. 5.19h). Thus, two to three layers of h-BN [9] were used to increase the time spent within the monolayer TMDC by es^- and hs^+, allowing for radiative recombination. The devices efficiency established on WSe_2 was observed to be exceptional: Their QE has been established to improve when temperature and injection current is increased, reaching upto 20% at ambient temperature [140]. Similarly, by stacking many layers of TMDCs in succession, the QE of such devices is expected to improve further (Fig. 5.19i) [9].

5.4.3 Plasmonic Devices

Plasmons in continuous form to GR garner considerable interest based on the fact that their frequency may be tuned by altering the carrier concentration and, consequently, the plasmonic frequency [141]. Simultaneously, additional 2DMs have been investigated for their plasmonic and phonon-polaritonic characteristics. For example, because h-BN possesses polar dielectric characteristics, it is capable of supporting surface phonon polaritons with extremely low optical losses [142]. In vdW HSs, a variety of novel polaritonic phenomenons are visible. By encapsulating GR in h-BN, one may prevent contaminants from scattering GR plasmons, increasing their inverse damping ratio by a factor of five over naked GR [143]. By sandwiching multiple GR layers separated by h-BN spacers, hybridizing plasmonic modes in such MLs is conceivable, and they can then be controlled by an external V_g. It is conceivable to reach a zone in which the plasmon polaritons in GR and the phonon polaritons in h-BN coexist in such HSs (Fig. 5.19j–l) [139]. The strong interaction between the two produces plasmon–phonon polaritons, which are new collective modes. Gating GR enables for adjustment of the new collective modes' amplitude and wavelength. The development of a moiré pattern further modifies the GR plasmonic spectrum in aligned GR/h-BN HSs. Zone folding produces secondary Dirac points [144], enabling a novel sort of V-optical transition. The modified damping factor, which peaks at such Fermi energies, reflects these changes instantaneously [145]. In addition, novel plasmonic modes with carrier density dependency, as seen in parabolic electronic bands, have been anticipated to arise in the region of the reconstructed spectrum's van Hove singularities [145]. Aksimsek et al. [146] theoretically predicted the existence of GR-MoS_2-metal hybrid structures with significantly increased sensitivity when a few atomic thin layers of MoS_2 are integrated into the sensor structure. In a further study using the transfer matrix approach on the suggested hybrids with GR augmentation, GR and MoS_2 were then contrasted in terms of sensitivity improvement, and it was reported that both atomic layers boost the biosensor's sensing

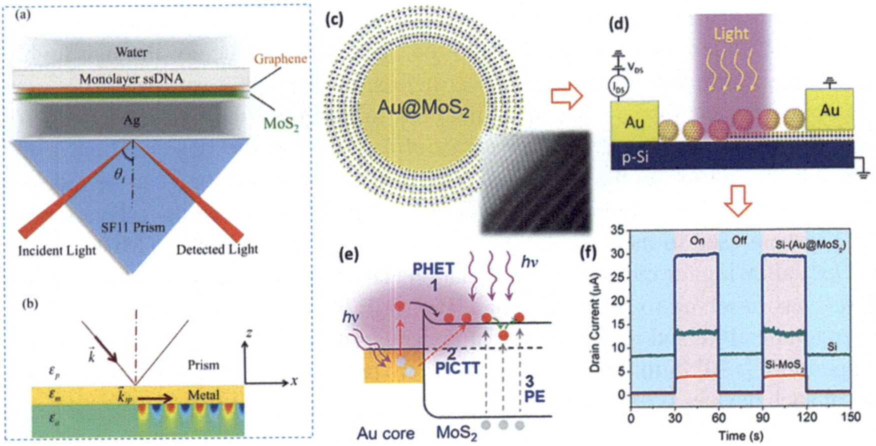

Fig. 5.20 **a** GR-MoS_2-metal hybrid biosensor schematic diagram, prism focuses polarized light onto metal, which is sensed by the photodetectors. **b** Surface plasmons excitation (Reproduced with permission from [146]), **c** The Au core ML-MoS_2 shell structure is shown schematically, **d** Au@MoS_2 HSs-based field-effect phototransistors, **e** photocarrier production pathways in Au@MoS_2 HSs, **f** I-t curves of various devices when illuminated periodically (Reproduced with permission from [147])

capability. Considering the most challenging sensing circumstances, for example, as biomolecule layer comprised monolayer single-stranded DNA with a thickness of 3.2 nm (Fig. 5.20a, b). Photovoltaic applications might benefit from the incorporation of plasmonic materials into semiconductor media. The plasmonic nanostructures that arise improve matter–light interactions, create unique charge trap states, and enable effective charge transfer routes for light-harvesting devices, particularly when a tight contact between the semiconductor and the plasmonic nanostructure is achieved. Plasmonic photodetectors are being developed based on Au@MoS_2 HSs comprise of an Au NP core encased in a CVD-produced ML-MoS_2 shell that perfectly accomplishes Au and MoS_2 interface [147]. The first is to create a field-effect phototransistor with a huge interdigitated area with tenfolds the photoresponsivity of planar MoS_2 transistors. An HJ Au@MoS_2 gateless photodiode anchored on Si is an alternative device shape. Its outstanding recovery capacity and photoresponse, with a photoresponsivity of up to 22.3 AW^{-1}, greatly exceeds the highest notable figures hitherto recorded for comparable gateless photodetectors. Improved photosensing capacity might be aided by increased light absorption, the generation of additional trap states, and perhaps the formation of a charge transfer transition that is interfacial aided by the close contact between Au and MoS_2 (Fig. 5.20c–f).

References

1. Novoselov KS, Mishchenko A, Carvalho A (2016) Castro Neto AH Sci 353:aac9439
2. Lyu J, Pei J, Guo Y, Gong J, Li H (2020) Adv Mater 32:1906000
3. Liao W, Huang Y, Wang H, Zhang H (2019) Appl Mater Today 16:435
4. Azadmanjiri J, Srivastava VK, Kumar VK, Sofer Z, Min J (2020) J Gong Appl Mater Today 19:100600
5. Paul KK, Kim J-H, Lee YH (2021) Nat. Rev. Phys. 3:178
6. Yankowitz M, Ma Q, Jarillo-Herrero P, LeRoy BJ (2019) Nat. Rev. Phys. 1:112
7. Gong C, Zhang H, Wang W, Colombo L, Wallace RM, Cho K (2015) Appl Phys Lett 107:139904
8. Liu Y, Zhang S, He J, Wang ZM, Liu Z (2019) Nano-Micro Lett. 11:13
9. Withers F et al (2015) Nat Mater 14:301
10. Ricciardulli G, Blom PWM (2020) Adv. Mater. Technolo. 5:1900972
11. Hattori Y, Taniguchi T, Watanabe K, Nagashio K (2018) ACS Appl Mater Interfaces 10:11732
12. Bampoulis P, van Bremen R, Yao Q, Poelsema B, Zandvliet HJW, Sotthewes K (2017) ACS Appl Mater Interfaces 9:19278
13. Le Quang T, Cherkez V, Nogajewski K, Potemski M, Dau MT, Jamet M, Mallet P, Veuillen JY (2017) 2D Mater 4:035019
14. Woods CR et al (2014) Nat Phys 10:451
15. Li B, Huang L, Zhong M, Li Y, Wang Y, Li J, Wei Z (2016) Adv. Electron. Mater. 2:1600298
16. Liu K et al (2014) Nat Commun 5:4966
17. Dankert SP (2017) Dash Nat Commun 8:16093
18. Wang J, Fang H, Wang X, Chen X, Lu W, Hu W (2017) Small 13:1700894
19. Tan H et al (2017) Adv Mater 29:1702917
20. Gong F et al (2016) Adv Func Mater 26:6084
21. Liu YP, Goolaup S, Lew WS, Purnama I, Chandra Sekhar M, Zhou TJ, Wong SK (2013) Appl Phys Lett 103:163108
22. Bao X, Ou Q, Xu Z-Q, Zhang Y, Bao Q, Zhang H (2018) Adv. Mater. Technol. 3:1800072
23. Hossain MA, Yu J, van der Zande AM (2020) ACS Appl Mater Interfaces 12:48910
24. Rahimi K (2021) Mater Lett 282:128680
25. Guan Y, Yao H, Zhan H, Wang H, Zhou Y, Kang J (2021) RSC Adv 11:14085
26. Zhang L et al (2021) Mater Today
27. Jang CW, Kim H, Nazeeruddin MK, Shin DH, Choi S-H (2021) Nano Energy 84:105899
28. Li M, Esseni D, Snider G, Jena D, Grace Xing H (2014) J Appl Phys 115:074508
29. Yan X, Liu C, Li C, Bao W, Ding S, Zhang DW, Zhou P (2017) Small 13:1701478
30. Yamaoka T, Lim HE, Koirala S, Wang X, Shinokita K, Maruyama M, Okada S, Miyauchi Y, Matsuda K (2018) Adv Func Mater 28:1801021
31. Cai Z, Cao M, Jin Z, Yi K, Chen X, Wei D (2018) NPJ 2D Mater Appl 2:21
32. Vu QA et al (2016) Nat Commun 7:12725
33. Zhang Y, Yuan Y, Huang J (2017) Adv Mater 29:1603969
34. Hong X et al (2014) Nat Nanotechnol 9:682
35. Chen H et al (2016) Nat Commun 7:12512
36. Wen X et al (2018) Nat Commun 9:1859
37. Hanbicki T, Chuang H-J, Rosenberger MR, Hellberg CS, Sivaram SV, McCreary KM, Mazin II, Jonker BT (2018) ACS Nano 12:4719
38. Gong Y et al (2014) Nat Mater 13:1135
39. Liu Y, Lew WS, Sun L (2011) Phys Chem Chem Phys 13:20208
40. Liu Y, Goolaup S, Murapaka C, Lew WS, Wong SK (2010) ACS Nano 4:7087
41. Zhou B, Ji S, Tian Z, Cheng W, Wang X, Mi W (2018) Carbon 132:25
42. Wei P et al (2016) Nat Mater 15:711
43. Yang HX, Hallal A, Terrade D, Waintal X, Roche S, Chshiev M (2013) Phys Rev Lett 110:046603

44. Seyler KL et al (2018) Nano Lett 18:3823
45. Huang B et al (2017) Nature 546:270
46. Gong C, Zhang X (2019) Science 363:eaav4450
47. Lee J-U, Lee S, Ryoo JH, Kang S, Kim TY, Kim P, Park C-H, Park J-G, Cheong H (2016) Nano Lett 16:7433
48. Du J, Xia C, Xiong W, Wang T, Jia Y, Li J (2017) Nanoscale 9:17585
49. Liu Y et al (2018) Nano Lett 18:3377
50. Lotsch BV (2015) Annu Rev Mater Res 45:85
51. Caldwell JD et al (2010) ACS Nano 4:1108
52. Lim H, Yoon SI, Kim G, Jang AR, Shin HS (2014) Chem Mater 26:4891
53. Schneider GF, Calado VE, Zandbergen H, Vandersypen LMK, Dekker C (2010) Nano Lett 10:1912
54. Wang L et al (2013) Science 342:614
55. Li H, Wu J, Huang X, Yin Z, Liu J, Zhang H (2014) ACS Nano 8:6563
56. Song J, Kam F-Y, Png R-Q, Seah W-L, Zhuo J-M, Lim G-K, Ho PKH, Chua L-L (2013) Nat Nanotechnol 8:356
57. Fang H et al (2014) Proc Natl Acad Sci 111:6198
58. Haigh SJ et al (2012) Nat Mater 11:764
59. Hsu W-T, Zhao Z-A, Li L-J, Chen C-H, Chiu M-H, Chang P-S, Chou Y-C, Chang W-H (2014) ACS Nano 8:2951
60. Chiu M-H, Li M-Y, Zhang W, Hsu W-T, Chang W-H, Terrones M, Terrones H, Li L-J (2014) ACS Nano 8:9649
61. Li X et al (2016) Sci Adv 2:e1501882
62. Fu L et al (2016) ACS Nano 10:2063
63. Zheng B et al (2018) J Am Chem Soc 140:11193
64. Sahoo PK, Memaran S, Xin Y, Balicas L, Gutiérrez HR (2018) Nature 553:63
65. Zhang Z, Chen P, Duan X, Zang K, Luo J, Duan X (2017) Science 357:788
66. Yang W et al (2013) Nat Mater 12:792
67. Shi Y et al (2012) Nano Lett 12:2784
68. Lin Y-C et al (2014) ACS Nano 8:3715
69. Lin Y-C et al (2015) Nat Commun 6:7311
70. Lee E, Lee SG, Lee WH, Lee HC, Nguyen NN, Yoo MS, Cho K (2020) Chem Mater 32:4544
71. Ma Z et al (2020) Small 16:2000596
72. Levendorf MP, Kim C-J, Brown L, Huang PY, Havener RW, Muller DA, Park J (2012) Nature 488:627
73. Lu J, Gomes LC, Nunes RW, Castro Neto AH, Loh KP (2014) Nano Lett 14:5133
74. Huang C et al (2014) Nat Mater 13:1096
75. Li M-Y et al (2015) Science 349:524
76. Gong Y et al (2015) Nano Lett 15:6135
77. Glavin NR, Rao R, Varshney V, Bianco E, Apte A, Roy A, Ringe E, Ajayan PM (2020) Adv Mater 32:1904302
78. Pagaduan JN, Hight-Huf N, Datar A, Nagar Y, Barnes M, Naveh D, Ramasubramaniam A, Katsumata R, Emrick T (2021) ACS Nano 15:2762
79. Bhimanapati GR et al (2015) ACS Nano 9:11509
80. Craciun MF, Khrapach I, Barnes MD, Russo J (2013) J Phys: Condensed Matt 25:423201
81. Kuila T, Bose S, Mishra AK, Khanra P, Kim NH, Lee JH (2012) Prog Mater Sci 57:1061
82. Georgakilas V, Otyepka M, Bourlinos AB, Chandra V, Kim N, Kemp KC, Hobza P, Zboril R, Kim KS (2012) Chem Rev 112:6156
83. Fang H, Chuang S, Chang TC, Takei K, Takahashi T, Javey A (2012) Nano Lett 12:3788
84. Tongay S et al (2013) Nano Lett 13:2831
85. Lin JD et al (2014) ACS Nano 8:5323
86. Mouri S, Miyauchi Y, Matsuda K (2013) Nano Lett 13:5944
87. Li Y, Xu C-Y, Hu P, Zhen L (2013) ACS Nano 7:7795
88. Wi S, Kim H, Chen M, Nam H, Guo LJ, Meyhofer E, Liang X (2014) ACS Nano 8:5270

89. Chen M, Nam H, Wi S, Ji L, Ren X, Bian L, Lu S, Liang X (2013) Appl Phys Lett 103:142110
90. Kiriya D, Tosun M, Zhao P, Kang JS, Javey A (2014) J Am Chem Soc 136:7853
91. Andleeb S, Kumar Singh A (2015) J Eom Sci Technol Adv Mater 16:03500992
92. Lee C-H et al (2014) Nat Nanotechnol 9:676
93. Cheng R et al (2014) Nano Lett 14:5590
94. Suh J et al (2014) Nano Lett 14:6976
95. Shi Y, Kim KK, Reina A, Hofmann M, Li L-J, Kong J (2010) ACS Nano 4:2689
96. Shin H-J et al (2010) J Am Chem Soc 132:15603
97. Du Y, Liu H, Neal AT, Si M, Ye PD (2013) IEEE Electron. Dev. Lett. 34:1328
98. Fang H, Tosun M, Seol G, Chang TC, Takei K, Guo J, Javey A (2013) Nano Lett 13:1991
99. Yang L et al (2014) Nano Lett 14:6275
100. Wu C-I, Lin C-T, Chen Y-H, Chen M-H, Lu Y-J, Wu C-C (2006) Appl Phys Lett 88:152104
101. Huang J, Xu Z, Yang Y (2007) Adv Func Mater 17:1966
102. Kang D-H et al (2015) ACS Nano 9:1099
103. Li H-M, Lee D, Qu D, Liu X, Ryu J, Seabaugh A, Yoo WJ (2015) Nat Commun 6:6564
104. Keller SW, Kim H-N, Mallouk TE (1994) J Am Chem Soc 116:8817
105. Decher G, Hong J-D (1991) Makromolekulare Chemie. Macromol Symp 46:321
106. Buck MR, Schaak RE (2013) Angew Chem Int Ed 52:6154
107. Geng F, Ma R, Ebina Y, Yamauchi Y, Miyamoto N, Sasaki T (2014) J Am Chem Soc 136:5491
108. Chalasani R, Gupta A, Vasudevan S (2013) Sci Rep 3:3498
109. Harper WR (1934) Trans Faraday Soc 30:636
110. Onoda M, Liu Z, Ebina Y, Takada K, Sasaki T (2011) J Phys Chem C 115:8555
111. Kotov NA, Meldrum FC, Fendler JH, Tombacz E, Dekany I (1994) Langmuir 10:3797
112. Taguchi Y et al (1998) Langmuir 14:6550
113. Muramatsu M, Akatsuka K, Ebina Y, Wang K, Sasaki T, Ishida T, Miyake K, Haga M-A (2005) Langmuir 21:6590
114. Akatsuka K, Haga M-A, Ebina Y, Osada M, Fukuda K, Sasaki T (2009) ACS Nano 3:1097
115. Li B-W, Osada M, Ozawa TC, Ebina Y, Akatsuka K, Ma R, Funakubo H, Sasaki T (2010) ACS Nano 4:6673
116. Lee G-H, Yu Y-J, Lee C, Dean C, Shepard KL, Kim P (2011) J Hone Appl Phys Lett 99:243114
117. Britnell L et al (2012) Science 335:947
118. Amet F, Williams JR, Garcia AGF, Yankowitz M, Watanabe K, Taniguchi T, Goldhaber-Gordon D (2012) Phys Rev B 85:073405
119. Mishchenko et al (2014) Nat Nanotechnol 9:808
120. Georgiou T et al (2013) Nat Nanotechnol 8:100
121. Sup Choi M, Lee G-H, Yu Y-J, Lee D-Y, Hwan Lee S, Kim P, Hone J, Jong Yoo W (2013) Nat Commun 4:1624
122. Yang H et al (2012) Science 336:1140
123. Bonaccorso F, Sun Z, Hasan T, Ferrari AC (2010) Nat Photonics 4:611
124. Wang QH, Kalantar-Zadeh K, Kis A, Coleman JN, Strano MS (2012) Nat Nanotechnol 7:699
125. Hu P et al (2013) Nano Lett 13:1649
126. Tamalampudi SR, Lu Y-Y, Sankar RKUR, Liao C-D, B KM, Cheng C-H, Chou FC, Chen Y-T (2014) Nano Lett 14:2800
127. Buscema M, Groenendijk DJ, Blanter SI, Steele GA, van der Zant HSJ, Castellanos-Gomez A (2014) Nano Lett 14:3347
128. Xia F, Wang H, Xiao D, Dubey M, Ramasubramaniam A (2014) Nat Photonics 8:899
129. Koppens FHL, Mueller T, Avouris P, Ferrari AC, Vitiello MS, Polini M (2014) Nat Nanotechnol 9:780
130. Baugher BWH, Churchill HOH, Yang Y, Jarillo-Herrero P (2014) Nat Nanotechnol 9:262
131. Mak KF, Lee C, Hone J, Shan J, Heinz TF (2010) Phys Rev Lett 105:136805
132. Chaves et al (2020) NPJ 2D Mater Appl 4:29
133. Roy K, Padmanabhan M, Goswami S, Sai TP, Ramalingam G, Raghavan S, Ghosh A (2013) Nat Nanotechnol 8:826
134. Rivera P et al (2015) Nat Commun 6:6242

135. Wang F, Wang Z, Xu K, Wang F, Wang Q, Huang Y, Yin L, He J (2015) Nano Lett 15:7558
136. Deng Y et al (2014) ACS Nano 8:8292
137. Britnell L et al (2013) Science 340:1311
138. Mudd GW et al (2015) Adv Mater 27:3760
139. Dai S et al (2015) Nat Nanotechnol 10:682
140. Withers F et al (2015) Nano Lett 15:8223
141. Grigorenko N, Polini M, Novoselov KS (2012) Nat Photonics 6:749
142. Dai S et al (2014) Science 343:1125
143. Woessner et al (2015) Nat Mater 14:421
144. Wallbank JR, Patel AA, Mucha-Kruczyński M, Geim AK, Fal'ko VI (2013) Phys Rev B 87:245408
145. Tomadin, F. Guinea, and M. Polini, Physical Review B 90, 161406 (2014).
146. Aksimsek S, Jussila H, Sun Z (2018) Opt Commun 428:233
147. Li Y, DiStefano JG, Murthy AA, Cain JD, Hanson ED, Li Q, Castro FC, Chen X, Dravid VP (2017) ACS Nano 11:10321

Chapter 6
Energy-Related Applications

From fundamental research to the development of next-generation technologies, considerable progress has been achieved in the field of ultrathin 2DMs research during the previous decade. With the rapid rise of smart devices, wearable electronics, and the Internet of Things, a lot of work has lately gone into developing practical, portable, and distributed energy sources for a new era of energy. Mechanical, thermal, solar, and chemical energy are just a few examples of environmentally friendly energy sources. Energy harvesting applications such as mechanical (piezoelectricity), solar (solar cell, hydrogen evolution, and CO_2 resection), and electrochemical (oxygen evolution) energy converters will be addressed, as well as storage applications such as batteries and supercapacitors will be addressed in the present chapter.

6.1 Energy Harvesting

With the fast growth of smart gadgets wearable electronics, and the Internet of Things, substantial effort has recently been put into building viable, portable, and distributed energy sources for a new era of energy (Fig. 6.1) [1–3]. Sustainable, self-sufficient power sources such as nanoenergy fields can help satisfy these rising energy demands [4, 5]. Numerous types of environment-friendly energy sources comprise mechanical, thermal, solar, and chemical energy. In the recent past, progress in nanotechnology and nanoscale materials has led to the successful unraveling of viable more efficient energy harvesting techniques. Harvesting solar and thermal energies, using novel materials and techniques has substantially increased device performance and efficiency [6, 7]. Since the first piezoelectric (PE) and triboelectric nanogenerators (NGs) were reported in 2006 and 2012, substantial progress has been made in mechanical energy harvesting utilizing nanomaterials [2, 8, 9]. New ways can now be utilized to capture different energy sources. Nanoscale 2DMs have a unique combination of electrical [10, 11], optical [12], mechanical [13], and thermal properties [14, 15]

M. Ikram et al., *2D-Materials for Energy Harvesting and Storage Applications*, Nanostructure Science and Technology,
https://doi.org/10.1007/978-3-030-96021-6_6

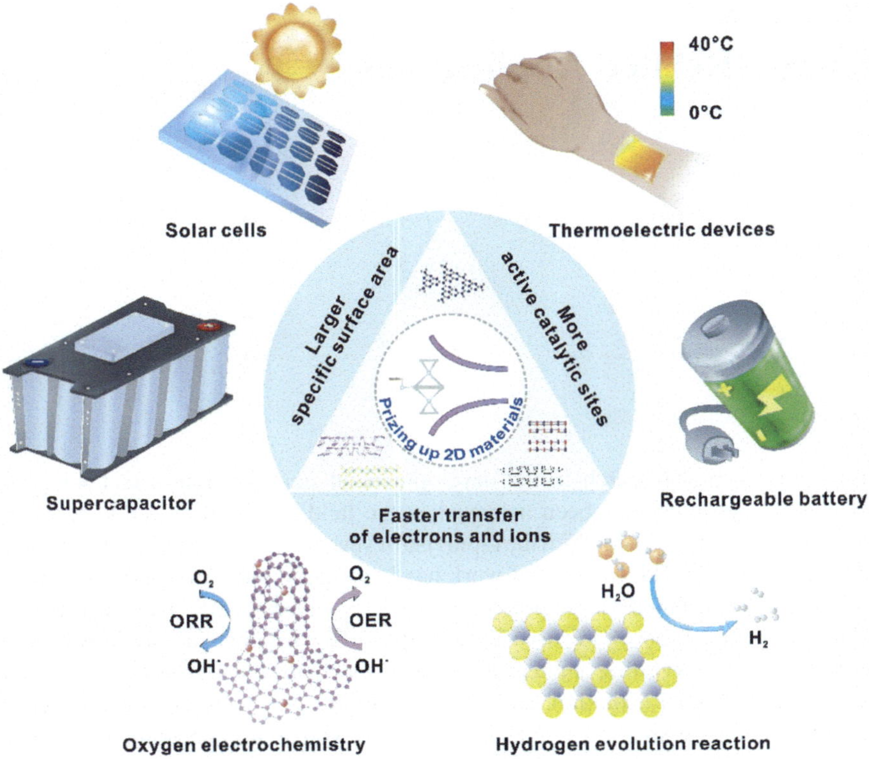

Fig. 6.1 Numerous uses of 2DMs in the field of energy (Reproduced with permission from [23])

that entice researchers to investigate innovative energy harvesting approaches [16, 17]. As a result of its exceptional conductivity and transparency, ML-GR may be utilized as transparent electrodes in different solar and piezo/triboelectric NGs [18, 19]. In addition, interactions with external species and other physical fields by all π-bonds accessible on the surface of the crystal are possible, making GR an ideal hydrovolcanic material [20, 21]. Other 2D ML materials, e.g., TMDCs and BP, show semiconducting behavior within visible light range E_g and are interesting for application as interlayers or active layers in solar energy systems [11, 22].

6.2 Mechanical Energy Harvesting

6.2.1 Piezoelectric Energy Harvesting

The comparatively recent topic of nanoenergy involves collecting energy from nanostructures and nanotechnology to power micro/nanosystems. It is currently either being used to prolong the life of batteries or replacing them outrightly. Nanoenergy idea is significantly different from megawatt or gigawatt grid-scale used for worldwide electricity necessity. Significantly, NG is considered novel technology type converting mechanical energy into electricity from a physical change on a small basis. NG has two types: (i) piezoelectric nanogenerator (PENG) and (ii) triboelectric nanogenerator (TENG).

The primary uses of nanogenerator are in three directions:

i. A self-powering source for small devices and gadgets.
ii. Active sensors of medical field, infrastructural, manpower interaction, environmental as well as security control.
iii. Fundamental network the collection of water movement energy with small frequency to blue energy.

The original self-powering idea emerged in 2006 when PENG was discovered [8] that uses ZnO nanowires to piezoelectrically transform mechanical energy into electricity. On the other, TENG was initially developed in 2012 [24] has its basic process can use electrostatic interactions produced by physical connection between two different materials to produce a potential drop that allows charge flowing among both electrodes established on upper and lower material sides. On comparing with PENG, TENG has been established to generate more output (o/p) power. However, TENG has its unique mechanism since its performance is reliant on the materials' air gap and surface potential, causing major problems during implantation. On the other hand, PENG is an excellent choice at this stage because of its better sensitivity and consistency in o/p performance over an extended length of time. Researchers have attempted to enhance PENGs by utilizing a variety of known active materials, including ZnO, PZT, PVDF, and other nanomaterials [25]. As a result, PENGs are capable of collecting mechanical energy and hence have significantly promising remote/wireless sensing, battery charging, and power generating devices [26, 27]. Owing to an increasing need for compact, portable, and lightweight electronic devices, atomically thin 2DMs with unique optoelectronics, mechanical, and thermal properties have gained significant scientific interest [28, 29]. The advantages of ultrathin, clarity, elasticity, high surface/volume ratio, and layer stacking have encouraged an almost exponential growth of low-weight and high-outcomes multipurpose applications [30]. Mostly, 2D-layered materials with an asymmetric structure show considerable promise toward mechanical energy applications. This study thus explores 2D layer-type material with non-centrosymmetry (NCS) structures, their energy harvesting potential and feasible PENG designs using 2D framework.

6.2.1.1 Piezoelectricity in 2D Materials

Piezoelectricity is described as a material's capacity to produce polarization charges due to adverse mechanical stress. A PE reaction has been proposed for crystals, polymers, and bi-molecules since 1880. Many of these PE materials also find wide uses in actuators, sensors, and energy harvesting technology [31, 32]. A 2D crystal's PE coupling for mechanical and electrical paths is given by $S_{ij} = d_{kij}E_k$ where *i*, *j*, *k* indicate 1, 2, 3, and *S*, *d*, and *E* for 2nd strain tensor, PE coefficient, electrical field. *i*, *j*, and *k* represent *x*, *y*, and *z* of Cartesian frame [28]. An applied strain tensor can alter PE polarization direction. The converse PE effect was frequently employed in sound wave generation and AFM systems. Besides, small scaling and diverse function device requirements are indeed very essential. Because of their ultrathin shape, outstanding electromechanical responsiveness, and other unique physical characteristics, 2D PE nanomaterials have become incredibly fascinating [33]. Emerging nanomechanical systems, self-adaptive nano- and optoelectronic systems, and advanced robotics are expected possibilities and platforms for innovative development and production.

In-Plane Piezoelectricity

Discussion of IP PE material (TMDC, h-BN, and GR nitride) will be addressed; family discussed in this section is that of TMDCs. Each TMDC layer is composed of a metal plane existing among two dichalcogenide planes, a trigonal prismatic unit cell containing metal atoms at its center (see the structure in Chap. 3) [34]. NCS analyses are a quick and easy approach to evaluate the presence of piezoelectricity in PE materials [28]. More precisely, ML-MoS_2 and its corresponding TMDC explain the piezoelectricity origin observed in Fig. 6.2a. To distinguish them from their bulk crystal counterparts, in 2012 that ML-TMDC materials are proven as PE candidates [35], two years later, the prediction with a flexible MoS_2 PENG has been true [33]. The top view structure (Fig. 6.2a) demonstrates that the ML-MoS_2 appears to lack an inversion center. This NCS arrangement demonstrates the ML-MoS_2's PE action directly. Notably, this PE characteristic is limited to IP (d_{11}) rather than OP (d_{33}) direction, as ML-MoS_2 maintains perfect symmetry along the vertical *z*-axis (Fig. 6.2a). ML-MoS_2 naturally forms electric dipoles when subjected to external strain because of Mo and S atoms› movement in the form of cation and anion, respectively. PE polarized charges lying on material's surfaces, whereas PE fields throughout the material are generated by these internal electric dipoles. Additionally, due to ML-MoS_2's unusual hexagonal structure, the IP-PE field vector has 3-equivalent directions parallel to semiconductor's threefold symmetry axis. The reported SHG signals and PE coupling strength show that asymmetry and PE responses are substantially angle-dependent, with maximum at a multiple of 60° (identifying the "armchair" orientation as 0°) [33, 36]. Excitingly, BL-MoS_2 maintains crystal centrosymmetry (CS) by providing symmetric plane atoms projection located at 6-vertex locations of hexagonal architecture, resulting in deficiency of PE action. A new device design was developed for measuring IP piezoelectricity in ML-MoS_2 in 2014 [36]. Through a multistep nanofabrication method, one can successfully produce a free suspended membrane along with 2-terminals clamping with metal electrodes. In addition, fundamental concept underlying this measurement is that the AFM probe converts the applied voltage induced IP strain to OP

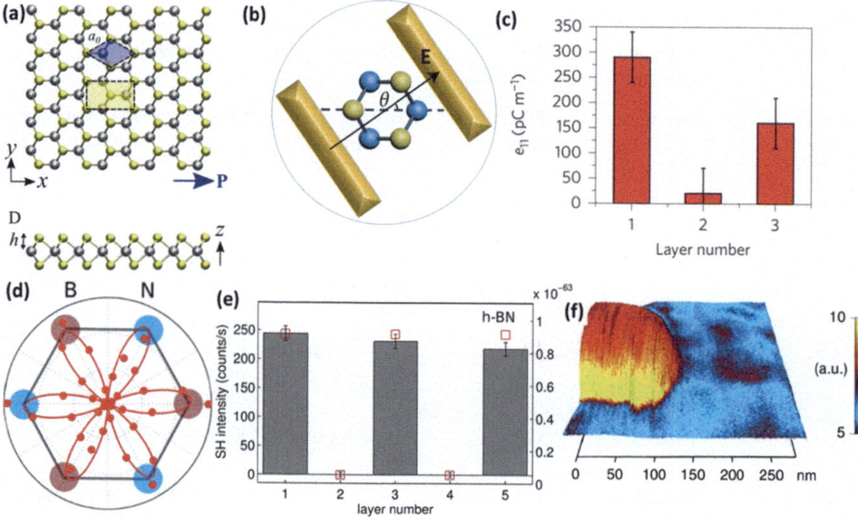

Fig. 6.2 Piezoelectricity in various 2DMs **a** geometries of a trigonal prismatic ML-MoS_2 from the top and side views (2H-MoS_2) (Reproduced with permission from [35]). **b** The electrodes were patterned at various angles θ to produce crystal rotation about the electric field and mechanical boundary, **c** with 1, 2, and 3-layer MoS_2 membranes, odd layers number demonstrated substantial coupling strength (Reproduced with permission from [36]). **d** Polar representation shows the SHG intensity in SL-BN, **e** surface nonlinear susceptibility and SHG intensity dependence for FL h-BN (Reproduced with permission from [40]). **f** g-C_3N_4 particle PFM response on conductive substrate with adjusted vertical amplitude (Reproduced with permission from [41])

strain (AFM tip closes to surface). Due to the D_{3h} group's symmetry, PE coupling of MoS_2 ML is caused by the crystal's azimuthal angle between mirror plane and electric field direction (Fig. 6.2b). Figure 6.2c shows the measured e_{11} PE coefficient, the odd-layer flakes show the most PE response, whereas even-layer flakes show no response. Above mentioned results are well-matched with NCS structural investigation for thickness-dependent piezoelectricity MoS_2 membranes. Inspired by these groundbreaking works on MoS_2 piezoelectricity, current research on piezoelectricity in TMDC has garnered considerable attention in the research community [37, 38]. In 2009, it was theoretically estimated that layered h-BN had planar PE characteristics [39]. The "B" and "N" substances in h-BN crystals are arranged hexagonally (Fig. 6.2d), and its structure is similar to GR also termed as white GR. Unlike GR, by alternating "B" and "N" are hexagonal vertex sites, h-BN ML displays substantial IP piezoelectricity. An optical SHG study in 2013 revealed the NCS structure of h-BN [40]; therefore, indirectly demonstrating the presence of h-BN piezoelectricity (Fig. 6.2e). Owing to the unique h-BN crystal structure, across threefold symmetry axis, it always has unique SHG intensity and signals exhibit layer-dependent effects. The g-C_3N_4 is graphitic form of C_3N_4 and more energetically stable tri-s-triazine or heptazine arrays comprise an SL of g-C_3N_4 (see Chap. 3). The heptazine g-C_3N_4 layer is fundamentally NCS, resulting in IP piezoelectricity [41]. In comparison to

Table 6.1 Summary of experimentally confirmed PE 2D materials

2D Material	Crystal structure	PE direction	Piezocoefficient	Comments
Monolayer MoS_2	Hexagonal	IP Angle dependence	d_{11} = 2.5–4 pm/V [35] e_{11} = 250–400 pC/m [35, 45]	Odd–even effect with the thickness increased
Monolayer h-BN	Hexagonal	IP Angle dependence	e_{11} = 100–400 pC/m [35]	Odd–even effect with the thickness increased [40]
g-C_3N_4		IP	e_{11} = 218 pC/m [41]	Existence of piezoelectricity regardless of thickness
Doped GR	Hexagonal	OP	d_{33} = 1.4 nm/V [46]	Extrinsic piezoelectricity
α-In_2Se_3	Rhombohedral	OP and IP		Indirectly confirmed by ferroelectricity [47]
Janus MoSSe	Hexagonal	OP and IP	d_{33} = 0.1 pm/V [48]	

TMDC, its ML equivalent exhibits significant IP piezoelectricity even in bulk as a result of unique atom-stacking pattern. The right side of Fig. 6.2f shows piezoelectric force microscopy (PFM) amplitude recording for g-C_3N_4 atom (30 nm) and findings validity for piezoelectricity of ML g-C_3N_4. Numerous 2D semiconductors should theoretically have an IP-PE impact. ML group-IV monochalcogenides such as MX, M = Sn/Ge, X = Se/S) have their own distinctive piezoelectricity computed, as demonstrated in Table 6.1 summarizes the d_{11} and d_{12} PE coefficients [42]. Surprisingly, the IP-PE coefficient d_{11} of this set of semiconductors varies between 75 and 250 pmV^{-1} as 2-orders of magnitude strongly compared to typical PE materials like α-quartz, ZnO, GaN, respectively. On the other hand, ML state for group III containing monochalcogenides, viz. GaS, GaSe, and InSe are considered as PE with piezocoefficients of 0.06, 2.30, and 1.46 pmV^{-1}, respectively [43]. Blonsky et al. [44] proved the IP (d_{11}) piezoelectricity existence in a variety of 2DMs (monolayer) including metal dichalcogenides family such as MX_2, M = Cr, Mo, W, Nb, Ta, and X = S, Se, Te), whereas group IIA as well as IIB metal oxides containing M = Be, Mg, Ca, Zn, Cd, Pb, and Pb), and finally group III–V comprising AX, A = B, Al, Ca, In and X = N, P, As, Sb.

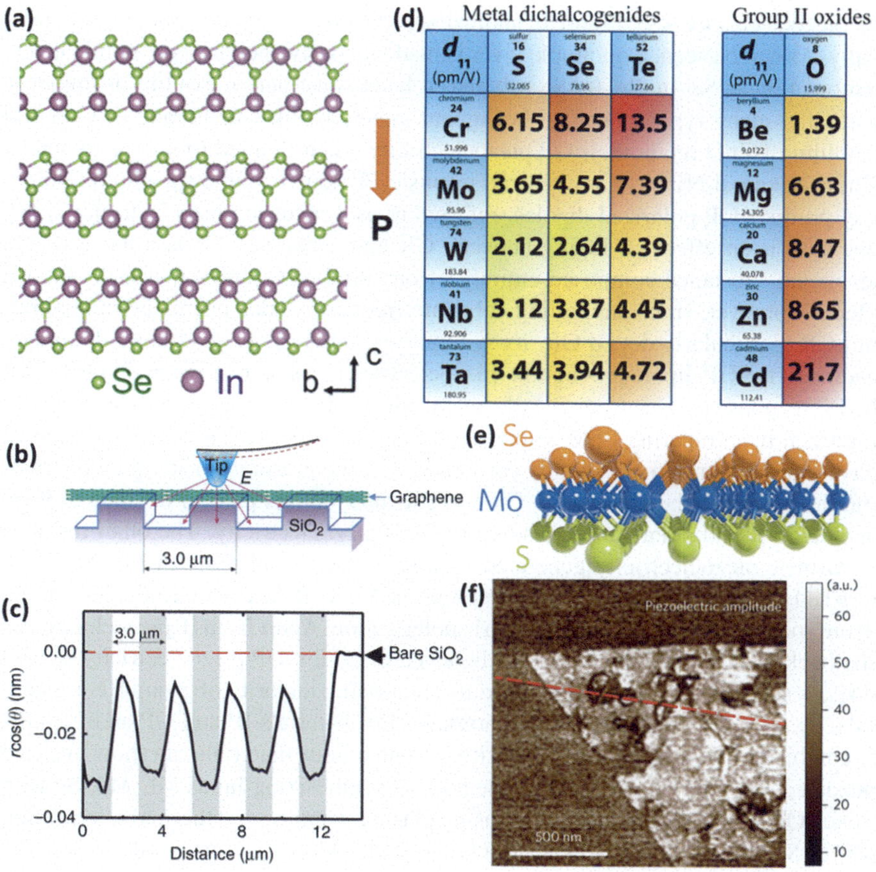

Fig. 6.3 **a** The arrows indicate side crystal structure for α-In_2Se_3 into R3m space group (Reproduced with permission from [49]). **b** PFM measurements on SL-GR adsorbed on TGZ_4 grating substrate are depicted schematically, **c** piezoresponse cross-section over grating structure indicated by shaded areas for supported GR (Reproduced under the terms of CC BY 4.0 license [46], https://creativecommons.org/licenses/by/4.0/). **d** Periodic location of d_{11} into metal dichalcogenides, metallic oxides, group III–V semiconductors along with d_{31} in group III–V semiconductors (Reproduced with permission from [44]) **e** MoSSe structure (front view), **f** PE amplitude of an ML of Janus MoSSe grown directly on highly oriented pyrolytic graphite later was determined using resonance-enhanced PFM (Reproduced with permission from [48])

Out-of-Plane Piezoelectricity

Following that, we will address OP-2D PE materials that have been experimentally proven. At various temperatures, α-In_2Se_3, a typical III–VI semiconductor, is claimed that among the five phases has a stable and layered structure. Based on TEM and SHG structural investigation, it has been proven that ML rhombohedral α-In_2Se_3 exhibits

OP piezoelectricity [49]. Figure 6.3a illustrates the precise crystal structure schematically. As can be seen, a quintuple layer is made up of five neighbored atomic layers combination as Se–In–Se–In–Se, whereas related construction exhibits asymmetry. Naturally, many typical buckled hexagonal materials, like membrane of CdS and nanofilm of ZnO retain their OP piezoelectricity when thinned to few nanometers. While structural NCS is a recognized intrinsic PE source, some specific interfaces may produce OP polarized dipoles, referred to as "extrinsic piezoelectricity" [28]. One case is an effect of doping between GR and substrate, GR is a flawless, CS hexagonal substance composed entirely of one kind of C-atoms that lacks the PE effect. Moreover, silica-supporting substrate frequently adds chemical compounds into GR, particularly doped GR, its symmetry structure is inadvertently disrupted, leading to the OP dipole production (piezoelectricity), as seen in Fig. 6.3b [46]. That the PE response derives from GR placed on silica substrate is proven to be crucially important by comparing PFM response on bare SiO_2 to GR. The GR's mechanical tensile strain distribution is highly correlated with the piezoresponse signal distribution in Fig. 6.3c. The C-atoms in GR react with O_2-atoms present in silica to form an electric μ and polarization, leading to GR's piezoelectricity. The other instance is extrinsic piezoelectricity generated by a ferroelectric substrate in ML MoS_2 on a $PbTiO_3$ sheet [50]. This OP dipole effect in MoS_2 flakes is more evident in the 19 nm specimen induced by ferroelectric polarization. Apart from IP piezoelectricity, ML-buckled hexagonal materials are believed to exhibit OP piezoelectricity as well [44]. As seen in Fig. 6.3d, these materials comprising majority of group III–V materials, i.e., AX are considered. Intercorrelated PE materials IP and OP additionally, by combining IP and OP piezoelectricity, certain unique materials can show intricate piezoelectricity. Moreover, a novel method for synthesizing Janus ML-MoSSe with broken OP symmetry was developed using plasma stripping and thermal selenization [48]. A Se atom layer replaces S atom layer in ML-MoS_2, as illustrated in Fig. 6.3e, resulting in the appearance of OP piezoelectricity not seen in pure ML-MoS_2. As seen in Fig. 6.3f, a PFM signal was utilized to quantify the vertical piezoelectricity. Such a Janus ML should exhibit an IP-PE effect because of the comparable atom planar projections to those found in TMDC family. Notably, CVD-produced rhombohedral α-In_2Se_3 flakes were proven to be ferroelectric IP and OP, implying this material retaining IP and OP piezoelectricity (ferroelectric materials must have PE effect). The PE characteristics of the typical PE 2DMs that have been experimentally shown are discussed in Table 6.1. In general, NCS is a necessary condition for PE materials. By doing symmetry analysis on a 2DM, one may determine if it is PE or not [28]. Although several 2DMs were theoretically proposed to be PE, the majority of them require more investigation or confirmation through experiments.

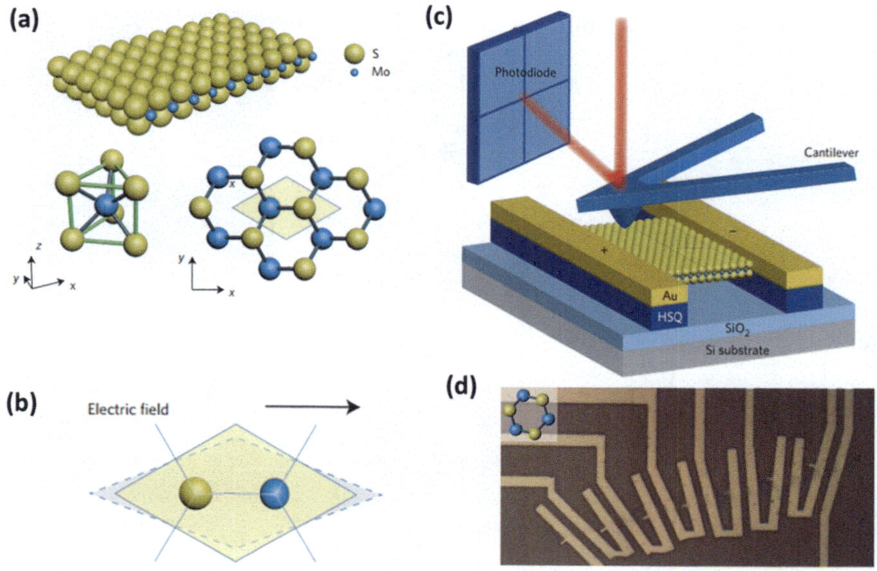

Fig. 6.4 **a** ML-MoS_2 is composed of stacks of S–Mo–S having 0.6 nm total thickness, **b** stress creation under *E*, **c** PFM designing: two hydrogen silsesquioxane electrodes were used to suspend the MoS_2 film now mounted as well as clamped under twins Au electrodes. A scanning AFM probe had been used to indent the film. The inducing stress altered voltage applied on cantilever, as seen via laser beam deflection, **d** multiple electrode pairing combined on MoS_2 long-striped-dashed-outline with each segment oriented through 10° (Reproduced with permission from [36])

6.2.1.2 Piezoelectric Nanogenerators

MoS_2-Based Energy Harvesters

A monolayer 2D MoS_2 membrane piezoelectricity was detected experimentally in 2014 [33, 36]. Wu et al. [33] were the first to employ and harvest for mechanical energy and piezotronics sensing. Keep in mind that MoS_2 associated with crystal D_{3h} regularity exhibits IP piezoelectricity (e_{11}) than OP piezoelectricity (e_{33}). A Mo atomic layer sandwiching two similar S layers packing into hexagonal lattice makes up the MoS_2 membrane. Two S atoms surround each rhombic prismatic unit cell on the left side, whereas one Mo atom is at the right as in Fig. 6.4a. On the other hand, an E_{ext} extending from S to Mo sites into hexagonal lattice toward an armchair direction, E_1 will change shape of unit cell while extending Mo-S bond leading to internal PE stress as in Fig. 6.4b. For measuring piezoelectrically produced membrane stress, AFM findings with a lateral electric field and nanoindentation are shown in Fig. 6.4c [36]. An angular dependency from 0 to 60° was investigated by designing devices that are fabricated by same single-crystal flake (Fig. 6.4d).

Mechanical energy is converted into electricity through stretching and releasing odd layers of MoS_2 flakes repeatedly. Strain-inducing-polarized careers in SL-MoS_2

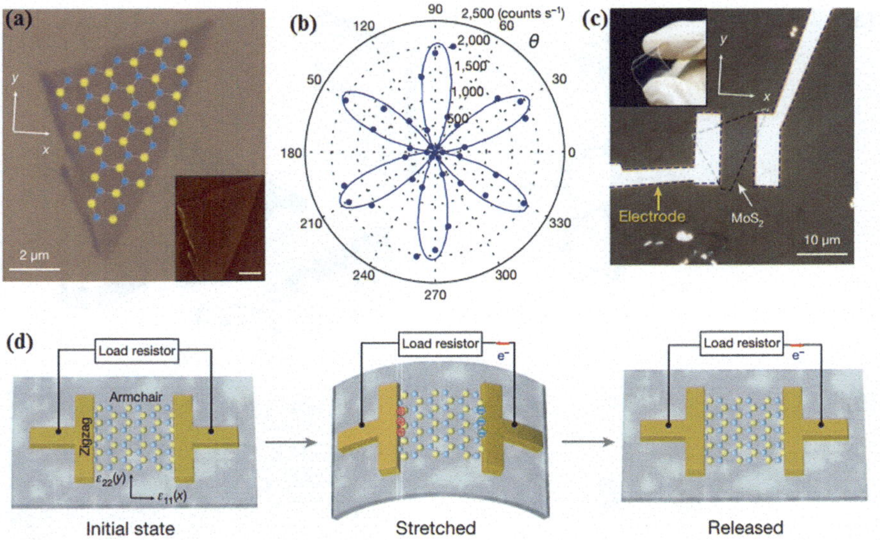

Fig. 6.5 **a** Superimposed lattice orientation determined from SHG findings on an optical image of SL-MoS_2 flake blue (Mo atom) and yellow (S atom) spheres. *Inset* AFM image of the flake, **b** polar plot indicates SH intensity in SL-MoS_2 depending on θ, **c** flexible device associated with SL-MoS_2 flakes as well as electrodes along zigzag boundary. The flexible device's optical image is included in an inset, **d** the functioning of an SL-MoS_2 PE device under stretched and released conditions is illustrated (Reproduced with permission from [33])

may boost strain detection by affecting charge carrier transportation across MoS_2-metal interface. A high piezoresistivity of approximately 230 for the BL material was reported in even-layer MoS_2, indicating a potential strain-inducing alteration into band structure. An SL-MoS_2 flake is being placed over flexible polyethylene terephthalate (PET) substrate for developing flexible device. AFM assessed the layer thickness (Fig. 6.5a, inset) and Raman spectroscopy verified it (extended Data Fig. 6.5a). The crystallographic orientation of MoS_2 flakes was then determined using SHG. The polar plot in Fig. 6.5b illustrates the strength of the 2nd-harmonic signal originated from SL-MoS_2 flake, depending on crystal's azimuthal angle. As illustrated in Fig. 6.5c, the electrical contacts have been formed using Cr (1 nm)/Pd (20 nm)/Au (50 nm) deposition. With mechanical bending of PET substrate, MoS_2 receives a uniaxial strain proportional to bending radius (Fig. 6.5d, bending figure). To avoid sample sliding, the applied strain was restricted to 0.8%. The PE response of a device is linked to load resistance (R_L) by applying strain to it (Fig. 6.5d). In this scheme, strain-inducing-polarized careers lying at sample edges may facilitate electron flowing into an external circuit. After releasing substrate, negative charges travel in opposite directions.

The PE (I and V) responses of an SL-MoS_2 are depicted in Fig. 6.6a. Upon applying strain toward x-direction ("armchair") then increasing strain results in positive V and I o/p, while decreasing strain results in negative o/p. Both responses were

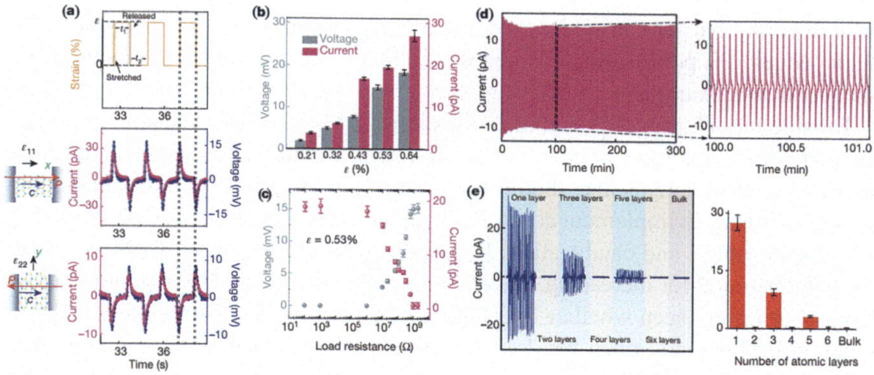

Fig. 6.6 **a** Top: applied strain versus time. Middle: PE o/p from SL-MoS_2 (armchair direction). Bottom: corresponding PE o/p of the device at *y*-direction strain application (zigzag orientation), **b** PE o/p from an SL-MoS_2 device based on applied strain, **c** dependence of *I*–*V* o/p from an SL-MoS_2 device showing 0.53% strain depending on R_L, **d** cyclic test demonstrating device stability over an extended period, and **e** The PE o/p of MoS_2 flakes increase in proportion to atomic layers number (*n*) (Reproduced with permission from [33])

more pronounced as the applied strain magnitude has increased. The highest V_{oc} of SL-MoS_2 was 18 mV, while the peak I_{sc} was 27 pA (Fig. 6.6b). By examining the *V* and *I* o/p as a function of R_L, the o/p power was calculated (Fig. 6.6c). The highest instant power supplying to load at 0.53% strain was 55.3 fW (10–14 W) for R_L of 220 M, equivalent to a P_D of 2 mWm^{-2}. The coulombic efficiency of SL-MoS_2 NG may thus be calculated to be 5.08%, defined as the ratio belonging electric power supplied to load in respect of entire mechanical distortion energy harvested into SL-MoS_2 after stretching. This energy conversion remained constant throughout time, as demonstrated during 300 min of cyclic load ~ 0.43% strain influencing 0.5 Hz (Fig. 6.6d). It was discovered that the PE signal decreases as the atomic layers (*n*) number increases, with virtually no discernible o/p of bulky flake as well as even layers sample (Fig. 6.6e). The strain-inducing alteration in devices relating to direct current electrical transportation features was studied using a 2-terminal arrangement along with applied voltage determined relativity toward drain electrode. The metal-semiconductor-metal device has been prepared to utilize with back-to-back Schottky barriers, and thereby current flow has been restricted by transport over reverse-biased Pd-MoS_2. In the mentioned configuration, alteration into transport performance was primarily caused by two belongings, the piezotronic effect showing strain-induced charge modulation of Schottky barriers asymmetrically, and secondly, piezoresistive effect indicates strain-inducing E_g alteration for modulation of device's as overall resistance. Under tensile strain, the current–voltage characteristics changed left side, i.e., toward negative drain biasing, whereas compressive strain shifted toward the right side. Under negative drain bias, the reverse trend was found, demonstrating the piezotronic consequence. In BL and bulky MoS_2 responsive is merely piezoresistive and "*I*" increases perfectly symmetrically with applied strain, whereas gap region lying in *I*–*V* curve narrows, precisely proportional to decrement into source and drain Schottky barrier caused by a decrease in E_g or change in carrier density. The zigzag edges' PE polarization charges were able to directly impact the metal MoS_2 contacts in this example.

Tunable modulation of p–n connections into 2DMs via strain-inducing polarization can permit a unique strategy not feasible with traditional strategies that utilize electrical control signals and should not require modification for interfacial structure or chemistry to implement controllable optoelectronics, improved photovoltaic, hybrid spintronics, and catalyzation. Lastly, array incorporation of SL-MoS_2 flakes was demonstrated for increasing PE o/p for energy conversion. Consistent increases in o/p V or I were seen whether four flakes were connected constructively in series or parallel. This could open the door for growth of useful technologies on a much wider scale using 2D PE nanomaterials to empower nanodevices, solid imaging, and wearable electronics. It was then examined how MoS_2 NGs for seven atomic layers performed theoretical study [51]. The MoS_2 PENG structure is shown in Fig. 6.7a. This device's left and right electrodes each have a MoS_2 flake with 3-atomic layers. R_{In} indicates MoS_2-PENG's internal resistance, and C_{NG} represents MoS_2-PENG's capacitance, and V is the voltage source that drives electron flow across the circuit through the piezopotential. Figure 6.7b illustrates the analogous circuit for the MoS_2 PENG with R_{Ex}. Under a periodic tensile strain (0.5%) and a square-wave frequency indicating 0.5 GHz (Fig. 6.7c), MoS_2 PENGs for various layers demonstrated a considerable out I_{sc} that diminishes as MoS_2 layers number (n) increases (Fig. 6.8d). Furthermore, at a frequency of 5 GHz, the MoS_2 PENGs were responsible for harvesting optimum frequency square-wave application of strain (Fig. 6.7e). In comparison to the applied strain (0.5 GHz), as seen in Fig. 6.7c, d, optimum current for applied strain (5 GHz) barely eased median period (0.1 ns) arising before post-peak current emerged, resulting in diverse effects on energy conversion efficiency. The energy conversion efficiency of the SL-MoS_2 PENG is illustrated in Fig. 6.7f for various square-wave applied strain frequencies.

The lateral electrode arrangement of ML-MoS_2 flexible PENG associated with armchair/zigzag atomic directions, respectively, is illustrated in Fig. 6.8a, b, and d. Non-conductive AFM reveals tip in the form of mechanical sensor in detecting lateral piezoresponse performed by Kim et al. [37]. Figure 6.8c illustrates the manufactured flexible PENG on PET substrate. Optical pictures of mechanical energy to harvest dynamic areas caused with armchair/zigzag atomic directions owing to PENGs are shown in Fig. 6.8e, f. Mechanical strain was used to examine the PE o/p performance. The "V" and "I" o/p illustrated in Fig. 6.8g–j have been attained at a frequency (0.5 Hz) with a bending strain (0.48%). The observed o/p voltage from PENG with an armchair orientation of MoS_2 neared 20 mV (Fig. 6.8g) and measured o/p "I" exceeded 30 pA (Fig. 6.8i). However, the o/p "V" (Fig. 6.8h) and "I" (Fig. 6.8j) from PENG containing zigzag orientation for MoS_2 are measured less than 10 mV and 20 pA. This finding implies that manipulation of MoS_2 atomic direction into an armchair orientation on large-scale basis is important for significantly improving the PE power o/p performance of single-crystalline MoS_2 ML-based PENGs [37]. Dai et al. [52] recently published a study in 2020 on the atomic assembly of GBs on ML-MoS_2 flakes, focusing on increased PE effect caused by the presence of

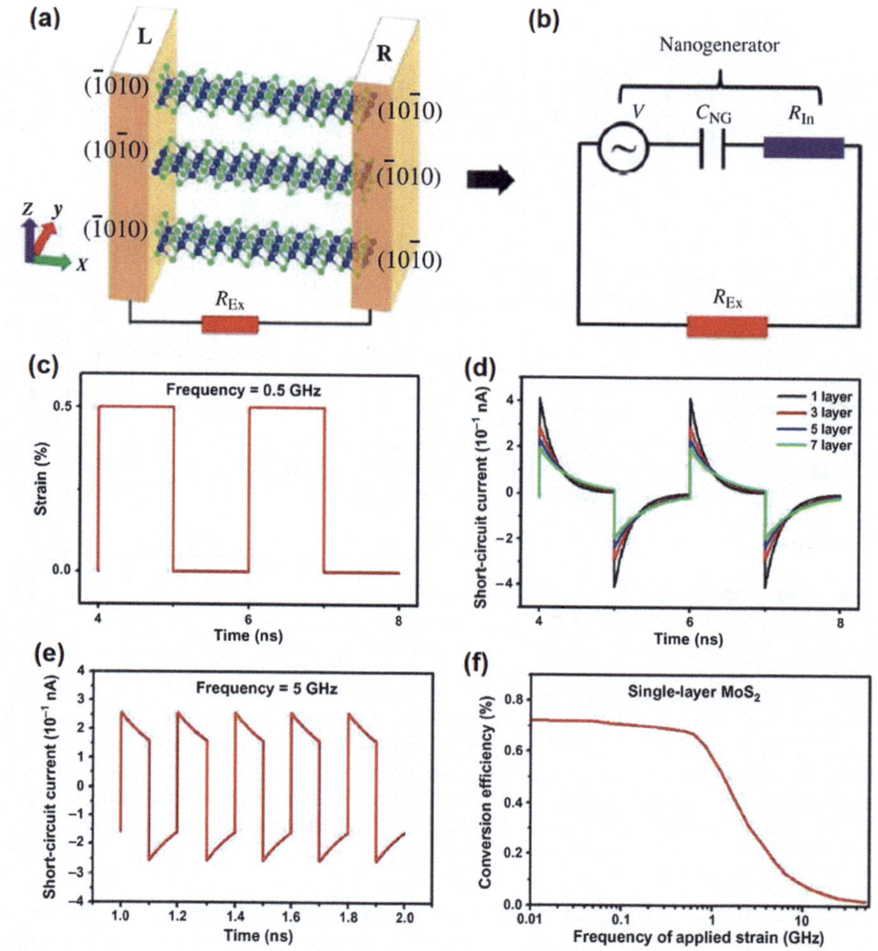

Fig. 6.7 **a** A 3-layer MoS_2 PENG connected with R_{ex}, as shown schematically **b** 3-layer MoS_2 NG (equivalent circuit), **c** square-wave outside strain applying on and releasing from NG, **d** attributing J_{sc} for odd layers MoS_2 NGs influenced by applied strain, **e** J_{sc} of SL-MoS_2 PENG influencing 5 GHz square-wave applying strain, and **f** energy conversion efficiency for SL-MoS_2 PENG versus frequency oriented square-wave strain (Reproduced with permission from [51])

deformable grain boundaries (GBs). Based on preliminary principle evaluations and PFM tests were established as GB increased for PE behavior. Furthermore, the PE o/p of MoS_2 showing GB-MoS_2 as 50% is large than monocrystal MoS_2 was observed. This finding enables to use of the butterfly-like GB-MoS_2 in self-powering force sensors, demonstrating the material's enormous potential for use in flexible/wearable electronics.

The device's PE response has been measured by applying external strain, and strain-induced polarized charges employed to control electrons flow into an external

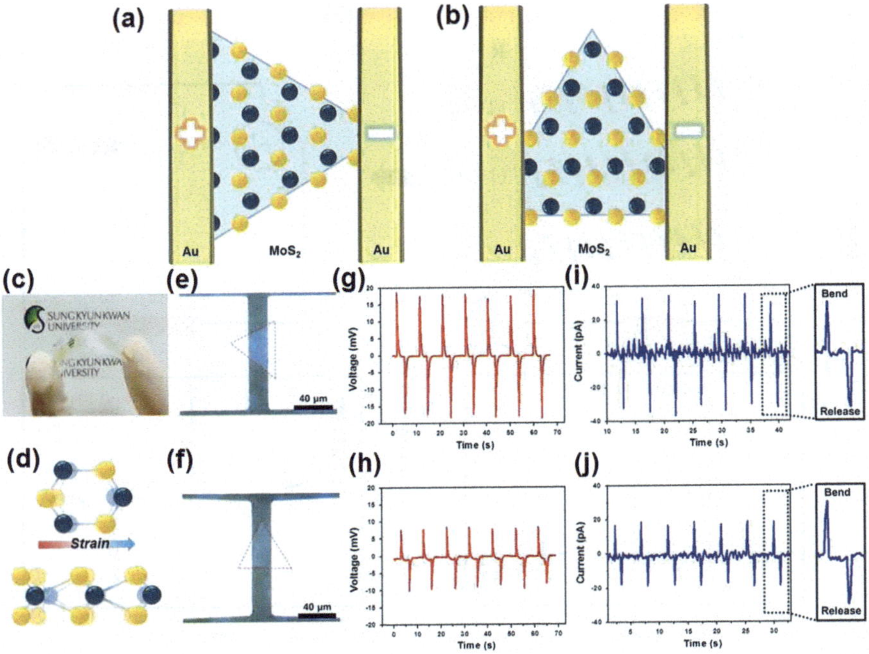

Fig. 6.8 Configurations of lateral electrodes for the measurement of the PE coefficient **a** armchair **b** MoS_2 zigzag orientations into horizontal electrode formation, **c** digital image owing to CVD developed ML-MoS_2-based flexible PENG, **d** top and lower cross-sectional atomic structures of ML-MoS_2 strained by external stress, **e, f** images of energy to harvest active areas associated with armchair/zigzag PNG atomic directions. PNG power o/p depending on MoS_2 atomic directions for an armchair or zigzag: **g, h** NG "V" o/p **i, j** NG o/p "I" (Reproduced with permission from [37])

circuit (Fig. 6.9a, b). Using an AFM, the thickness of ML-MoS_2 was determined. As seen in Fig. 6.9c, the ML-MoS_2 flake has a thickness (1 nm). As previously observed, when MoS_2 is ML thick, the structure turns out to be NCS [33]. Furthermore, because the SHG is extremely sensitive to structural symmetry, anisotropic polarization patterns predict the armchair/zigzag orientations of ML-MoS_2 [53]. The NCS crystal structure was studied using SHG measurements. As demonstrated in Fig. 6.9d, when an 800 nm laser is used to excite the ML-MoS_2 flake, a 400 nm SHG peak is observed, but the bare substrate shows no signal that confirms the ML-MoS_2 flakes' NCS structure [53, 54]. A six-folding rotational regularity calculated by $I = I_0 \cos^2(3\theta)$ is observed for the angle depending on SHG intensity measuring toward parallel (XX) or vertical (XY) corroborating the threefold rotational symmetry of ML-MoS_2 (Fig. 6.9e). The armchair orientation is confirmed as being perpendicular to edge using SHG patterns [55]. The consistent intensity demonstrated by polarized SHG mapping demonstrates the monocrystalline nature of the triangle-like MoS_2 ML (Fig. 6.9f). These observations indicate that such GBs in ML-MoS_2 are complex than

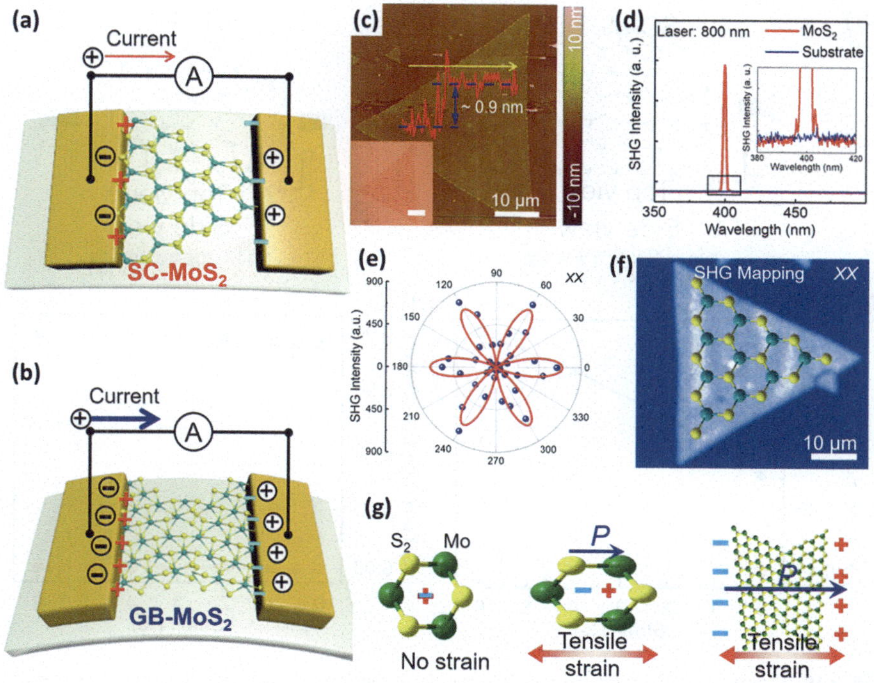

Fig. 6.9 Characterization of MoS_2 flakes **a** triangle ML-MoS_2 flake (AFM image). *Inset* Corresponding optical image **b** SHG spectrum of MoS_2 flake with substrate, **c** SHG intensity polarization angle θ-dependent in parallel (*XX*). With a six-folding rotational regularity by $I = I_0 \cos^2(3\theta)$, **d** triangular MoS_2 flake deposited on SiO_2/Si substrate SHG mapping, **e** polarization angle θ dependent SHG intensity under parallel conditions (XX). It exhibits a clear six-fold rotational symmetry ($I = I_0 \cos^2 3\theta$), **f** SHG mapping of a triangle MoS_2 flake on SiO_2/Si substrate, **g** schematic diagram of in-plane polarization under tensile strain in GB-MoS_2 monolayers. (Reproduced with permission from [52])

theoretically modeled "periodical" structures [56, 57]. When a tensile strain is introduced to GB-MoS_2 ML, the NCS structure of GB results has increased polarization (Fig. 6.9g).

Superior Piezoelectricity from Grain Boundary in MoS_2 Monolayers

Optimal learning about how GBs affect structural, dipole moment (μ), and electronic structures of the ML-MoS_2 were performed. Figure 6.10a, b shows atomic assemblies of monocrystal armchair edges for MoS_2 ribbon, and GB-MoS_2 ribbon associated with uniaxial tensile stressing [52]. To stabilize the formations, hydrogen atoms passivate all dangling bonds. A Mo atom is ended at the edges by two atoms of hydrogen, whereas a sulfur atom is finished by a hydrogen atom [58]. In Fig. 6.10b, two MoS_2 grains are linked by a seven-atom and a five-atom ring shown by GBs. The element configuration of the GB unit cell is $Mo_{17}S_{34}H_{13}$. Thus, 2–3 coordination of Mo atoms are making Mo–Mo bond at the border, thereby terminating by one-bridged H atoms. As a result, the Mo–S–Mo sequence is broken at GB. A single addition

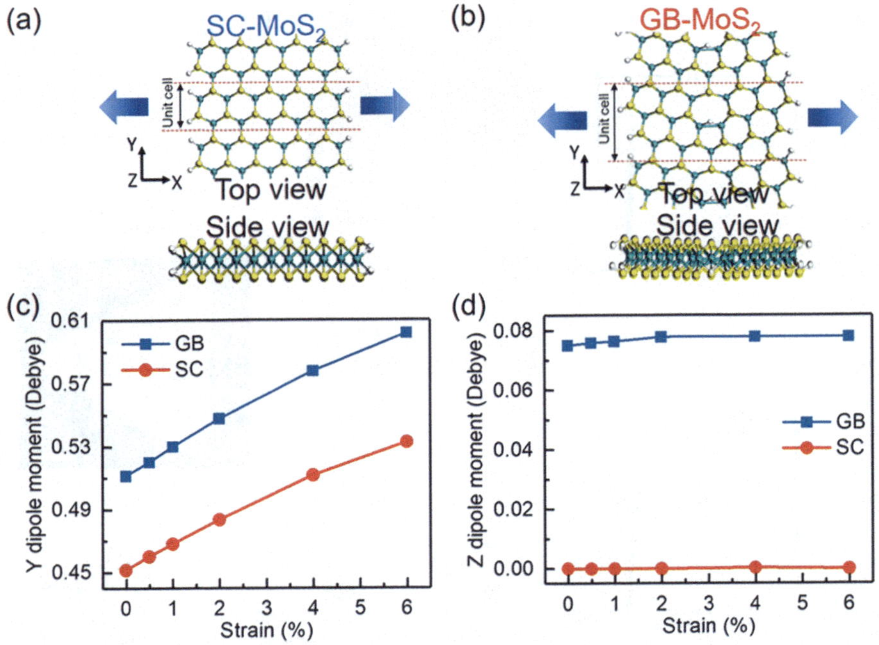

Fig. 6.10 Origins of GB increased piezoelectricity: a theoretical and experimental investigation: **a**, **b** MoS_2-ribbon model completed with hydrogen atoms as white balls toward DFT calculations, upper views of crystalline left side with domain boundaries of right side, and corresponding bottom-side-view, **c**, **d** the dipole moment related to MoS_2 ribbons subjected to uniaxial tensile straining was determined. (Reproduced with permission from [52])

of Mo atom produces hole inside the scheme. The boundary edging demonstrates a joint armchair to zigzag with a ratio of 1:2. The DFT-derived μ, such as μ_y and μ_z components are presented in Fig. 6.10c, d for a uniaxial tensile strain along the x-axis. In comparison to the semiconductor ribbon, the GB ribbon has higher μ_y and μ_z values [52].

WSe_2-Based Energy Harvesters

The WSe_2 BLs produced by turbostratic stacking exhibited dependable PE characteristics that are not acquired from mechanically exfoliated WSe_2 BLs stacked Bernal method [38]. In comparison to ML-WSe_2 degrades the performance of device above 0.63% strain, the flexible PE WSe_2 BLs demonstrate high mechanical durability around (0.95%) strain and a dependable energy harvesting performance sufficient while driving little liquid crystal view in the absence of external energy foundations. Figure 6.11a shows a geometric depiction for monolayer WSe_2 (m-WSe_2)-depending NG, while Fig. 6.11b depicts an optical photograph for Cr/Au electrode installed on PET substrate employed to fabricate the device. The PE coefficient d_{11} for m-WSe_2 was calculated as 3.26 ± 0.3 pmV^{-1}, an acceptable measurement in comparison

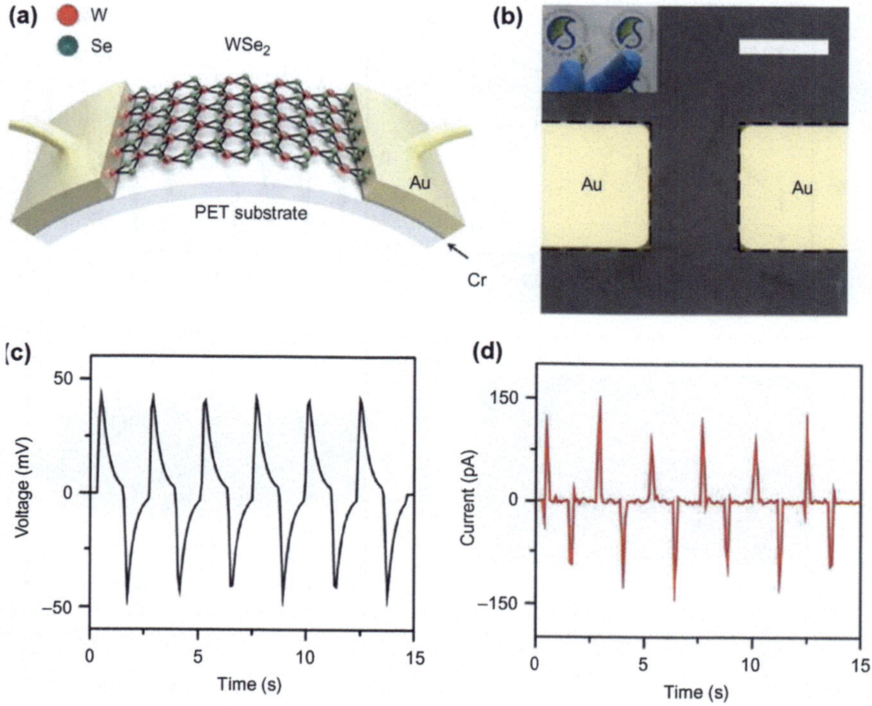

Fig. 6.11 **a** m-WSe_2 depending PENG schematic, **b** the electrode configuration optical photograph with original PENG (inset), **c** o/p voltage, and **d** J_{sc} from PENG under periodic tensile strain (Reproduced with permission from [38])

with earlier published simulation results of (2.79 pmV^{-1}) [35]. The PE o/p V and I responsiveness of the m-WSe_2 were also measured. With R_L ~ 1 GΩ, the peak V and I_{sc} were produced ~ 45 mV and 100 pA, respectively, as depicted in Fig. 6.11c, d for 0.39% strain and a 40 mms^{-1} straining rate. The highest prompt power o/p was 2.54 pW for $R_L = 500$ MΩ and flexible m-WSe_2-based PENG converted at a rate of 2.41% for over 1000 cycles.

A direct PE response of bilayered WSe_2 (b-WSe_2) was studied by the preparation of two different kinds of b-WSe_2. One is b-WSe_2 producing direct with sapphire substrate through CVD (db-WSe_2), while other is m-WSe_2 producing through CVD, and thereby moving toward other m-WSe_2, leading to tb-WSe_2. The mechanical durability of tb-WSe_2 has been shown to be large than m-WSe_2. The m-WSe_2 may withstand 0.64% straining while creating a peak o/p (V ~ 90 mV) (Fig. 6.12a), whereas the tb-WSe_2 can withstand up to 0.95% of strain and provide an o/p voltage of 85 mV (Fig. 6.12b). Showing tb-WSe_2 of 0.89% straining, the greatest instant power production was 4.05 pW and remained constant for over 1000 cycles (Fig. 6.12c). Furthermore, m-WSe_2 produced an optimum power of 2.05 pW at 0.89% strain and then power dropped considerably with repeated strain (Fig. 6.12d). Thus, despite its lower PE coefficient than m-WSe_2, tb-WSe_2 has been shown as an appropriate

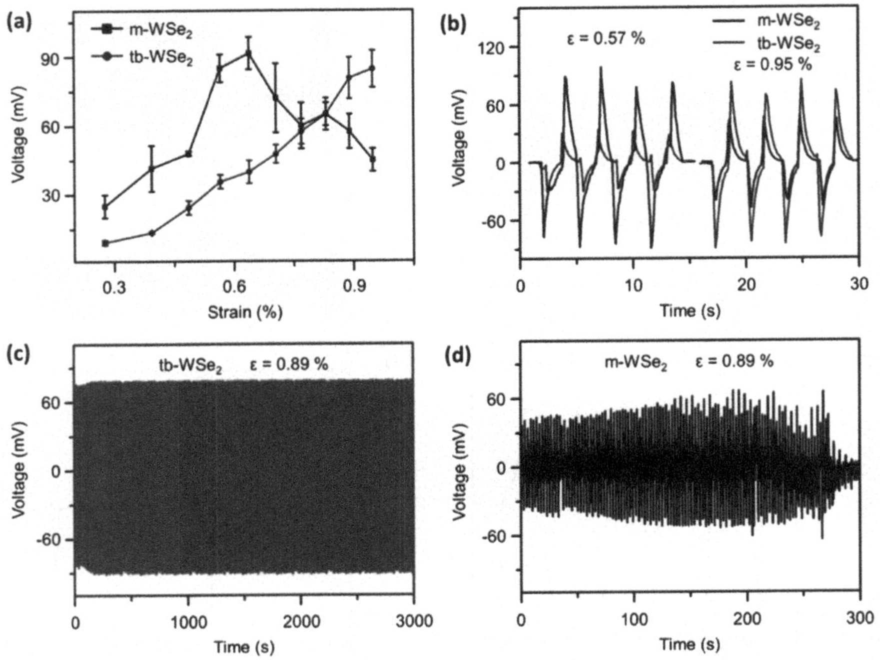

Fig. 6.12 Mechanical stability of WSe_2 mono/BLs investigated: **a** PE peak o/p "*V*" attributed to m-WSe_2 as well as tb-WSe_2 vs straining with $R_L = 1$ GΩ, **b** PE o/p "*V*" for m-WSe_2 indicated with black line whereas tb-WSe_2 as red line showing low straining of 0.57% and high straining of 0.95%, **c, d** durable testing was assessed for knowing the tb-WSe_2 and m-WSe_2 mechanical stability depending on PE energy harvester of 0.89% straining during a 1000 cycle period (Reproduced with permission from [38])

mechanical energy harvesting. To enable real-world applications with increased o/p power, multielectrode patterning was used located at large-area tb-WSe_2, whereas NGs were combined into a bundle. On a single substrate, 5-integrated NGs were produced. The 14 operational PENGs have been effectively merged with the twenty operational PENGs and measured o/p "*I*" rose linearly (1.4 nA) as parallel connections number improved. The o/p "*V*" of the paralleled PENGs are virtually identical to those of single PE energy harvesters. The PENGs used a very tiny, commercially accessible liquid crystal display (LCD). The o/p for the LCD screen was set to "1", and the LCD was directly connected with PENGs.

α-In_2Se_3-Based Energy Harvesters

The coexistence of OP and IP piezoelectricity into ML toward bulky α-In_2Se_3 was ascribed to their NCS extracted with hexagonal stacking [59]. Consequently, the associated d_{33} PE coefficient for α-In_2Se_3 rises without any odd–even impact from 0.34 pmV^{-1} (ML) to 5.6 pmV^{-1} (bulk). A flexible PENG depending α-In_2Se_3

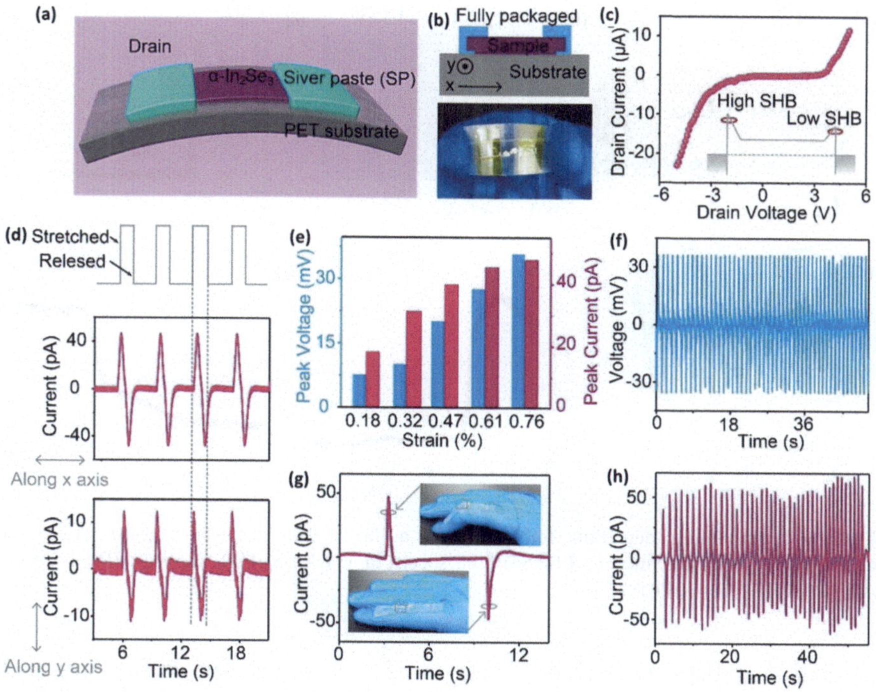

Fig. 6.13 **a** Flexible PENG is based on α-In_2Se_3 that are completely encapsulated in Ag-paste, **b** sketched devices side view in (**a**), **c** *I*–*V* curve, **d** J_{sc} response of NG for periodic straining of 0.76% toward two opposite directions, **e** straining dependence for PE o/p, **f** V_{oc} o/p with strain of 0.76%, **g** NG PE combination with index finger reclaiming mechanical energy from finger movements, **h** "*I*" o/p (Reproduced with permission from [59])

was exhibited as an energy cell with electronic skin. While producing flexible α-In_2Se_3 PENG, and ML specimens of thicknesses ranging from 100 to 200 nm were moved to PET substrates and two metal electrodes were formed using silver paste (Fig. 6.13a). Because Ag-paste is fluid, two terminals of the ML α-In_2Se_3 have been entirely encased as seen at the top of Fig. 6.13b that allowed for PE o/p to benefit from vertically as well as horizontally polarized charges. The *x*-axis represents the substrate's length and the *y*-axis its width direction. The bottom panel of Fig. 6.13b shows an optical photograph for flexible along with transparent PENG. An *I*–*V* measurement in Fig. 6.13c demonstrates PENG's metal-semiconductor junctions are Schottky contacts. Figure 6.13d illustrates the PE "*I*" retorts for as-fabricated PENG. Figure 6.13e illustrates relationship between PE o/p peak (*V* and *I*) and applied strain magnitude. In Fig. 6.13f, a consistent "*V*" signal is shown under 0.76% and short (1 s) straining cycle, exhibiting high o/p repeatability. Maximum *V* and *I* production as 35.7 mV and 47.3 pA were measured as 0.76% straining. By embedding PENG into human flesh, including back of index finger, the PENG has been exhibited as e-skin. This was accomplished by replacing the PET substrate with very supple and

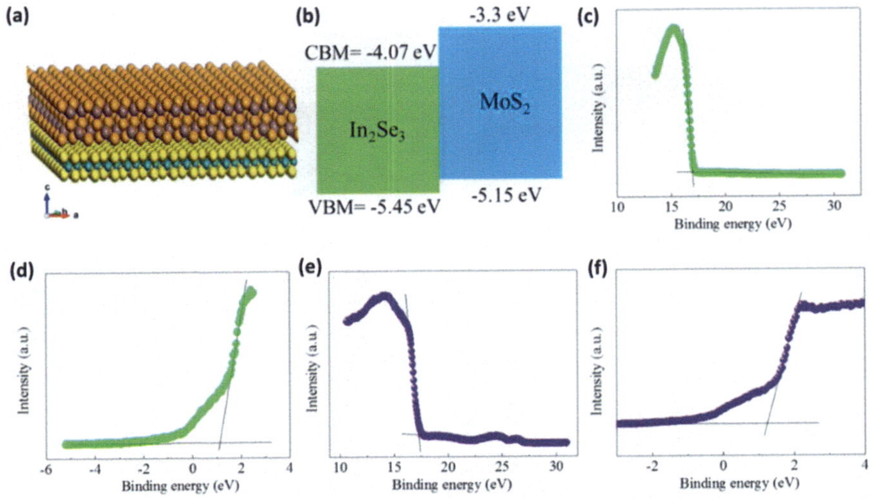

Fig. 6.14 Energy-band depiction for vdW-HS: **a** In_2Se_3/MoS_2 (atomic construction), **b** In_2Se_3/MoS_2 (band structure), **c**, **d** In_2Se_3 UPS spectrum, **e**, **f** MoS_2 UPS spectrum (Reproduced with permission from [60])

transparent ordinary "scotch tape". When index finger is attached with as-fabricated PENG, that device quickly responds with a peak "*I*" (Fig. 6.13g) to repeat curved or extended motions as illustrated in Fig. 6.13h. This illustrates the device's enormous potential for gathering biomechanical energy or serving in the form of electronic skin.

In_2Se_3-Based Heterostructures for Piezoelectricity

Very recently, in 2020, Hao et al. [60] developed a layered In_2Se_3/MoS_2-based vdW-HSs with an enhanced PE response. The extraordinary PE response was attained in a 2D vdW-HS system using In_2Se_3/MoS_2, in contrast to the IP piezoelectricity of MoS_2 that is constrained by edge effect along with odd and even layers. The physical mechanism behind vdW-HS's PE characteristics was investigated. A high-precision PE actuator at the 2D atomic scale is shown, paving the way for innovative applications in nanoelectronics and atomic scale coupling.

Physical Mechanism

It was possible to produce improved PE characteristics in the In_2Se_3/MoS_2 vdW HS ascribed to the large band offset [61]. Interfacial polarization charges modulate the PE effect, affecting band alignment with offsets. Figure 6.14a depicts the In_2Se_3/MoS_2 atomic structure. As given in Fig. 6.14b, substantial band offset between CBM and VBM of In_2Se_3 and MoS_2 layers results in significant electrical polarization. The ultraviolet photoelectron spectroscopy (UPS) was employed to infer band energy of

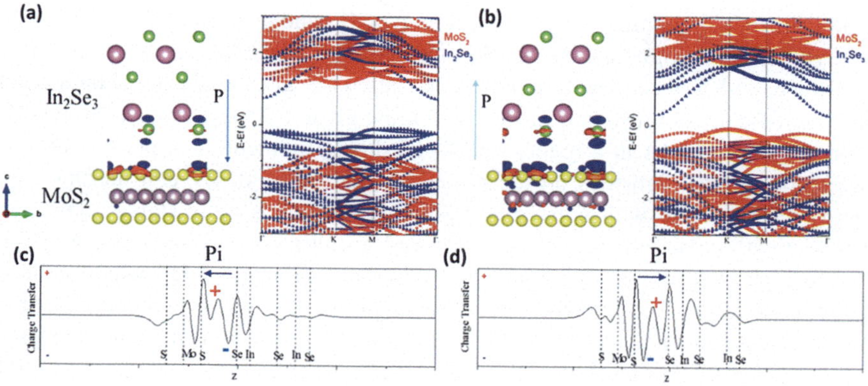

Fig. 6.15 DFT calculations: **a** atomic structure and band structure of the In_2Se_3/MoS_2 heterostructure when the polarization of In_2Se_3 is pointing downward. **b** Atomic structure and band structure of the In_2Se_3/MoS_2 heterostructure with polarization pointing upward, **c**, **d** plane-averaged differential charge density of (**a**) and (**b**), respectively (Reproduced with permission from [60])

In_2Se_3 and MoS_2, as shown in Fig. 6.14c–f. Secondary electron energy cut-off-edge of UPS spectra is determined to be ~ 16.9 eV by linear extrapolation to the baseline. According to equation $\varphi = hv - E_{\text{cutoff}}$ connected with an incoming photon energy of $hv = 21.2$ eV, work function (φ) of In_2Se_3 is 4.3 eV. Additionally, by linearly extending the UPS spectrum's onset edge with baseline, and variance among Fermi energy and valence band is determined as ~ 1.15 eV. On the same pattern, work function and VBM for MoS_2 are measured as 3.9 eV and 1.25 eV, respectively. The In_2Se_3 and MoS_2 possess significant band counterbalance, according to their E_g.

It should be prominent that while work function and VBM values above are for isolated In_2Se_3 (or TMDC), charge transfer at the In_2Se_3/MoS_2-HS interface may cause band alignment discrepancies when In_2Se_3 and MoS_2 are combined [62, 63]. It is important to compute the band structure with disparity of charge density for In_2Se_3/MoS_2-HS to determine the effect on interface charge transfer as well as potential screening in band alignment. The predicted E_g of In_2Se_3 (0.83 eV) and MoS_2 (1.83 eV) are somewhat smaller than experimentally observed E_g of 1.38 and 1.86 eV, respectively. Because of self-interacting error for complex that arises in fully occupying states in conventional DFT; E_g computed using PBE is frequently underestimated. Relevant with multielectron quasi-particle gigawatt computations, a more precise E_g may be achieved. But, in supercell HS system, computation is prohibitively expensive. According to estimated band structure, when polarization is oriented toward MoS_2 layer, the HS's band structure conforms to type I band alignment[1] (Fig. 6.15a), which means that In_2Se_3's CBM and VBM are entirely contained inside the E_g of MoS_2. When In_2Se_3 polarization is directed upward away

[1] There are two sorts of band lineups: type I with a spike, where the smaller gap entirely encircles the bigger, and type II with a cliff, where both bandgaps only partially overlap. A big spike can hinder carrier flow in the band, while a tiny spike can be overcome by carrier thermal excitation. In a cliff band lineup, the interface has the smallest gap, accelerating interface recombination.

from the MoS_2, in contrast, the HS exhibits type II band alignment (Fig. 6.15b). In this type of band alignment, In_2Se_3 layer contributes to CBM and the MoS_2 layer contributes to VBM can increase total polarization because of additional interfacial dipole moment. As illustrated in Fig. 6.15d, the greater charge transfers in type II band alignment result in a higher interfacial dipole moment. The plane-averaged differential charge densities are described in Fig. 6.15c, d. By considering total energy, the structure with polarization pointing upward is more stable relative to structure with polarization pointing downward, resulting in a 70 meV energy gain. Due to high energy gain, it is predicted that the experimental HS would exhibit type II band alignment. In comparison to a single In_2Se_3, type II band alignment is considered advantageous to vdW-HS′ interfacial charge transfer. If vacuum level is utilized as a reference level, the DFT calculation agrees with the experimental UPS band alignment.

h-BN Energy Harvesters

The h-BN is a broad E_g energy around 5.9 eV 2DM [64], with remarkable features, including high IP thermal conductivity and its chemical inertness, superior mechanical capacities with non-toxicity. Occurrence of piezoelectricity is FL-h-BN that was anticipated theoretically using a PE stress coefficient of $e_{1,11} = 0.731\ \text{cm}^{-2}/N_{\text{odd}}$ (N is the layer number) [65, 66]. From a recent experimental study, piezoelectricity of an FL-h-BN is demonstrating potential applications to harvest energy and sensor/actuator [67]. Since then FL h-BN has attracted considerable interest from scientific community and industry, despite the fact that its study is still in its infancy [65]. The h-BN and MoS_2 share a point group (D_{3h}), containing B and N atoms in h-BN crystals arranged hexagonally. Unlike GR, ML of h-BN exhibits significant IP piezoelectricity as a result of alternating order for B and N atoms at hexagonal vertex area. The layer and thickness dependency of IP piezoelectricity by SHG extracted from atomically thin samples for h-BN comprising 1–5 layers has been investigated [68]. The SHG signal is only detectable for layers with an odd layers number and not for layers with an even layers number. This implies that IP piezoelectricity arises for h-BN with odd layers, as demonstrated earlier for 2D-MoS_2. Therefore, h-BN has no thickness dependency, but MoS_2 was shown to decrease in thickness with increasing thickness. The latest first-principles simulations have indicated h-BN with a B/N atomic vacancy to exhibit enhanced piezoelectricity with magnetic moment [69]. Lee et al. [67] made the first effort to quantify the PE voltage coefficient attributed to single BN nanoflakes (NF) to assess energy harvesting capability for flexible PE energy devices based on 2D h-BNNFs. Produced h-BNNFs with a high yield and efficiency using a simple ball-milling method aided by sodium hydroxide (Fig. 6.16a). A PE single h-BNNF has been linked to Au electrodes on a flexible substrate using simple methods including transfer and focused ion beam-Pt deposition to accurately examine its energy conversion efficiency during mechanical agitation as depicted in Fig. 6.16b. Figure 6.16c, d show SEM images and a single h-BNNF photograph on soft substrate ($5 \times 5\ \text{cm}^2$) containing edges for single NF (adjacent size as 1 μm)

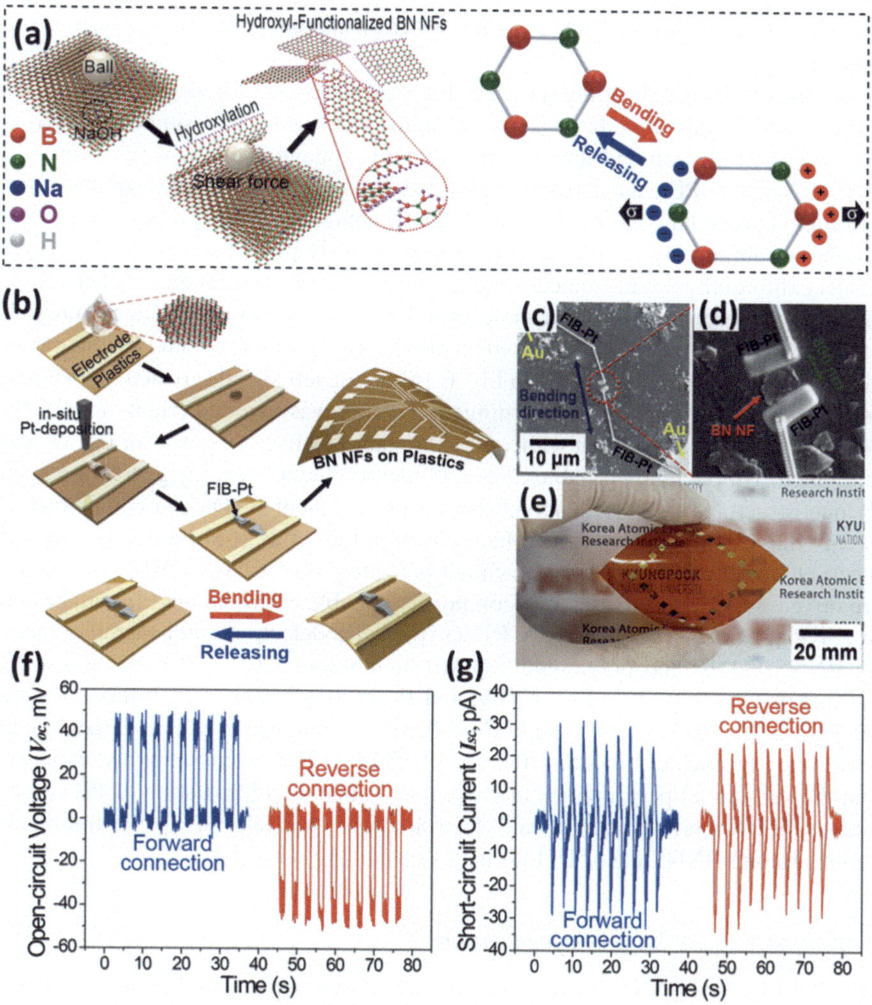

Fig. 6.16 **a** Schematic of serial procedure for producing 2D h-BNNFs (left side); the scheme also illustrates the PE effect for 2D h-BN at deforming time mechanically (right side). **b** The preparation procedure for showing the h-BNNF on an electrode pattern for flexible soft substrate is schematically depicted. **c**, **d** SEM images of single h-BNNF unified with Au lines with 800 nm electrode space employing focused ion beam-Pt deposition process, **e** soft substrate image with a PE single-NF deforming of human fingers, **f** V_{OC}, and **g** J_{sc} from a unit h-BNNF under application of external stress (Reproduced with permission from [67])

have been connected to electrode lines through focused ion beam-Pt deposition. Figure 6.16e illustrates the deformation of a soft substrate by even a PE single NF of human fingers. Single h-BNNF produced 50 mV and 30 pA "*V*" signals and "*I*" pulses respectively, after repetitive mechanical deformation of the supple substrate. At $R_{\mathrm{ex}} \sim 200\ \mathrm{M}\Omega$, h-BNNFs-based flexible PE energy harvesters having electrode area (3 × 3 cm^2) with ~ 9 V (V_{oc}), ~ 200 nA (I_{sc}), and ~ 0.3 μW (instantaneous o/p power).

During mechanical bending showing strain rate 2.32% s^{-1}, these generated values (50 mV and 30 pA) have been observed on the top surface of a single h-BNNF at 800 nm electrode gap and may be altered as PE voltage coefficient (g_{11}) for 2.35 × 10^{-3} VmN^{-1}. PE characteristics to BNNFs are enhanced by optimization of the exfoliation procedures, as the PE o/p of 2D PE nanomaterials is proportional to the number of atomic layers (n). According to previously published results, odd-layer sheets exhibit inherent piezoelectricity because of reversal symmetry; rather piezoelectricity is absent in even-layer samples and bulk 3D forms with a CS structure [33, 35, 66]. To confirm that the observed small electricity was generated by the piezoelectricity of the single h-BNNF, In Fig. 6.16f, g, the left signals created by bending as well as releasing toward forwarding connection was inverted (right-signal). By combining load V and I data with various R_L, an effective o/p power of 0.3 W was computed. This technique enables the exact measurement of energy generation from PE nanostructures and represents substantial advancement in 2DM-based flexible PE energy harvester's development. Much recently, Luo et al. [65] produced a hybrid piezo NG with good performance. Small percentages of BNNSs (BN nanosheets) are mixed with PDMS to generate composites capable of producing PE voltages up to 5.4 V with d_{33} ~ 12 pCN^{-1}. A PTEG (piezo/triboelectric nanogenerator) made by BNNS-PDMS and polyamide-6 (PA6) membranes (20 × 20 mm^2) has a J_{sc} 230 mAm^{-2}, an o/p "V" of ~ 1870 V, and P_D of 103.7 Wm^{-2}, each thrice timing the PDMS/PA6 NG control. The BNNS-PDMS membranes may be polarized electrically to significantly increase the o/p of PTEGs. The extensive study indicates that the substantial performance improvement is due to collaboration of PE consequence for BNNSs and the improved electron-affinity for BNNS-PDMS membranes, demonstrating BNNS-based PTEGs have outstanding potential.

Performance of BN-Based Nanogenerator

The PDMS-based BNNS PENG device architecture is depicted in Fig. 6.17a; PENG is made up of a BNNS-PDMS membrane of 20 × 20 mm^2 inserted between two "Al" tapes serving as top and bottom electrodes. This membrane has a thickness of 200 μm, and BNNS content in the composites has been varied from 0 to 10% by weight. Prior to device characterization, the BNNS-PDMS membranes have been electrically poled in a variety of fields. The electric potential distribution has been demonstrated using the finite element technique after BNNSs into PDMS composite have been embraced via vertical force, as illustrated in Fig. 6.17b. All BNNSs of varied sizes have been considered to be oriented parallel to the membrane's thickness direction. In all cases, irrespective of size or tilting angle, the BNNS-PDMS composite has a distinct vertical potential gradient and a considerable potential difference among top and bottom margins for each BNNS. Several examinations were repeated to investigate the PE retort of the Al/BNNS-PDMS/Al NGs, findings are presented in Fig. 6.17c, d. The devices were subjected to a 50 N loading force at a frequency of 2 Hz. Figure 6.17c shows a comparison of voltage o/p of a BNNS-PDMS PENG doped at 2% with a

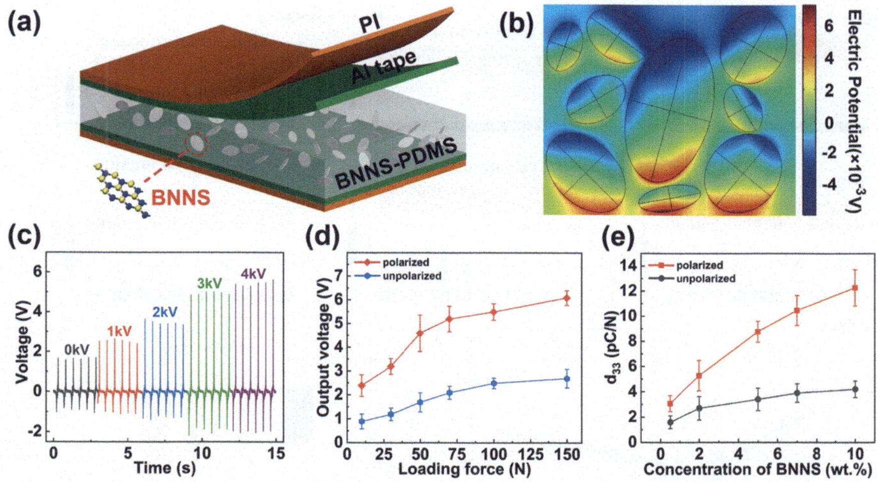

Fig. 6.17 **a** Device architecture for BNNS-PDMS PENG, **b** BNNSs PE response for composite *i* **c** o/p voltage of 2 wt% BNNS-PDMS PENG for various electric polarizations, **d** with varying loading forces, the PE response for polarized/unpolarized BNNS-PDMS membranes was investigated, **e** d_{33} for polarized/unpolarized BNNS-PDMS membranes due to varying BNNS concentration (Reproduced with permission from [65])

poling electric "*V*" ranging from 0 to 4 kV with electric fields of 0–20 kVmm^{-1} during 12 h at ambient temperature. When poling electric voltage is raised from 0 to 4 kV, the peak o/p voltage (V_p) rises from ~ 1.6 to 5.4 V. Without electric poling, the BNNS-PDMS device exhibits a rather weak PE response with a V_p of around 1.6 V is consistent with findings [67]. This is due to small concentrations with random dispersal of BNNSs have negligible PE impact. Polarization considerably increases the PE sensitivity for BNNS-PDMS films of ~ 1.6–5.4 V, more than thrice that of unpolarized membranes. As shown in Fig. 6.17d, the PE responses of both polarized (20 kVmm^{-1}) and unpolarized BNNS-PDMS membranes rise as the loading force increases; this is a characteristic of the PE effect. The PE constant d_{33} of the BNNS-PDMS membranes was measured before/after electric polarization (20 kVmm^{-1}) with various BNNS contents. Figure 6.17e shows the influence of electric polarization on PE responses for BNNS-PDMS membranes. The d_{33} unpolarized as well as polarized BNNS-PDMS membrane raises BNNS concentration, reaching 4.2 and 12.3 pCN^{-1} at 10 wt%, respectively.

Mechanism

Figure 6.18a illustrates the working of a BNNS-PDMS/PA6 PTEG device. For the beginning, PTEG is a particular parallel plate capacitor linked to R_{ex}. When the PTEG's two plates are repeatedly pushed and separated, electrostatic induction causes "*I*" in flowing to and fro among bottom and top electrodes, creating a current (or voltage) pulse. There is no charge produced prior to the connection for

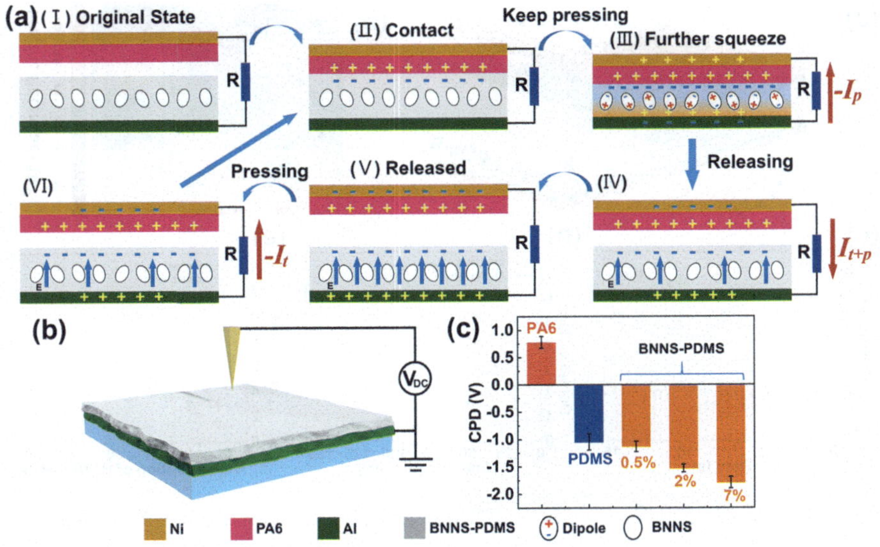

Fig. 6.18 **a** Graphic representations for BNNS-PDMS/PA6 PTEG's functioning while demonstrating the joint outcome related to PE and triboelectric effects, **b** KPFM setup shown schematically, **c** contact potential differences of KPFM measurements on PA6, PDMS, and BNNS-PDMS membranes with variable amounts of BNNS (Reproduced with permission from [65])

two polymer layers, and therefore no difference in electric potential (Fig. 6.18aI). Then, by applying vertical pressure to the two PTEG plates, opposing kinds of charges are produced as a result of connection electrification influence. Electrons are transported from the PA6 membrane toward BNNS-PDMS membrane forming positive and negative careers on PA6 and BNNS-PDMS membrane surfaces. This is because of two polymer tribomaterials' insulation properties produced charges are not readily transported and neutralized (Fig. 6.18aII). With additional stretching, a single BNNS's inversion symmetry is disrupted, leading to a nonzero μ in the strain direction [67]. The PE action occurs and results in the generation of extra surface charges, therefore enhancing the triboelectric effect [70]. Toward vertical direction for BNNS-PDMS membrane, already illustrated, a piezoelectric potential (ΔU_{P}) among upper and lower planes for BNNS-PDMS membrane drives negative charge career transport originated from top to bottom electrodes and producing $-I_{\mathrm{P}}$, as shown in Fig. 6.18aIII. Following that, at the time of twin friction layers start for separating, an o/p "I" may be split into two phases. To begin at the time contact force between twin friction layers gradually decreases to zero, inversion regularity for BNNS is returned; the contact force-induced dipole moment of BNNS and ΔU_{P} also vanishes; and the electrons are driven by the ΔU_{P} flow back, resulting in a reverse o/p "I" of the PE (I_{P}). Secondly, when twin friction layers isolate, electric potential difference (ΔU_{t}) among two electrodes is produced through friction-induced surface careers, causing electrons to flow from bottom to top electrodes via R_{L}, leading to an o/p current (I_{t}) of triboelectric generator. When the PTEG reverts to its initial condition and ceases to separate, the ΔU_t is completely counterbalanced by electrons and charge carriers transmitted from an external circuit that has collected into electrodes (Fig. 6.18aV). At the time PENG is repeatedly pushed, the reduction in ΔU_t causes an electron backflow and produces a triboelectric reverse current ($-I_t$) until zero potential difference exists among two layers (Fig. 6.18aVI). Additionally, prior research has shown that a strong electric field can reorient BNNSs parallel to the field direction [71, 72]. The charges collected on surface of BNNS-PDMS as well as bottom electrode may produce a perpendicular electric field directing upward. The electric field intensity becomes optimal at momentary separation for maximum twin friction layers. When V_{oc} of BNNS-PDMS/PA6 PENG becomes greater than 1000 V, the electric field intensity for BNNS-PDMS membrane approaches 8 kV/mm is near or equal to value utilizing the pole of BNNSs previously published [67, 73, 74]. Thus, the high electric field inside the BNNS-PDMS membrane makes reorientation of BNNSs, enhancing their PE action macroscopically and resulting in to increase in BNNS-PDMS/PA6 PENGs' electric o/p. When compared to the pure PDMS/PA6 TENG and BNNS-PDMS/PA6 PTEG exhibited significantly improved triboelectric o/p performance. KPFM was subsequently used to evaluate the surface potentials of BNNS-PDMS membranes accompanied with varied BNNS contents (wt% ratios) Fig. 6.18b. The outcomes of the KPFM experiments are presented in Fig. 6.18c.

Other 2D Materials-Based Energy Harvesters

In the recent past, it has been uncovered that 2DMs like filler is desirable for PE energy harvesting. For instance, it was discovered that 2D SnO_2 NS [75], MoS_2 [76], rGO [58], and GO [77] may be utilized as filler-material to enhance PE energy harvesting efficiency for PE polymer material (PVDF along with copolymer). Because 2DMs have an optimal surface: volume ratio, capacitance for microcapacitor embedded within a material is effectively increased. In addition to interfacial contacts among 2DMs connected with polymer sequence efficiently organize chain, culminating into composite material with high piezoelectricity. Developing composites from the aforementioned components may be a viable strategy for obtaining a high-quality flexible PE material. The 2D-SnO_2 NS/PVDF composites of PENG could indeed produce higher o/p "V" (42 V), an appropriate current density (6.25 μcm^{-2}), and a high P_D (4900 Wm^{-3}) capable of directly lighting 85 LEDs along with charging of capacitors without any electrical poling [75]. Integrated 2D-MoS_2 on PVDF fiber results in the fabrication of extremely sensitive PENG (Fig. 6.19) [76]. The as-prepared NG has been assumed extremely sensitive, capable of transducing extremely light leaf of 0.68 g whereas matchsticks of 0.09 mg dropping from various heights. This sensitivity was also utilized to recognize speech signals and capture acoustic energy. Consequently, the PENG correctly identifies numerous English alphabets of A, B, C, D, and E to display an o/p response. The frequency profile spectrum of the acoustic waves was identified using the fast Fourier transform connected with frequency spectra. Furthermore, the PENG was utilized in identifying and distinguishing acoustic signals from a variety of musical instruments. Another research

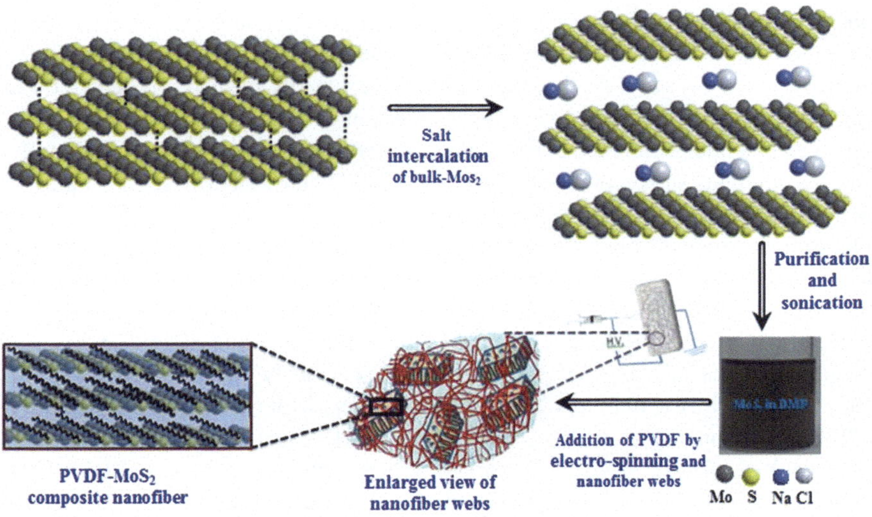

Fig. 6.19 MoS_2 NS merged with PVDF for preparing PVDF-MoS_2 composites nanofibers shown schematically (Reproduced with permission from [76])

examined the function of rGO to enhance mechanical energy collecting capability for P(VDF-TrFE) matrix in the absence of standard polling technique [58]. Five distinct PE sheets are produced by changing rGO concentration into P(VDF-TrFE) matrix for attaining optimal concentration. The rGO impact was achieved via increasing the amount of electro-active polar-beta (β) phase into P(VDF-TrFE). One of many manufactured PENGs contains 0.1% rGO exhibiting optimal V_{oc} of 2.4 V and a peak I_{sc} of about 0.8 μA when a force of 2 N is applied. Additionally, it has greatest o/p power showing 3.2 μW at R_L for 1.8 MΩ. Interestingly, GO is a promising 2DM for enhancing the sensitivity of PVDF nanofiber-based PENG [77]. A self-powered wearable sensor capable of detecting and discriminating pressure and temperature was constructed. For instance, with a high degree of sensitivity, the pressure sensor reliably and promptly identified as 10 Pa (4.3 $VkPa^{-1}$). The sensor has been used to monitor human physiological signals by connecting it to various areas of the human body; it can detect joint motions, coughing sounds, severe swallowing, whereas tracking throat muscle action in exact timely duration continuous speaking for distinct alphabetic letters, as well as pyroelectric breathing sensor. This implies that such produced sensors may be employed in a number of functions by wearable and portable electronics toward biomedical e-skin sensing.

6.3 Solar Energy Harvesting

6.3.1 Solar Cells

2DMs are demonstrated excessive potential alternatives for several applications in next-generation electrical devices due to their intrinsic energy and varied crystal structures [78, 79]. Figure 6.20 illustrates a diverse range of 2DMs with variable E_g that are used in a variety of applications across a range of wavelength regimes. Especially, research and development of TMDCs [80, 81] and BP [82, 83] have accelerated due to their superior mechanical properties that were influential for designing flexible as well as stretchable future electronic devices [13, 84]. It is now well established that Si became the most preferred semiconductor for photovoltaics due to its narrow E_g, accessibility, affordability, low toxicity, and well-established manufacturing processes. The achievement of Si-based photovoltaic is potentially contingent on two aspects: (i) the power conversion efficiency (PCE) is solar energy conversion to electrical energy and (ii) the cost/watt of electricity generated. Si sun cells have dominated the market for the last 50 years, with record PCE of 24.7% for <

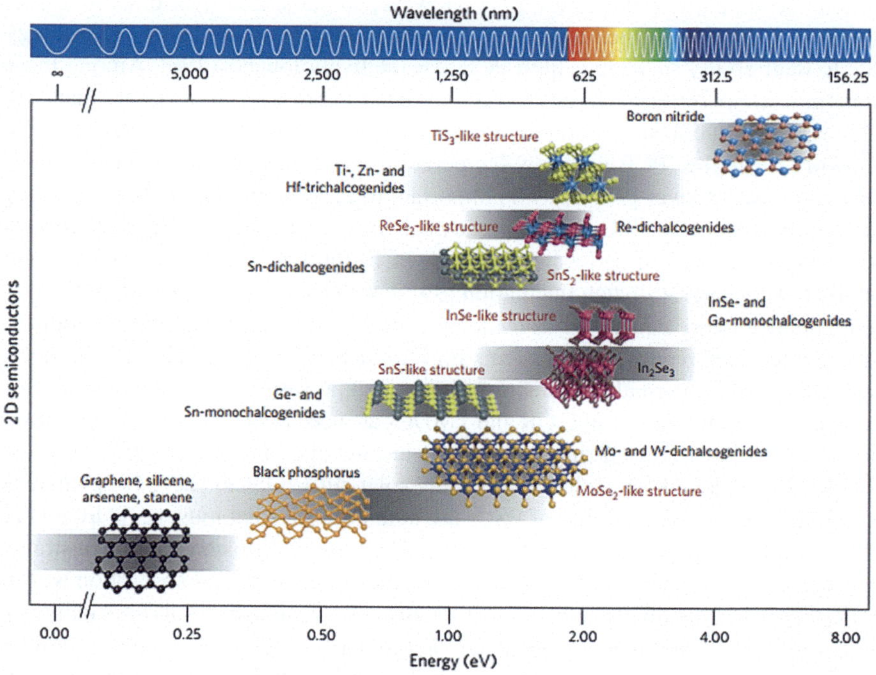

Fig. 6.20 E_g values for several 2D semiconductor material families are compared. Furthermore, the crystal structure shows for demonstration of similarities/dissimilarities across various families (Reproduced with permission from [89])

100 μm thick low-temperature-fabricated Si semiconductors and 25.6% for comparable back-contact semiconductors more recently [85, 86]. What is the next "big" thing in Si photovoltaics, even so? In the battle to lower the cost/watt and increase watts/gram usage, it is critical to minimize the active absorber thickness of Si [87, 88].

The Si-based photovoltaic technology roadmap states clearly that substrate thickness becomes small than 25 μm [79]. Currently, academics and industry have concentrated their efforts on overcoming the Shockley–Queisser limit using thin monocrystalline Si substrates [90–92]. With the requirement for developing semiconductors to be bendable and capable toward high speed in roll by roll manufacturing, Si is progressively progressing to "flexible" future with thickness less than ~ 50 μm [93, 94]. Flexible Si substrates might result in more effective Si utilization, but they would require advancements in wafer production, efficient sawing technologies, reasonable kerf loss reduction, novel handling ideas, and alternative encapsulation processes. As a result, the Si photovoltaic sector is on the verge of a paradigm shift during coming decade. The issue now is about 2DMs associated with numerous unusual characteristics and will be a viable choice for next-generation Si semiconductors. The exceptional properties of scalability, modularity, and roll-to-roll manufacturing enable 2DMs to be a basic player toward advancement of lightweight photovoltaic systems techniques in the future [95]. However, organic and hybrid semiconductors have recently emerged as an effective alternative for photovoltaic technology due to their ability to achieve high PCE at a relatively low cost [96]. Among these materials, 2D perovskites have recently attracted interest as photovoltaic material. This is partly due to possession of longer excitons diffusion length relative to Si, allowing for higher PCE with low-temperature processing and a less costly light-harvesting active layer [97, 98]. It is important to keep in mind that high-efficiency perovskite solar cells (PSCs) were already achieved, and work is being done toward achieving great area of cells showing high PCE, throughput, device stability, and finally lowers batch-to-batch fluctuation [99]. For instance, semitransparent flexible semiconductors are generating considerable attraction owing to appealing application in integrated photovoltaic systems for buildings [97]. Atomically thin TMDCs with enhanced light–matter interaction and optical performance have potential uses in organic and photonics. Thus, GR and TMDCs are used as anode/cathode, in extraction of e^-/h^+, and as interfacial layers in organic and photonic devices. However, the short diffusion length of excitons in organic solar cells (OSCs) [100, 101] whereas environmental unpredictability of PSCs are notable issues that must be addressed in order to enhance performance considerably. The limited diffusion length constrains the organic layer thickness, therefore limiting the effective optical absorption within material's particular thickness [102]. Although many organic materials have naturally high absorption coefficients, inadequate photon harvesting occurs owing to limited absorption window for organic materials [103, 104]. The 2DMs were used in the form of third component into ternary-based OSCs by employing materials with complementary absorption ranges being second donor or accepter for harvesting high-

and low-energy photons independently (to enhance optical absorption in photoactive layer) and stabilizer material for photovoltaic cells (enhance the PSC stability) [105, 106].

6.3.1.1 TMDCs in Si-Based Solar Cells

A unique sub-nanometer ML-TMDC may theoretically absorb sunlight of 50 nm Si or 15 nm of GaAs ML, resulting in $J_{sc} = 4.5\ \text{mAcm}^{-2}$ [107]. Figure 6.21 shows the absorbance estimated for MoS_2, $MoSe_2$, and WS_2 ML with GR [107]. GR and other TMDC absorbance (A) is determined using (Eq. 6.1):

$$A = 1 - e^{-\alpha L} \tag{6.1}$$

Assume a flat layer of thickness $L = 1$ nm, and an absorption coefficient α transformed to absorbance derived from the material literature (Eq. 6.2) [79, 107]

$$J_{\text{abs}} = e \int_{E_g}^{\infty} A(E) J_{\text{ph}}(E) \mathrm{d}E \tag{6.2}$$

where E denotes photonic energy, $J_{\text{ph}}(E)$ denotes input photon flux, E_g denotes the absorbing material's optical energy, and e denotes the elementary charge. With a thickness of only 0.65 nm, ML-MoS_2 absorbs 5–10% of incoming light. The work function of TMDCs may also be changed significantly depending on the TM's thickness employed (in case of ML; WS_2: ~ 4.6, WSe_2: ~ 4.3, and MoS_2: ~ 5.1, in case of bulk; WS_2: ~ 4.2, WSe_2: ~ 3.9, and MoS_2: ~ 4.5 eV) [108]. With p-Si, ML-MoS_2 produces a Schottky connection showing direct E_g 1.9 eV [109]. A Si/MoS_2 p–n junction semiconductor showing V_{oc} 0.41 V exhibits PCE of 5.23%. The device structure for MoS_2/p-Si HJ semiconductor is depicted in Fig. 6.22a, along with the

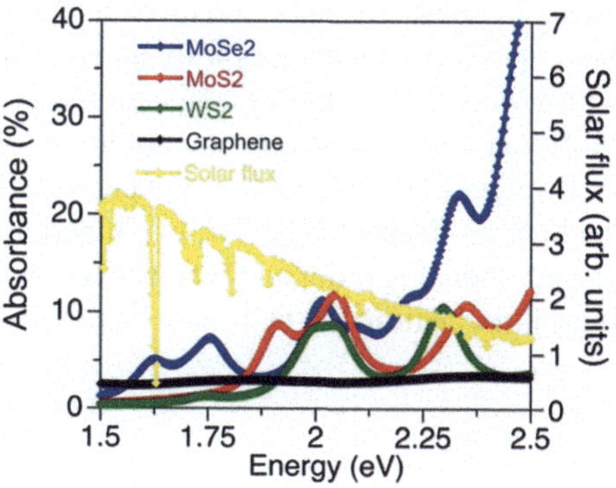

Fig. 6.21 Three TMDC MLs and GR absorbance were examined to the incident AM1.5G solar flux (Reproduced with permission from [107])

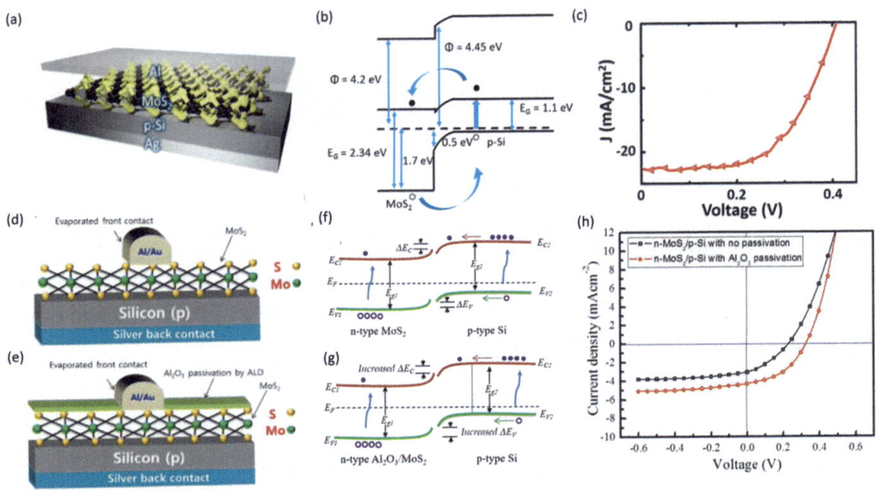

Fig. 6.22 **a** MoS_2/p-Si HJ semiconductor device structure, **b** band depiction for MoS_2/p-Si HJ observed by UPS measurements, **c** *J–V* characteristics (Reproduced with permission from [110]). Demonstration for n-MoS_2/p-Si semiconductors schematically: **d** non-passivated layer and **e** 10-nm-thick Al_2O_3 passivated layer on MoS_2 surface. Energy level diagrams of **f** MoS_2/p-Si material, and **g** MoS_2/p-Si decorated with Al_2O_3 passivation, **h** *J–V* characteristics of semiconductor appliances with/without Al_2O_3 passivation (Reproduced with permission from [111])

Si/MoS_2 HJ's energy-band diagram (Fig. 6.22b) [110]. It is worth noting that MoS_2 forms a type II HJ with p-Si in this situation, resulting in 5.23% PCE (Fig. 6.22c).

The built-in potential for n-MoS_2/p-Si interface was observed higher through the addition of an Al_2O_3 passivation layer [111]. The photovoltaic efficiency for bulky semiconductors, such as MoS_2/Si, has been seen to increase as a result of Al_2O_3 depending on passivation on the MoS_2 surface. The design of the device without and with an Al_2O_3 passivation layer is shown in Fig. 6.22d, e. Upon passivation of MoS_2 surface with atomic layer deposition of Al_2O_3, the built-in field for MoS_2/Si contact is increased (Fig. 6.22f, g). By decreasing surface trapping density for MoS_2 owing to relevant passivation layer, a PCE (5.6%) is produced (relative to 2.21% for MoS_2/p-Si semiconductors without passivation) (Fig. 6.22h) [111]. A semiconductor with a MoS_2/ITO Schottky connection was constructed by transferring a CVD-grown MoS_2 nanomembrane onto an indium tin oxide (ITO) coating glass substrate [112]. PCE values of (0.7%, 1.8%) were published in a stack of MoS_2 with 110 and 220 nm thicknesses, respectively. The p–n junction is established inside the MoS_2 layers by plasma-inducing p-doped ITO electrode/ML MoS_2/ITO stack [113]. Under AM1.5G solar illumination, the PCE may be raised to 2.8% in this construction.

6.3.1.2 TMDCs in Organic Solar Cells

In the fabrication of OSCs, TMDCs were utilized as the electron and hole transporting layer (HTL, ETL), interfacial and protective layers. Under usual AM1.5 illumination circumstances, an OSC associated with stacking architecture for ITO/TiO_2/MoS_2/P_3HT/Au exhibits 1.3% PCE [114]. MoS_x anode-buffer-layer development in situ via thermal breakdown for $(NH_4)_2MoS_4$ improved the efficiency of P_3HT: PCBM with the maximum PCE was achieved from 2.9 to 3.9%. MoS_x buffer layers demonstrated a range of device performance characteristics based on processing annealing, with the highest performance at 300 °C. MoS_3 and MoS_2 were shown to be helpful to increase the performance of the device [115]. Ultraviolet ozone (UVO) treatment of exfoliated MoS_2 sheets changes the sheets' local crystalline structures, introducing O_2 atoms into the lattice and so enhancing the device PCE of OSCs. This treatment minimizes the HTL's surface flaws, resulting in improved contact with the photoactive layer. MoS_2 NS treated with UVO treated and placed between the PEDOT: PSS and photoactive layer boost the device's stability and efficiency to 2.81% [116]. Yang et al. [117] found 7.64% efficiency for UVO treatment of MoS_2 is 53% more efficient than the untreated sample. By adjusting the layer thickness and sputtering deposition temperature, a diffusion length of MoO_3/MoS_2 film treated with UVO may function as an efficient HTL with a 4.15% PCE. It is discovered that tantalum disulfide treated with UVO functions in the form of hole and electron extracted layer in OSCs. The results indicate that normal and inverted structures have improved significantly. It is demonstrated that by integrating WSe_2 flakes with an optimal thickness of 6.1 nm into organic layer, efficiency for OSC rises from 8.1 to 9.28% [118]. MoS_2 sheets were included as semitransparent electrodes for fabrication of semitransparent polymer solar cells (Fig. 6.23a). The OSCs without ITO dependent on glass/AgNW-MoS_2/n-MoS_2/organic layer/p-MoS_2/Ag obtained high 8% PCE. The transmission for spin coating AgNW (silver nanowire) as well as AgNW-MoS_2 over glass coated with various spinning speeds is depicted in Fig. 6.23b and showed electrodes and interfacial layer energy diagram is shown in Fig. 6.23c [119]. The cathode and anode of this structure are made of new AgNW-MoS_2/n-MoS_2-NS complex with p-MoS_2/thin-Ag-metal, respectively. MoS_2 layers were utilized in plasmonic OSCs. A mixture HTL for gold NPs (20 nm) over 2D MoS_2 with optimal coverage resulted in 7.25% PCE that becomes 17.3% higher than the device in the absence of Au NPs [120]. Additionally, an active layer composed of hybrid P3HT: PCBM/MoS_2 on Ag-NPs arrays enables rapid charge transfer and increased absorption inside the active layer.

6.3.1.3 2D Materials in Perovskites Solar Cells

Device Architecture

The exponential growth of PSC research over the last decade has led to the development of several device architectures, particularly mesoscopic and planar (Fig. 6.24).

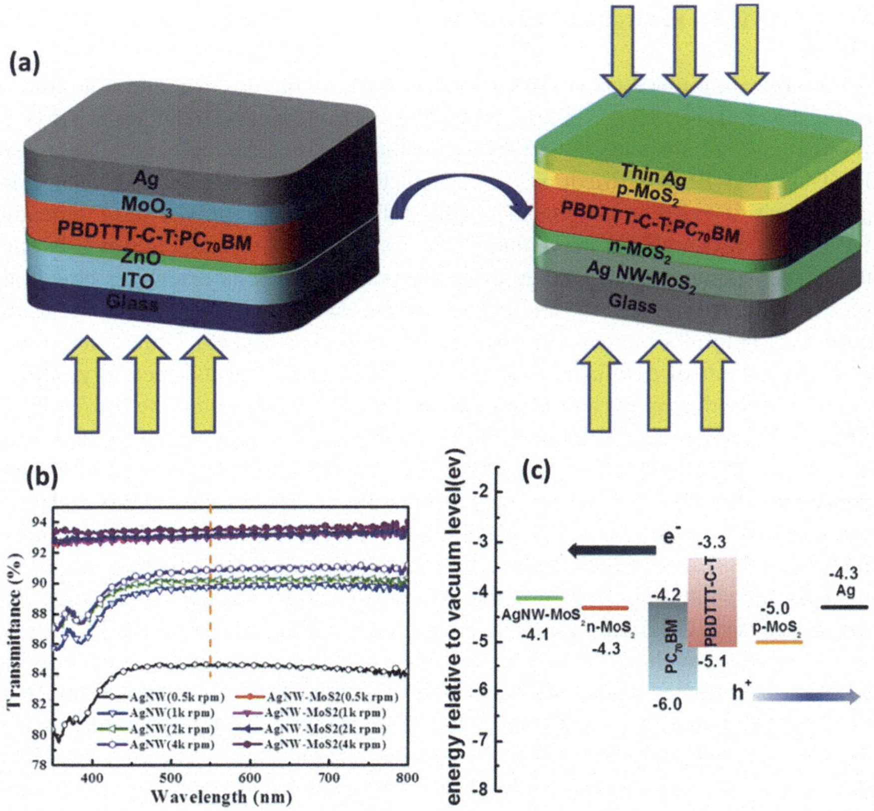

Fig. 6.23 **a** Common inverted PSC and MoS_2 NS modified ITO-free semitransparent polymer solar cell device configuration, **b** transmittance of AgNW and AgNW-MoS_2, **c** semitransparent polymer solar cell electrodes and interfacial layers schematic energy diagram (Reproduced under the terms of CC BY 4.0 license [119], https://creativecommons.org/licenses/by/4.0/)

In comparison to planar PSCs, [121] mesoscopic PSCs incorporate an extra porous layer that serves in the form of scaffold of perovskite absorber, therefore increment in charge collecting performance [122]. The mesoscopic structure is more tempting for high-efficiency PSCs, [123], but planar PSCs exhibit talent since they need low annealing due to compatibility with flexible substrates [124, 125]. The progression for charge-selective layer may additionally separate the device design into conventional and inverted PSCs as (n–i–p) and (p–i–n), respectively. PSCs are composed of transparent conductive oxide significantly perovskite absorber resided among transport layers and metal contact. Regardless of device design, the PSC working mechanisms follow the same principles. The exciton (electron–hole pair) is produced by light absorption by the perovskite photoactive layer. These generated charges are then collected at their respective electrodes (Fig. 6.25a). These two elements work together to extract and transport charge carriers, as well as govern perovskite crystal

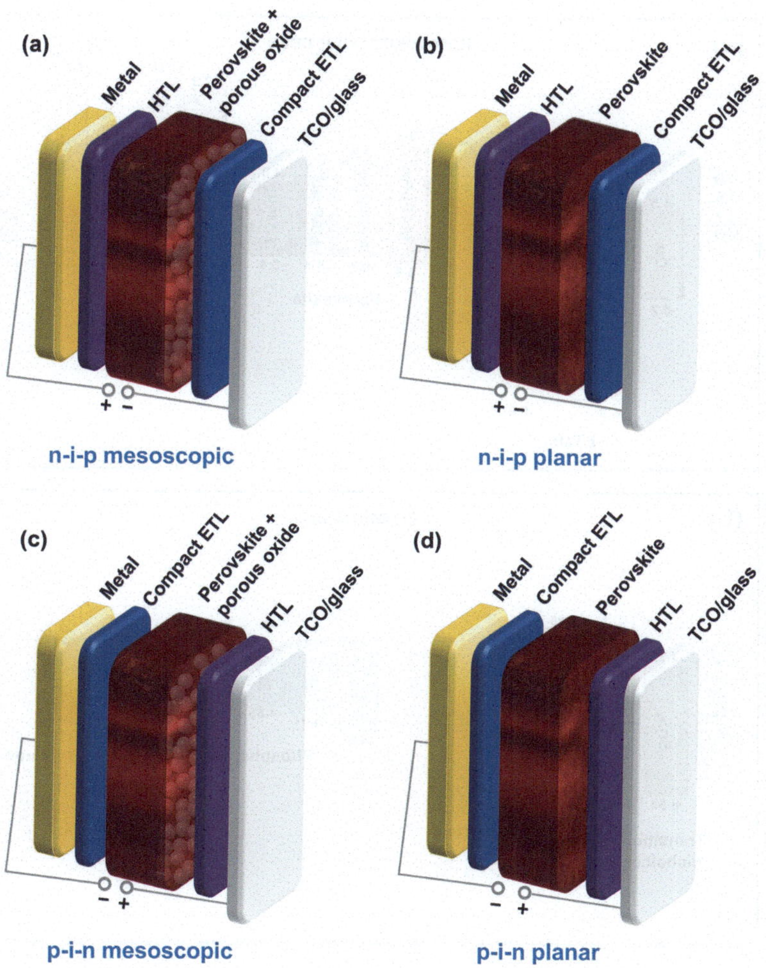

Fig. 6.24 Four typical PSC's device structures: **a** n-i-p mesoscopic, **b** n-i-p planar, **c** p-i-n mesoscopic, **d** p-i-n planar. (Reproduced with permission from [128])

formation. These charge-carrying layers are also vital for device stability. In addition, the hysteresis phenomena are highly connected with the charge selective layers' characteristics and interface to the perovskite [126, 127].

Certainly, all of these criteria are vital for achieving great photovoltaic performance and stability. Ideal ETL and HTL materials should have these features: great charge mobility, good energy alignment with other components, inherent stability, high transparency, and low cost. For regular n–i–p PSCs, TiO_2 and Spiro-OMeTAD seem to be the most commonly employed electron and hole transporting material (ETM and HTM, respectively), whereas inverted devices, PCBM (ETM) and PEDOT: PSS (HTM), are commonly used [129, 130]. Despite the efforts made to use these

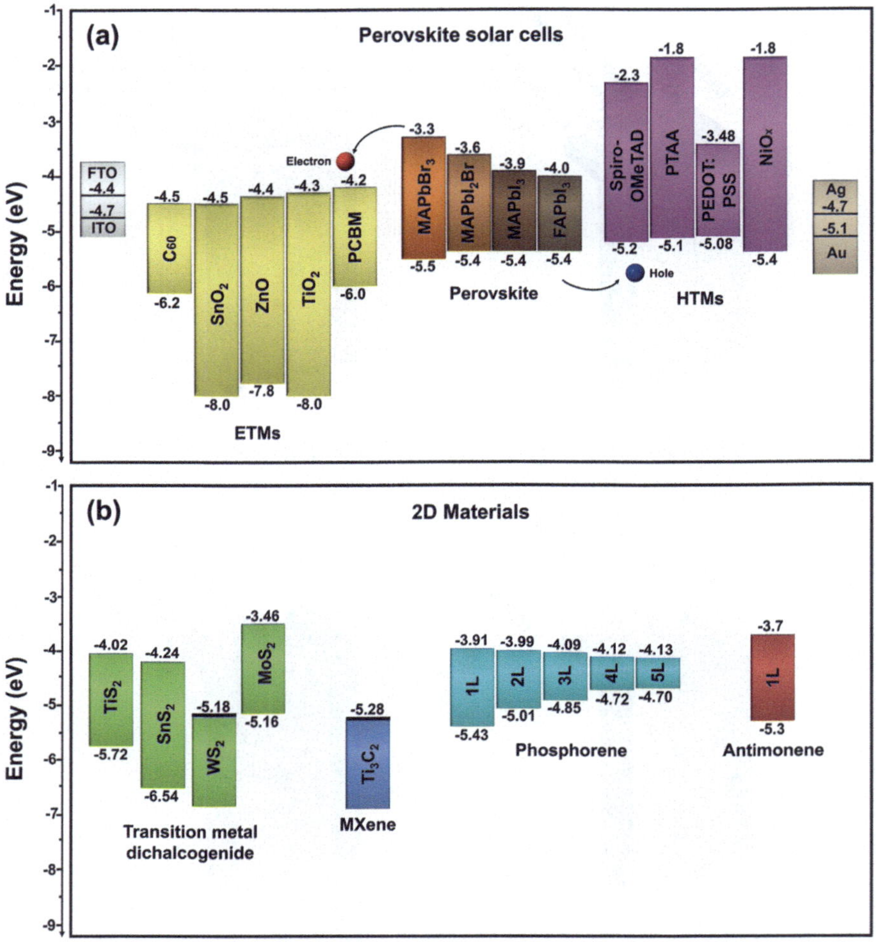

Fig. 6.25 **a** Energy level diagrams of typical PSC materials, **b** illustration of 2DMs energy alignment schematic (Reproduced with permission from [128]). The energy levels values are sourced from: Antimonene [137], Phosphorene [138], MXene [139], MoS_2 [140], WS_2 [141], SnS_2 [142], and TiS_2 [143]

materials, have certain inherent flaws. The poor conductivity, high annealing temperature (above 450 °C), a wide GB associated with UV irradiation instability of TiO_2 are still serious problems [131, 132]. Similarly, poor charge mobility for pure Spiro-OMeTAD along with excessive acidity of PEDOT: PSS are major problems affecting device PCE with stability [133, 134]. Fortunately, many innovative ETMs and HTMs are produced and proved to be viable alternatives, as recently reviewed [130, 135]. In addition to ETLs and HTLs, regulating perovskite crystal formation with additives has been a recurrent topic in dramatically improving device performance [136]. Different solution-processed ultrathin 2D-layered materials are being investigated

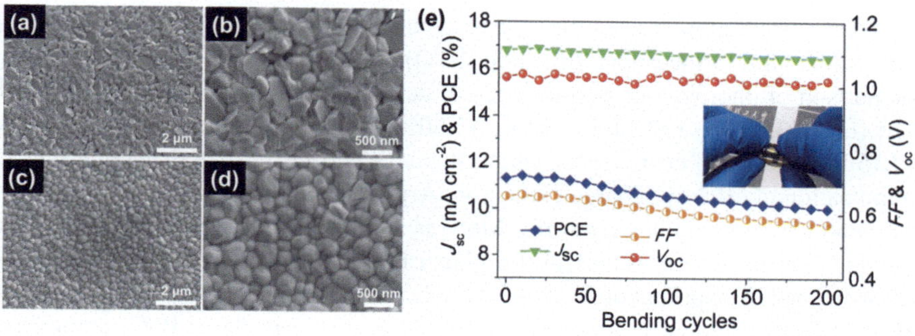

Fig. 6.26 **a** Low-magnification SEM image **b** high magnified SEM photographs for perovskite films on ITO/PEN substrate, **c** low- and **d** high-magnified SEM photographs for perovskite films on BPQD-coated ITO/PEN substrate, **e** photovoltaics constraints for BPQDs-depending stretchy PSC (Reproduced with permission from [145])

as PSC possibilities. 2D-layered materials are gaining popularity as low-cost PSC additions because of their unique characteristics, compatibility with perovskites, and energy alignment with other components (Fig. 6.25a).

2D Material-Based Conventional Perovskites Solar Cells

Electron Transport Layer

Aside from proper band alignment, high ETL carrier mobility is important for high-performance PSCs. Anisotropic (orientation-dependent) characteristics of BP make it a suitable candidate for layered materials in electrical and optoelectronic devices [144]. The bulky BP E_g gives 0.3 eV that may be adjusted via layering (see Fig. 6.25b). Black phosphorus quantum dots (BPQDs) have recently replaced ETL in PSCs [145]. The exfoliated BPQDs were 3–10 nm in diameter. On comparing ETL-free PSCs with that of 3.58% PCE, BPQD ETLs significantly enhanced photovoltaic performance. After optimized layer thickness, 11.26% efficiency was obtained in favor of devices built on soft substrates (ITO/PEN) and 14.6% for devices built on rigid glass. This significant improvement in device efficiency is due to compatibility, uniformity, and pinhole-free perovskite thin films; however, effective electron extraction is because of energy-band alignment among BPQDs as well as perovskite (Fig. 6.26a–d). These flexible devices exhibited modest variation into short-circuit current–density connected with V_{oc} even after 200 cycles of bending (Fig. 6.26e). This may be owing to higher shunt resistance (R_{sh}) of ITO/PEN substrate from ITO film cracking. A lot of work has gone into the solution processing of FL-BP NS. Producing good quality, consistent, and repeatable FL-BP has proven difficult [146, 147].

The TMDC MLs with a thickness < 1 nm can absorb 5–10% of incoming sunlight which are higher than GaAs and Si [107]. Based on these amazing characteristics, TMDCs have lately attracted considerable interest as PSC components. For example, microwave irradiation is used directly to synthesize thin, clear, and homogeneous MoS_2 films on fluorine-doped tin oxide (FTO)-coated glass for PSC ETL [148]. A comparison of photovoltaic performance of devices manufactured on FTO showed PCE (6.11%), MoS_2 (13.14%), SnO_2 (15.80%), and TiO_2-coated FTO substrates (17.15%). The MoS_2 occupies the smallest charge carrier lifespan than other ETLs, an indication of effective charge extraction. Arising from its high electrical conductivity and low-temperature synthesis, MoS_2 is an ideal ETL for efficient PSCs. Yin et al. [143] utilized TiS_2 as an ETL for planar PSCs. Using eight to nine-layer TiS_2 NS suspended in isopropanol alcohol for PSCs, a PCE was 17.37%, equivalent to the TiO_2-dependent reference device (17.07%). Also, unencapsulated PSCs connected with TiS_2 ETL were very robust during UV irradiated 10 $mWcm^{-2}$ in air environment after 50 h, retained by 90% of original efficiency. In the same conditions, TiO_2-dependent perovskite devices only retained 56% of their starting value. These devices› excellent efficiency and UV stability are ascribed to TiS_2's superior electron transport and collecting characteristics, as well as its better energy-band position and small photocatalytic properties than TiO_2. TiS_2 as a transport layer of electrons in planar PSCs reduced the efficiency of the electron transport layer-free device from 13.42 to 9.44% [149]. The scientists discovered that when TiS_2 is deposited on ITO, the work function increases from 4.75 to 4.79 eV well below the perovskite's CBM. Work function shifts to 4.64 eV when TiS_2-coated ITO film exposed to UVO for fast electron transfer. This update improved device performance by 18.79%. The UVO-treated TiS_2 ETL device PCE retained 95.8% after 816 h of storage in the absence of encapsulation in normal settings showing relative humidity (RH) of ~ 10% (Fig. 6.27a). After 100 h of exposure to high humidity (45–60 RH), this gadget retained over 80% of its initial PCE. Huang et al. [150] recently achieved a PCE of 21.73% employing diffusion length of SnO_2 as well as 2D-TiS_2 being effective ETLs. This outperformed the SnO_2 for ETL, 19.65%.of reference device. Upon film, transparency remained unchanged following the addition of TiS_2 to SnO_2, film conductivity increased because of TiS_2's higher electrical conductivity. A greater energy level position among ETL and perovskite, and passivated-SnO_2 defects will trap at the time TiS_2 is present, leading to improving PCE (Fig. 6.27b). The SnO_2/TiS_2 double structure-based device showed reduced hysteresis and increased stability. TiS_2 has significant potential in producing extremely efficient and stable PSCs. With its appropriate E_g and strong carrier mobility (50 cm^{-2} V^{-1} s^{-1}), SnS_2 was employed in the form of potential ETM into PSCs [151]. Zhao et al. [152] utilized hydrothermal 2D-SnS_2 NS in the PSC ETL, after improving reaction time and acquired 13.63% PCE. A self-assembly stacking deposition technique may greatly enhance the PCE for SnS_2-ETL-dependent devices, exhibiting 20.12% of PCE is considered greater than the SnO_2-based control device (17.72%). This technique improves the density as well as homogeneity for ETL film, resulting in better perovskite crystallinity. With SnS_2 as an ETM, the charge transferring along with collection is facilitated, whereas charge buildup at ETL/perovskite boundary is passive. The SnS_2-based devices have low hysteresis and high stability and collection are facilitated, and the charge buildup at the ETL/perovskite interface is passive. The SnS_2-based gadget also has low hysteresis and high stability (Fig. 6.27c). After 600 h of storage at 30–40% RH

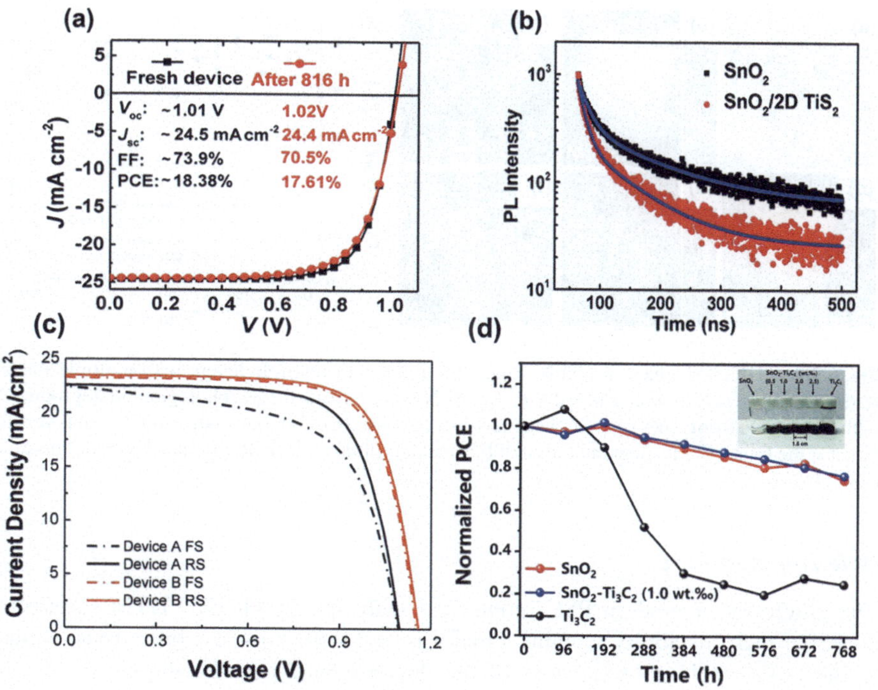

Fig. 6.27 **a** Current–voltage characteristics of PSCs measured on preliminary day after 816 h with an RH of ~ 10% (Reproduced with permission from [149]), **b** transient PL spectrum for SnO_2 and SnO_2/2D TiS_2 in the form of ETL into PSCs (Reproduced with permission from [150]), **c** electrical parameters toward forward as well as reverse scanning orientation for devices A and B (Reproduced with permission from [142]), **d** stabilities of PSC devices dependent on SnO_2 and SnO_2–Ti_3C_2, whereas Ti_3C_2 subjected to air environment without encapsulation. As shown in the inset, SnO_2 hydrocolloid with varied Ti_3C_2 concentrations (0.05, 0.5, 1.0, 2.0 and 2.5 wt%) and Ti_3C_2 dispersion (Reproduced with permission from [155])

at room temperature, the encapsulated SnS_2 ETL device preserved 90% preliminary efficiency, whereas control device declined by 60%. The PCE dropped 9.8% when exposed to continuous UV radiation (15 mWcm^{-2}) for 80 h, suggesting the SnS_2-based devices are more stable than the reference device (46%). In 2011, novelty in 2D transition metal carbide/nitride family member named "MXenes" was described, attracting much attention [153]. Thus far, over 70 MAX phase members have been recorded [153, 154]. The SnO_2–Ti_3C_2-dependent device was also stable in air environment of RH ~ 20%, keeping around 80% of its original performance after 700 h (Fig. 6.27d), indicating that using MXene may not affect device stability at the time of boosting efficiency.

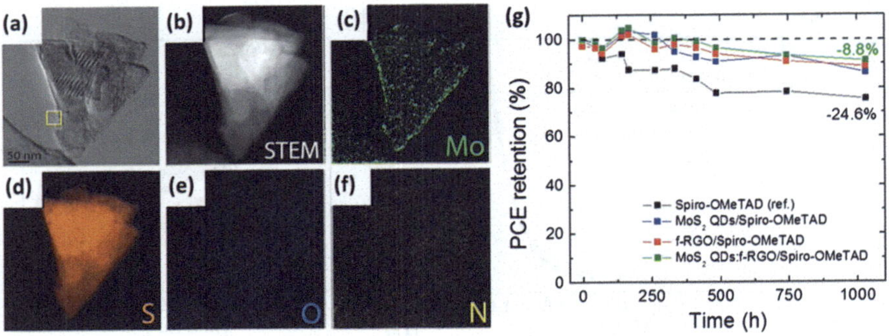

Fig. 6.28 **a** HRTEM image **b** STEM image for flake with multiple layers with multiple flakes stacked on top of one another, **c** Mo, **d** S, **e** O, and **f** N energy-cleaned TEM photographs (Reproduced with permission from [160]), **g** *J–V* characteristics were used to derive normalized PCE trends versus time for the four PSCs throughout the shelf life analysis of ISOS-D-1 (Reproduced with permission from [161])

Hole Transport Layer

The efficiency of modern PSCs connected with Spiro-OMeTAD HTM surpassed 20%, but their instability remains a major issue [156]. This instability issue is due to the unfavorable consequences of the dopants employed for doping of Spiro-OMeTAD for controlling its poor electrical conductivity. Numerous investigations have also demonstrated about substantial iodine species movement originated from perovskite layer through Spiro-OMeTAD toward metal electrode occurring during working, affecting device performance and stability [157, 158]. A buffer layer among perovskite layer where HTL can help in solving the issue. Being an active buffer layer among perovskite as well as Spiro-OMeTAD HTL, above-mentioned alteration addresses two requisites: (i) preventing metal electrode movement and (ii) facilitating hole collection and injection. *N*-methyl and 2-pyrrolidone was the best solvent for exfoliating MoS_2 [159]. To add a perovskite-friendly solvent like isopropanol alcohol, a solvent-exchange procedure was required. The flakes in the purified isopropanol alcohol sample had minimal O and N content (Fig. 6.28a–f).

To get a smooth and pinhole-free HTL containing full surface treatment located on perovskite layer, MoS_2-QDs were fabricated in hybrid shape connected with functionalized reduced graphene oxide (f-rGO) flakes that are modified with MPTMS. The PSCs along with FTO/c-TiO_2/mp-TiO_2/$MAPbI_3$/MoS_2 QDs or f-rGO or MoS_2QDs: f-rGO/Au device structure were prepared as given by Fig. 6.29a. The device utilizing MoS_2 QDs: f-rGO in the form of HTL possesses high PCE as compared with HTL-free devices exhibiting 7.60% versus 3.01% efficiency. Adopting this test, f-rGO with MoS_2 QDs: f-rGO has been utilized in the form of ABLs for manufacturing PSCs by making Spiro-OMeTAD in the shape of HTM (Fig. 6.29b). The MoS_2 and QDs: f-rGO active buffer layer-dependent cell present PCE of 20%, whereas the reference, MoS_2, and f-rGO-based PSCs had PCEs of 16.85, 14.40, and 18.64%. This increase in PCE was attributed to better charge

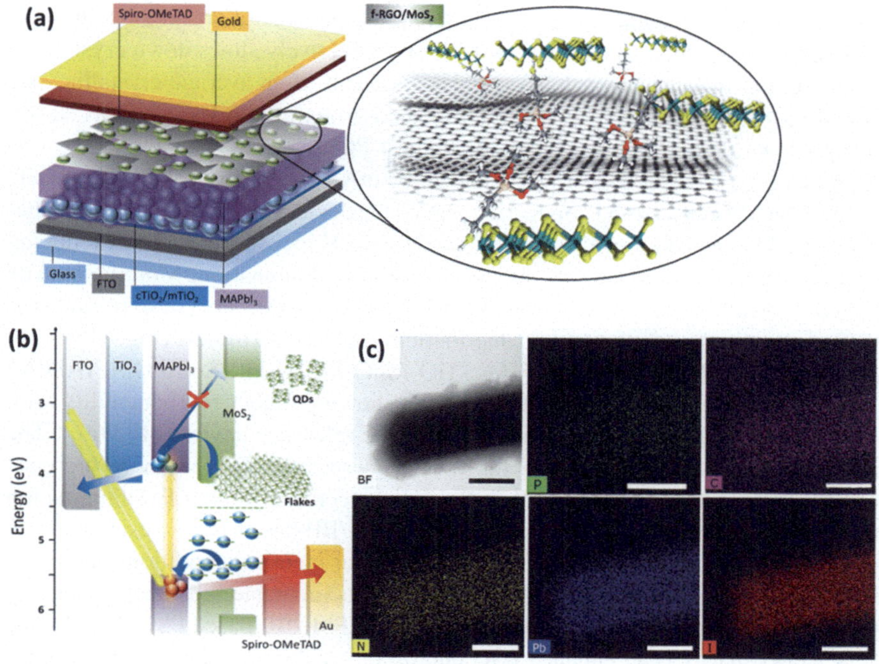

Fig. 6.29 **a** Schematic of mesoscopic $MAPbI_3$-based PSC with MoS_2 QDs: f-RGO hybrid in the form of HTL and ABL, **b** energy level diagram of materials for $MAPbI_3$-based PSC (Reproduced with permission from [161]), **c** $MAPbI_3$/BP (TEM image) and the corresponding elemental distributions. Scale bars ~ 500 nm (Reproduced with permission from [163])

collecting efficiency. The MoS_2 QDs: f-rGO active buffer layer-based device also showed significant hysteresis suppression and increased stability (see Fig. 6.28g). FL-BP NS was utilized as an HTL replacement and as a buffer layer residing among perovskites and/or Spiro-OMeTADs for PSCs [162]. The device had a PCE of 7.88% at the time FL-BP was employed as HTL was greater than the HTL-free 4.0% device, whereas lower than Spiro-OMeTAD-dependent 13.1% device. Its lower efficiency was due to multiple reasons: high R_s, poor FL-BP surface coverage on perovskite layer, whereas uneven FL-BP surface is caused by poor contact to Au electrode. Surprisingly, adding Spiro-OMeTAD to the FL-BP layer increased charge origination and extraction along with lower R_s and R_{sh}, increasing PCE to 16.4%. The boost came from higher J_{sc} and fill factor (FF). When kept in an Ar-filled glove box, an FL-BP-based unencapsulated device performed well. However, when kept in humid settings (65% RH), the devices lost roughly 60% of their initial PCE after 96 h.

Active Layer

Additive engineering has been shown to increase perovskite film crystallization and surface coverage [164, 165]. In addition to 2D-$Ti_3C_2T_x$ to boost crystal size, defect density, and photogenerated electrons efficiency, a perovskite precursor solution is being used [166]. MXene additive of 0.03 wt% effectively slows down crystallization, resulting in large perovskite crystals as a result of an average crystallite size increasing from 150 to 358 nm. The terminated grouping for $Ti_3C_2T_x$ was shown to slow down perovskite nucleation. Its highest PCE was 17.41%, and its J_{sc} was 22.26 $mAcm^{-2}$, compared to 15.54% and 20.67 $mAcm^{-2}$ for the pristine perovskite device. The MXene's strong conductivity aided the devices' electron transfer mechanism. The MXene addition also lowered the hysteresis loops for devices by reducing density of trap locations on GB, reducing charge recombination. More perovskite grains (> 500 nm) were obtained when FL-BP was added to $MAPbI_3$ films (400 nm). It verified the existence and consistent distribution of BP and perovskite (Fig. 6.29c). By using $MAPbI_3$/BP in PSCs with FTO/c-TiO_2/SnO_2/perovskite/Spiro-OMeTAD/Ag structure produced a prominent 20.65% PCE showing higher reproducibility with low hysteresis. This efficiency is greater than the reference devices (18.96%). The increased charge transport capacity for $MAPbI_3$ perovskite absorber in the presence of BP reduces charge carrier recombination. Under constant white LED illumination of 100 $mWcm^{-2}$ in an N_2 glovebox, the $MAPbI_3$/BP-based PSCs sustained 94% of the original PCE after 1000 h. When tested under identical conditions, the reference device's PCE dropped to 30%. The BP with its favorable self-healing capacity was shown to be related to the high stability of $MAPbI_3$/BP-based PSCs [167]. This work shows that BP can enhance PSC photostability that is required for commercial cutting-edge technology. In the perovskite active layer, 2D-layered promising materials are currently being researched. There is still plenty to learn about PSC efficiency. The applications of 2DMs in the perovskite light absorber might lead to a big advance in this cutting-edge scientific sector. For example, 2D-layered materials like phosphorene may efficiently absorb photons in the UV, visible, and NIR (near-infrared) ranges. But optimizing materials and devices will require significant research.

Inverted Perovskites Solar Cells

Electron Transport Layer

The PCE for p–i–n PSCs has improved recently, nearing 21% [168]. The substantial mismatch in the energy locations of device constituents results in low V_{oc}. The work function of hole transport layer PEDOT: PSS is 5 eV [169]. Showing non-optimal energy level alignment with $MAPbI_3$ HOMO level at 5.4 eV [170]. This is in contrast to the work function for Ag with 4.7 eV and Au (4.0 eV) (5.1 eV). Wang et al. [171] utilized GO as well as MoS_2 in the form of interfacial layers lying among PEDOT: PSS, $MAPbI_3$, PCBM, and Ag to enhance energy level matching (Fig. 6.30a, b). Using this architecture, the PCEs improved from 14.15 to 16.89% when 3 nm thin layer of MoS_2 was introduced as 19.14% when a 1 nm GO film was incorporated. This substantial improvement was attributed to better energy level matching, resulting in less non-radiative charge carrier recombination, as seen by

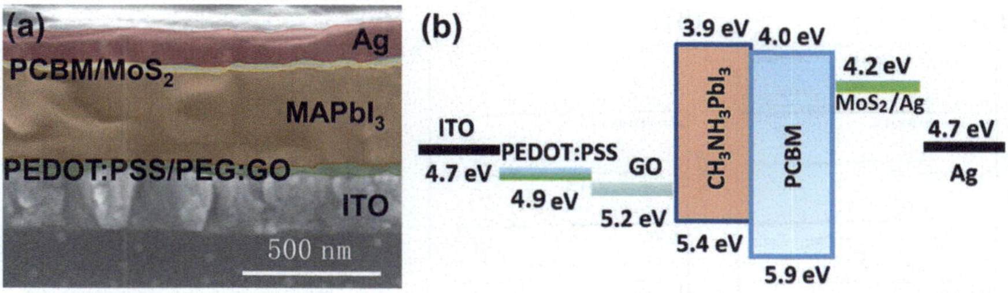

Fig. 6.30 **a** Cross-sectional image of SEM of devices with MoS_2 and PEG: GO interfaces, **b** energy level diagram schematic in PSCs (Reproduced with permission from [171])

increment in V_{oc} showing 0.962 to 1.135 V. The interfacial changes also reduced hysteresis and improved stability. In p–i–n devices, new 2DMs as TiS_2 and SnS_2 are predicted to dramatically improve device performance as previous 2DMs have in n–i–p devices. Logically, ETL of p–i–n devices is placed at optimum level of perovskites; using large-area 2DMs may assist to enhance device stability.

Hole Transport Layers

In planar p–i–n-structured PSCs with $CH_3NH_3PbI_{3-x}Cl_x$ as an active layer, Kim et al. [172] used MoS_2, WS_2, and GO instead of PEDOT: PSS HTL. MoS_2 and WS_2 were made via CVD, whereas GO was made by the Hummers technique. These materials' work functions were measured using UPS to be 5.25, 5.1, 5.0, and 4.95 eV, proving their suitability as HTLs. The PEDOT: PSS films possess surface root mean square value unevenness indicating 0.841, 1.391, 1.778, and 1.108 nm. On comparing, PEDOT: PSS PSCs represent PCEs percentage values as 9.93, 9.62, 9.53, and 8.02%. The efficiency of HTMs is significantly dependent on work function values because of direct impact on devices' for V_{oc}. The lowering PCE for WS_2-based cell is owned by lowering V_{oc} of 0.82 V because of lowering work function as compared with other materials showing 0.91, 0.96, and 0.92 V in favor of PEDOT: PSS, MoS_2, and GO, using water-soluble MoS_2 and WS_2 as the HTMs, which were prepared by lithium intercalation-based exfoliation. The AFM estimated the thickness of the obtained MoS_2 and WS_2 flakes to be 3.68.0 nm (4–9 layers) and 2.54.0 nm (3–5 layers). The MoS_2 and WS_2 have an average of six and three layers, respectively. The influence of MoS_2 and WS_2 crystalline phases (semiconducting 2H and metallic 1T phases) on PSC performance was studied. The presence of 1T phase in TMDC HTLs impacted device performance. The 1T percentage improved following exfoliation from their bulk counterparts but decreased after annealing at 90 and 200 °C (Fig. 6.31a, b). The devices were made from 11 nm thick MoS_2 and WS_2 HTL with high 1T content (> 50%) and low (20%) 1T content. Its photovoltaic performance declined considerably as 1T content decreased. 1T-rich MoS_2 and WS_2 HTLs had PCEs of 12.44, 14.35, and 15.00%, respectively. Improvements in V_{oc} and J_{sc} contributed significantly to the increased PCEs, possibly best alignment, and the beneficial effect of MoS_2 and WS_2 on charge dissociation and transportation. While the MoS_2 and WS_2 devices

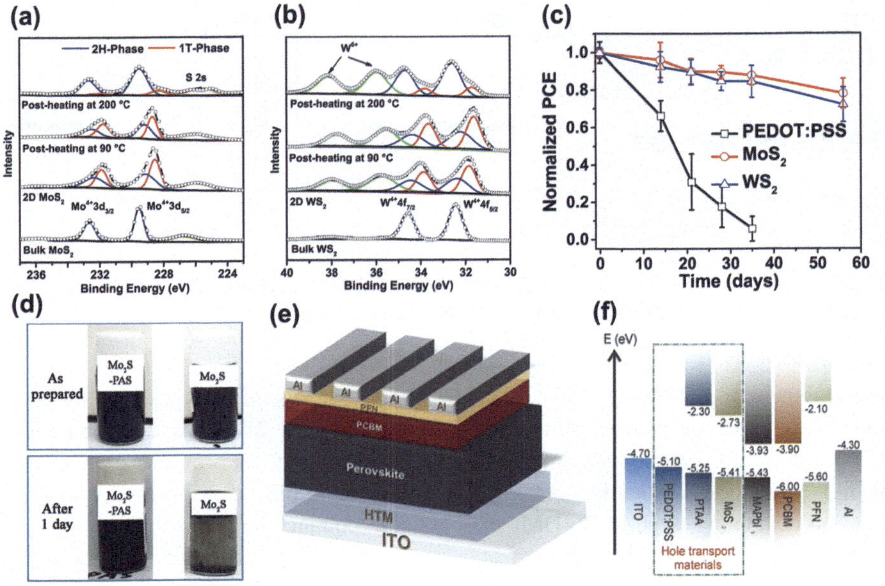

Fig. 6.31 XPS spectral study for **a** bulky MoS_2 with MoS_2 NSs, **b** WS_2 bulky with NS, **c** PCEs for PSC-dependent PEDOT: PSS, for MoS_2 along with WS_2 versus storage duration (Reproduced with permission from [141]), **d** images for various phases of solution for MoS_2-PAS into MoS_2 dispersed in water (Reproduced with permission from [173]), schematically **e** device sketch and **f** energy-band diagram for inverted PSCs (Reproduced with permission from [175])

retained 78 and 72% of their initial values after 56 days in an N_2-filled controlled glove box, the PEDOT: PSS-based PSC demonstrated virtually total failure after 35 days (Fig. 6.31c). The hygroscopic characteristic of PEDOT: PSS was blamed for the PSCs' low stability. Solution-processed is a facile approach o functionalize MoS_2 utilizing a metal-organic complex [173]. The PEDOT: PSS modified MoS_2 was subsequently used as the HTM in $MAPbI_3$-based PSCs. The functionalized MoS_2 concentration ranged from 0 to 5 $mgmL^{-1}$. The PCE improved from 14.69 to 16.47% as the concentration grew from 0 to 3 $mgmL^{-1}$, and then decreased as the MoS_2 level increased, resulting in a PCE of 10.43% at 5 $mgmL^{-1}$. The aggregation of highly concentrated MoS_2 produced poor quality, non-uniform film, lowering the FF value. After 10 days, PSCs with modified MoS_2 (3 $mgmL^{-1}$) retained around 87% of their initial efficiency, but PSCs without MoS_2 lost 52% of their initial efficiency. The devices with phenylacetylene silver (PAS)-functionalized MoS_2 and pure MoS_2-incorporated PEDOT: PSS solution was made to examine the role of PAS in the reported efficiency increase. The pure MoS_2 device reduced total cell performance from 14.69 to 11.68% compared to the PEDOT: PSS reference device. Incorporating PAS-functionalized MoS_2 increased PCE to 16.47%. Pure MoS_2 has a lower dispersibility than PAS-functionalized MoS_2 that may explain the decreased PCE (Fig. 6.31d). This can cause flakes to aggregate and impair the PEDOT: PSS/MoS_2-PAS mix film homogeneity. The UVO p-type doping of the MoS_2 films enabled efficient hole transmission while efficiently blocking electrons [174]. As a consequence, PSC had a 6.01% higher PCE than the HTL-free reference device (1.76%). The devices were made using MoS_2 concentrations of 5, 10, and 20 vol.% in hybrid HTMs. The top device would have a PCE of ~ 14.7% at 10 vol.% of MoS_2, whereas the PEDOT: PSS-dependent reference device keeps PCE of 12.0%. While enhancing charge withdrawal and transfer in devices, MoS_2 flakes are added. The contact resistance found at MoS_2-blend on HTL interfacial lowered by 10%, and the recombination resistance is raised to 50%. For 28 day period, the PEDOT: PSS/MoS_2 HTL-based PSCs retained 95% for the initial PCE, thereby reference device retained 60%. In inverting planar PSCs, MoS_2 adopted an HTL with hole transportation interlayer as given in Fig. 6.31e [175]. Exfoliation of MoS_2 flakes is performed in ethanol/water solution. The MoS_2 flakes occupy 240 nm lateral dimensions with 1.03 nm thickness. The structure of ITO/HTM/$MAPbI_{3-x}Cl_x$/PCBM/PFN/Al structure has shown 7.80% PCE more than 5.05% HTL-free PSC whereas smaller than 12.30%. PEDOT: PSS HTM-dependent cell and single spin-coated MoS_2 HTL-based PSCs exhibit small efficiency because of poor surface coverage and homogenous for MoS_2 HTL. Increasing the number of spin-coating cycles to remove this issue is practiced. The device efficiency was improved to be 12.46% with 6-spin coating, thereby decreasing to 12.09% with 7-spin coating. Regardless of efficiency-boosting, device repeatability was identified as poor. MoS_2 for hole extraction interlayer was also investigated for increasing device efficiency and repeatability. For compatibility connected with the underlying PEDOT: PSS, the non-aqueous processing of PTAA-HTM doped with F4-TCNQ HTM was adopted. Because MoS_2 VBM (− 5.41 eV) is among HOMO levels for PTAA (− 5.25 eV) whereas perovskite absorber (− 5.43 eV) and is regarded optimal toward device performance (Fig. 6.31f). By utilizing MoS_2 as an interlayer among PTAA as well as $MAPbI_3$ enhanced device efficiency as 16.42% of two spins than 15.51% of conventional devices. For 12 h of stress testing, the PSCs and MoS_2 in the form of an interfacial layer were preserved above 97% for original PCE, whereas reference device was almost totally failed. Without doping for PTAA and PTAA/MoS_2 HTLs becomes responsible to study outcome of dopants on device efficiency and stability. Both PCEs devices dropped to 13.06 and 13.95%. On comparing with doped PTAA, the undoped PTAA retained 70% of its initial PCE with 12-h stress testing. The dopants are requisite for excellent device performance, whereas they induce rapid device deterioration that is successfully reduced through addition of MoS_2. The PSC connected with MoS_2 interlayer demonstrated extraordinary stability, maintaining 80% initial performance for 568 h, whereas reference device attained 80% for ~ 171 h. Following these results, huge PSCs with active areas of 0.5 cm^2 are offered. The MoS_2 interlayer-based PSC with PCE of 13.17% than 10.64% for the reference device was presented to enhance efficiency.

Active Layer

Controlling crystallinity, GB, and defect passivation in perovskite films with effective additives is well known to improve PCE, stability, and repeatability of PSC devices. From this perspective, Yang et al. [176] added liquid phase-exfoliated BPQDs to the

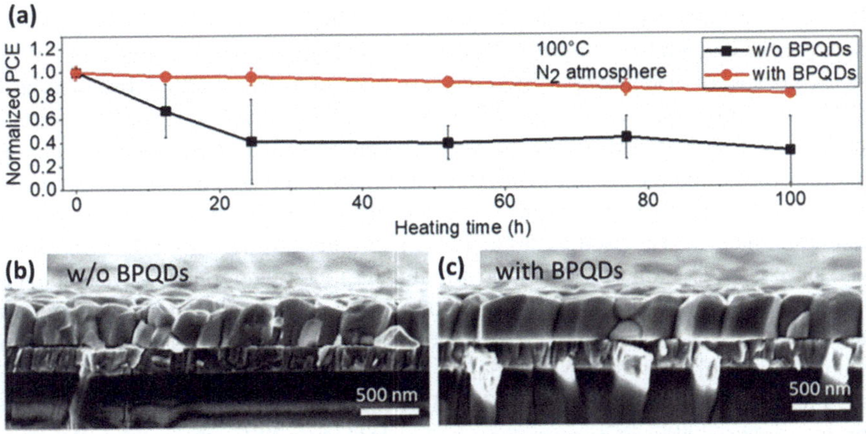

Fig. 6.32 **a** Thermal stability exploring of unsealed PSCs annealed at 100 °C into glove box, **b** BPQD-independent perovskite films, and **c** BPQDs doping in perovskite film at annealing of 100 °C for 50 h (Reproduced with permission from [177])

perovskite precursor to improve perovskite quality. These PSCs were made by adding BPQDs to a $MAPbI_3$ precursor solution in various quantities. Less non-radiative defects and greater crystallinity were observed in the BPQD-based perovskite films compared to the reference perovskite film without BPQDs. Using time-of-flight secondary ion mass spectrometry, the BPQDs were revealed to be heterogeneous nucleation sites for perovskite crystal formation. The additive-assisted perovskite film had a PCE of 20.0%, while the reference device had a PCE of 17.7%. However, when tested under comparable parameters, the unsealed BPQD additive-based device retained over 80% of its original PCE after 100 h of thermal aging at 100 °C within controlled glove box, whereas the conventional devices lost over 40% after 24 h (Fig. 6.32a). The devices' improved thermal stability is due to the $MAPbI_3$-BPQDs film's better crystallinity and reduced strain-induced grain break due to the BPQDs at the film's bottom (Fig. 6.32b, c).

6.3.2 *Hydrogen Evolution*

6.3.2.1 Transition Metal Dichalcogenide

TMDCs are a fascinating class of layered materials. Following a series of theoretical and practical demonstrations, as discussed earlier (see Chap. 3) it is determined that the active sites are the edges of S–Mo–S in the MoS_2 sheet. As a result, several studies detailing techniques for increasing the number of active sites have been published. The purpose of fabricating numerous nanostructures nanowire, nanorings, and NSs was to increase the structure's surface area [109, 178]. While the morphological modifications did not affect electronic property and found results in an increase in active sites number per unit area. Recently, the photocatalytic H_2 generation ability of a co-doped MoS_2/CdS composite was explained. Due to the rapid recombination, bare MoS_2 and CdS did not provide meaningful findings. Conversely, doping with a specified wt% of MoS_2 increased the yield (14.10 mmolh^{-1} g^{-1}). Combining the composite with Co increased the amount of H_2 produced [178]. The production of a CdS composite with a FL-rich TMDCs in a single-pot wet chemical process FLs of TMDCs developed on the CdS surface, and the nanohybrid among two structures boosted the surface active sites number, resulting in large production of WS_2-CdS 1984 mmolh^{-1} g^{-1} and MoS_2-CdS nanohybrids 1472 mmolh^{-1} g^{-1} [179]. 2D nanoplatelets or nanocrystals of Cd-dependent chalcogenides are ideal illustrations for chalcogenides in comparison with bulky counterparts because of their slower solar to hydrogen conversion [180, 181]. The higher recombination rate is ideal in this situation because 2D crystals' platelet features offer a high cross-section toward light absorption, whereas tight confinement permits rapid IP carrier movement [182]. Despite their various advantages, charged excitons possess a short lifetime and are prone to radiative and non-radiative recombination, thereby precluding their engagement in the form of light-collecting centers. This issue is clearly addressed through construction of a hetero-NS assembly involving CdSe/CdTe with core/crown and hetero-NS of type II (by showing CdTe crown lateral extension on CdSe NS core). According to PL measurements, the excitons were identified as domain-localized, with electrons into CdSe whereas holes into CdTe. This exciton localized production of electron with long lifetime and slow recombination process [183]. Similarly, in another research, the authors published the way of Ni-decorated CdS nanorods production for more hydrogen. The semiconductor surface provides location toward two steps hole transfer process taking usage for $^{\bullet}$OH/increment with OH's mobility along with effective interaction with the semiconductor as well as the sacrificial agent. Resultantly, H_2 generation possesses higher quantum yield (QY) [184]. Another study [185] analyzed for hydrogen efficiency generation by using CdSe capping with dihydrolipoic acid (DHLA) in the form of light absorber and soluble Ni^{2+}-DHLA catalyst to produce protons using ascorbic acid (electron donor). The QY exhibited a QY showing 36%, whereas material's endurance was rather extraordinary, offering an excellent choice toward water splitting systems. Zhukovskyi et al. [186] synthesized 2D CdS NS associated with lateral dimensions ranging 1–3 nm. However, photodeposition owing to Ni-NPs onto the NS was performed for H_2 efficiency test on the as-prepared materials indicating 25% increment in apparent quantum yield (AQY) into Ni-decorated sheets than pristine structures. Therefore, the greater efficiency attained in the semiconductor/metal HJ is because of the existence of large number of electrons. Similarly, 2D sheets production for NiOH/sulfide on CdS nanospheres by employing co-catalyst has been reported. The NiOH and sulfide sheets function into hole storage sites with surface reaction centers, thereby facilitating charge transfer. On comparing with CdS, the HS combination possesses 46 times fast rate toward evolution [187]. As with chalcogenides, novelty in 2DMs, their synthesis with hydrogen to generate efficiency has attracted increasing scientific

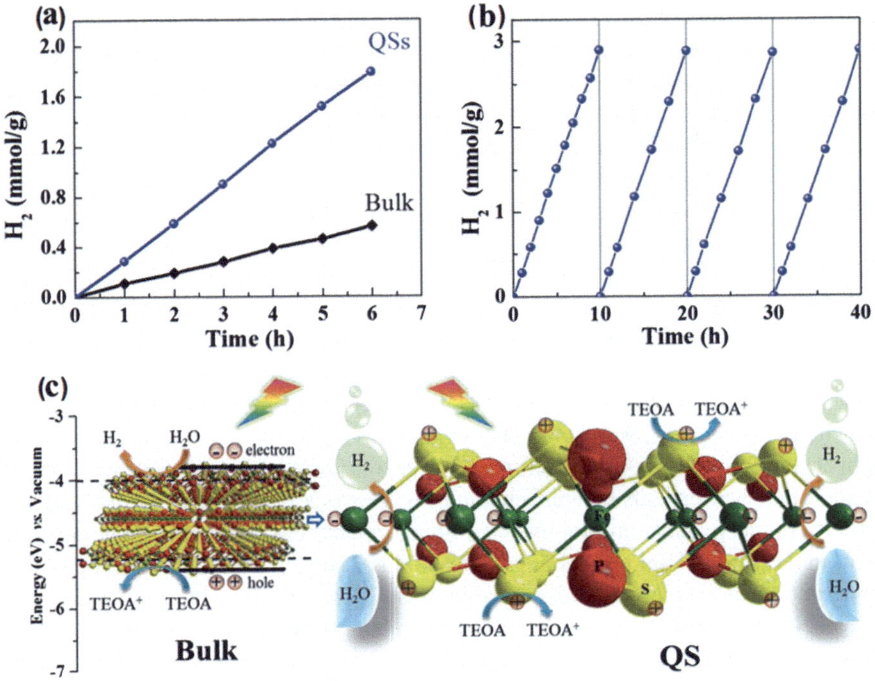

Fig. 6.33 Photocatalytic activity (PCA) for bulky $FePS_3$ along with QSs: **a** H_2 production photocatalytic graphs of bulky $FePS_3$ as well as QSs into 100 mL deionized water showing 10% triethanolamine, **b** cyclic evaluations of H_2 generation for $FePS_3$ QSs, **c** the PCA processing of bulky $FePS_3$ with QSs images (Reproduced with permission from [189])

attention in current era [188]. On comparing with graphite, the MPX_3 (metal phosphorus trichalcogenide) family requires less cleavage energy. $FePS_3$ quantum sheets (QSs) are formed via sonochemical exfoliation of the bulk material in hydrazine solution [189]. The material's E_g is 2.18 eV is substantially greater than E_g of bulk $FePS_3$ (1.6 eV) [189]. These 2D QSs of $FePS_3$ produced H_2 at a rate 3-times that of bulk material as seen in Fig. 6.33.

6.3.2.2 Graphene-Like Materials

Wulan et al. [190] recently described a simple calcination process for the production of nickel pyrophosphate-modified g-C_3N_4 reporting the catalyst's effectiveness in photocatalytic water splitting. The catalyst as produced exhibited a 37-fold increase in hydrogen generation rate (207 μmolh^{-1} g^{-1}). The tiny particle size, high visible light absorption, and capacity of excited electrons to perform fine reduction all contributed to the increased outcome. The BP acted as a breeding ground for holes. When compared to pristine g-C_3N_4 sheets, the changed 2D structure produced approximately 4-times more H_2 [191]. A similar 2-step thermal treatment on bulk sheets of

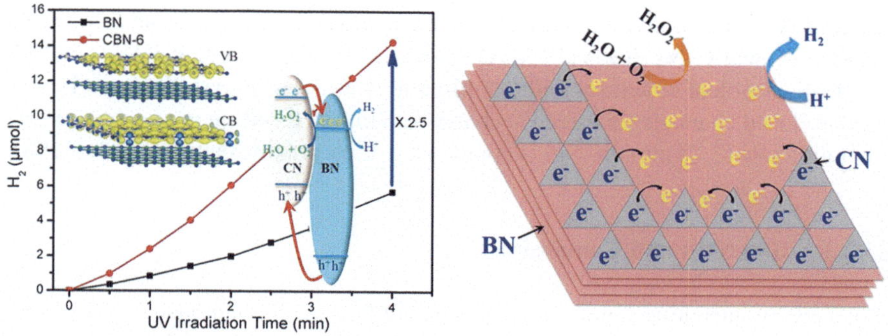

Fig. 6.34 PCA mechanism of 2D-HJ produced by loading C_3N_4 on h-BN NS (Reproduced with permission from [194])

C_3N_4 to achieve an atomically thin structure with abundant O_2 and amino groups is also employed [192]. The development of these groups resulted in a high concentration of active sites for surface reactions from an extremely rapid evolution rate of up to 20,948.6 $\mu molh^{-1}$ g^{-1} when Pt was used as a co-catalyst [192]. Recently, a research group [193] described the formation of new HJ of 0D-Ni_2P with 2D-g-C_3N_4 sheets, where a 3.5 wt% injection of Ni_2P NPs was adequate to increase hydrogen evolution rates, these new interfaces serve as excellent excitons sink. Another research developed a new hybrid photocatalyst composed of C_3N_4 and h-BN (Fig. 6.34). Due to the broad VB, BN often has a large E_g, when varied quantities of C_3N_4 were loaded, band edge mismatch occurred, effectively contributing to charge separation. The HS's total o/p was superior to that of Pt-loaded h-BN. As a result, this work established the principles of band edge tuning for catalytic applications [194]. A similar study demonstrated the integration of the 1,3,4,6,7,9b-heptaazaphenalene-2,5,8-triamine (melem) structure by annealing the C_3N_4 polymer structure. This annealed structure had a broad absorption spectrum in the visible range. After additional treatment with ammonia in this study, the annealed samples repaired the N vacancies and bridged the polymeric framework. These samples produced more H_2 as a result of their ability to efficiently harvest light and separate charge carriers [195]. The family of materials known as perovskite-based layered materials is an intriguing one. Their potential application as photocatalysts has lately received increased attention in scientific studies on hydrogen generation. Furthermore, the drawback of these ML structures is their poor electron mobility. However, these problems can be addressed by adjusting them using a suitable layered material such as C_3N_4. The hybridization of $La_2Ti_2O_7$ NS with C_3N_4 sheets results in the C_3N_4 sheets acting as a hole receptor and the perovskite NS acting as an electron sink. This effective charge transfer across the interface results in a significant amount of PCA [196]. Chen et al. [197] described the mechanochemical synthesis of the g-C_3N_4/C_{60} hybrid structure. This hybrid structure was formed by ball milling C_{60} in the presence of LiOH. The authors ruled out the use of any noble metals as co-catalysts and instead demonstrated visible-light-induced hydrogen generation using the hybrid structure in bare circumstances.

The results demonstrated a remarkable increase in H_2 generation of up to fourfold that of pristine g-C_3N_4 sheets [197]. After that, the synthesis of new 2D g-C_3N_4 NS @$ZnIn_2S_4$ nanoleaf structures without utilizing Pt as a co-catalyst, the nano-HJ has been produced channels for effective charge separation, resulting in an increased H_2 generation rate (2.78 mmolh^{-1} g^{-1}) compared to their pristine parent samples [198].

6.3.2.3 2D-MOFs and Composites

MOFs are an intriguing family of materials that have lately demonstrated a variety of uses in energy storage, with several studies demonstrating their PCA. The porous structure with a large surface area combined with the flexibility to customize the organic linkers or metal clusters make them an appealing class of materials for a broad variety of applications. Hybrids with 2DMs and other inorganic materials for photocatalytic hydrogen generation are an intriguing area of research. Gomes Silva et al. [199] were among the first to examine UiO-66's PCA; when light with a wavelength higher relative to 300 nm was irradiated in CH_3OH presence as a sacrificial agent, a significant AQY was observed. Another important study [200] utilized 2-amino-benzene dicarboxylic acid as an organic linker to establish amino-functionalized titanium (IV) MOF. Here, Pt was photo deposited in the hybrid to be used in research on H_2 production with triethanolamine used as a sacrificial electron donor in the aqueous solution. The resultant hybrid materials had intriguing properties; the organic linker enabled light harvesting, and the excited electrons have been transported to titanium oxo clusters. The preceding sections discussed the usage of 2D-chalcogenide materials for H_2 generation; chalcogenide is an attractive option due to the availability of suitable CB and VB edges, while the MOF offers sufficient space and surface sites for reaction time. The Pt photodeposition over the composite was tested for H_2 generation with the best possible outcomes obtained with a CdS loading of 10%, and subsequent tests up to 5 h demonstrated the material's outstanding stability [201]. Using triethanolamine as a sacrificial agent, the catalytic efficiency of metal (Au, Pd, and Pt)-doped MOFs for H_2 generation is investigated. The results indicate that the o/p is extremely low in Pt absence as a co-catalyst when compared to the doped materials. This demonstrates the critical role of co-catalysts and sacrificial agents [202]. Tilgner et al. [203] developed an intriguing work, constructed gold-doped anatase composite shell around a MOF core (MIL-101). Doping anatase TiO_2 with Au increases its absorption of visible light. MOF is utilized as a core that ensures structural stability while at the same time providing sufficient surface areas for work. In a water/methanol solution, the composite structure generates an astounding 903 μmolh^{-1} g^{-1} of H_2. Recently, explore the effect of $Co_xNi_{4-x}S_4$ on the surface of MIL-101 and the photocatalytic H_2 generation as shown in Fig. 6.35a. The authors established the critical role of surface atomic structures in facilitating electron transport as displayed in Fig. 6.35c. H^+ adsorption initiates the hydrogen production process, followed by H^+ reduction and hydrogen production by subsequent desorption. Since a result, adsorption becomes a critical stage in the H_2 evolution process, as it controls the adsorption process and driving force.

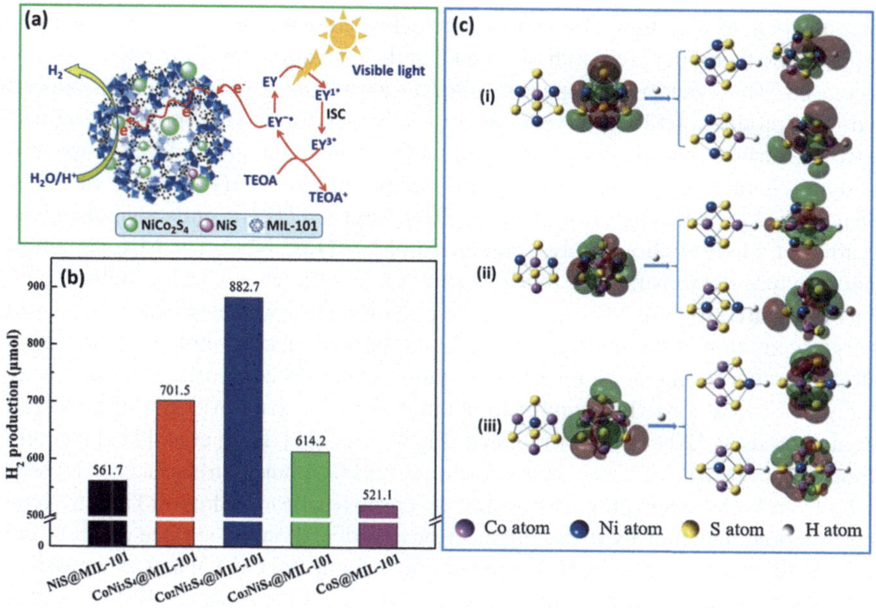

Fig. 6.35 **a** Schematic of photocatalytic mechanism for H_2 production on Eosin Y-sensitized $Co_2Ni_2S_4$@MIL-101 irradiated by UV light, **b** time progressions for H_2 production on NiS@MIL-101, $CoNi_3S_4$@MIL-101, $Co_2Ni_2S_4$@MIL-101, Co_3NiS_4@MIL-101 and CoS@MIL-101 photocatalysts of 10% into 100 mL (v/v). The triethanolamine aqueous solution of pH 9 by visible light for $\geq$ 420 nm, **c** surface architecture, H_2 adsorption with charge density to spread for HOMO (**i**) on $CoNi_3S_4$, (**ii**) on $Co_2Ni_2S_4$, (**iii**) on Co_3NiS_4 (Reproduced with permission from [204])

Earlier findings indicate that the difference in activation energy at various crystal facets is related to the intermediate structure's adsorption. As a result, it is critical to do research on structural orientations that exhibit minimal adsorption and overpotential energy. The authors noted in their research that the dual transition metal sulfide ($Co_2Ni_2S_4$@MIL-101) had most activity (Fig. 6.35b) [204].

6.3.2.4 MXenes and Composites

These 2DMs are architecture in careful etching, an elemental family of (Al/Ga/Si) in HF, NH_4HF_2, or LiF as well as HCl solutions [205]. These materials are considered robust with an increment of electrical conductivity as well as capacitance [206, 207]. Because of abundance in functional groups on MXenes surface are excellent candidates toward heteroatom doping. The advancement for unique technique to architect laminated, defect controlling C-supported/S-doped TiO_2 junction of photocatalyst for monument impregnation for Ti_3C_2 was recently performed. The authors significantly uncovered new materials to generate H_2 at higher rate (333 μmolh^{-1} g^{-1}).

Increments in visible light absorption were achieved using the new synthesis technique, while theoretical research revealed that the activation energy needed was also decreased. The synergistic impact resulted in an overall increase in light-induced hydrogen production [208]. MXenes are typically metallic and hydrophilic in nature; however, surface modification with terminal O_2 results in a material with improved catalytic characteristics. The as-prepared sample produced H_2 at a rate of 105% (88 μmolh^{-1} g^{-1}) than the clean samples [209]. Su et al. [210] recently announced the creation of a hybrid photocatalyst composed of Nb_2O_5/C/NbC. The MXene hybrid nanostructure demonstrated a fourfold increase in efficiency (7.81 μmolh^{-1} g^{-1}) when compared to pure Nb_2O_5. The increased efficiency was ascribed to improved charge separation at the interface has led to a reduced electron-hole recombination rate. In another recent work, researchers synthesized CdS composite structure using Ti_3C_2 as a co-catalyst. The combination emitted 14,342 μmolh^{-1} g^{-1} of hydrogen when photons at 420 nm were irradiated (Fig. 6.36) [211]. Peng et al. [212] recently reported the growth of TiO_2 sheets (exposed on 001 plane) on surface of layered $Ti_3C_2T_x$ and subsequent photo deposition of cuprous oxide nanodots on TiO_2. MXene acts as a hole mediator by tunneling electrons produced on the surface of TiO_2 and $Ti_3C_2T_x$ to the nanodots (Cu_2O) for reduction to elemental Cu. Cu acts as a possible co-catalyst for reduction in chemical process. The resulting photocatalyst is capable of producing hydrogen with an efficiency of 860 μmolg^{-1} h^{-1}.

6.3.2.5 Other 2D Materials

Recent research on new structures and their photocatalytic properties is discussed in this section. In comparison to their bulk counterparts, 2DMs have long been a source of fascination. Their adaptability and extensive applicability form them ideal candidates for a wide range of energy and environmental applications. The 2D NS enables the incoming photon flux to traverse less distance and hence enhance efficiency even at low intensities [213]. Additionally, reports of the production of p–n nano-HJs have surfaced, demonstrating the critical role of surface chemistry in HJ formation on the semiconductor surface [214]. Okamoto et al. [215] investigated the hydrogen generation efficiency of Rh-doped $CaNb_2O_6$ NS synthesized by exfoliation of their native material. In comparison to the parent material, the new synthesized 2D sheets demonstrated a 165-fold improvement in production efficiency. Another study also demonstrated the fabrication of a new HJ of $AgIn_5S_8/ZnIn_2S_4$. This 0D/2D HJ offered sufficient active sites number and charge separation for the photogenerated charge carriers. In comparison to its pristine parents, the composite had the highest production rate (949.9 μmolh^{-1} g^{-1}) [216]. Kong et al. [217] recently reported the fabrication of TiO_2 NS through plasma etching. Plasma etching established oxygen vacancies appear to decrease E_g to 2.88 eV by producing isolated bands. The o/p of H_2 has been increased to 160 μmolh^{-1} g^{-1}. Pilarski et al. [218] constructed layered cesium copper titanate structures and determined the material's structural efficiency through hydrogen generation experiments. Copper ions are deposited within cesium titanate voids, significantly increasing the sample's absorption of visible light. In this

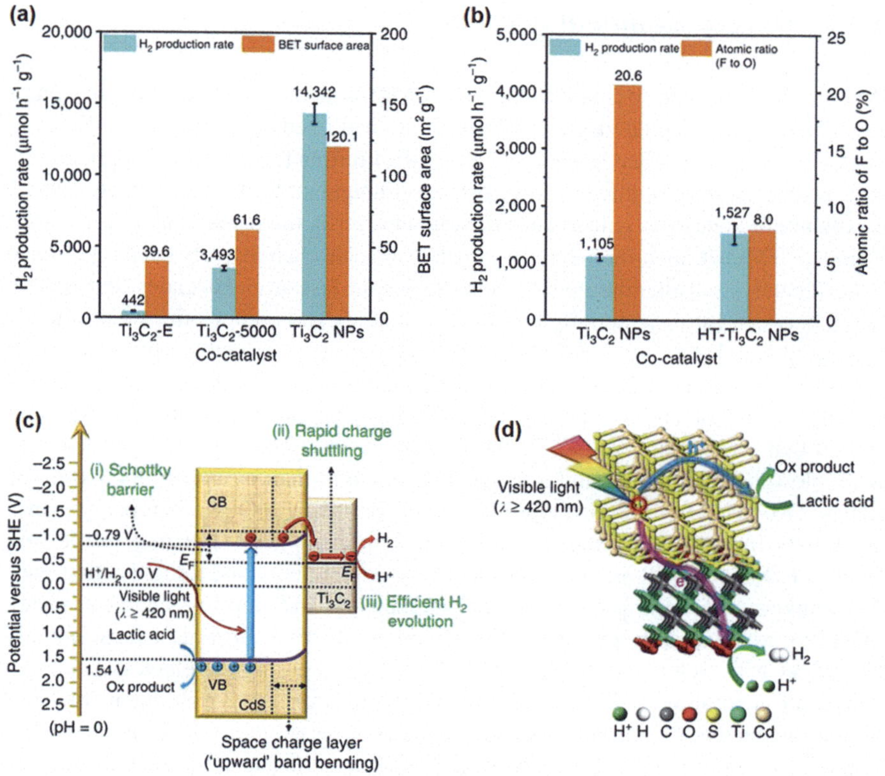

Fig. 6.36 **a** Plot of dominant effect for co-catalyst surface area of photocatalytic evolution of H_2 **b** the impact from F toward O atomic ratio found by co-surface catalyst's in favor of photocatalytic hydrogen production, **c** charge separation as well as transformation depiction of schematic in CdS/Ti_3C_2 irradiated with visible light. Red/blue spheres depicting photogenerated electrons and holes **d** credible H_2 evolution mechanism in CdS/Ti_3C_2 system toward visible light irradiation. Color scheme depicting green H^+ (Reproduced under the terms of CC BY 4.0 license [211], https://creativecommons.org/licenses/by/4.0/)

work, a continuous rise in the rate of H_2 generation was observed as attribute to copper atoms being reduced to their metallic form during the process and acting as a co-catalyst. Ajayan et al. [219] demonstrated the production of 2D QDs from exfoliated sheets of 2DMs simply by introducing a pretreatment in liquid nitrogen for a certain time period. This procedure represents a significant advancement over current LPE methods, as cryo treatment always results in the formation of pure 2D structures, primarily ML. There is no longer a need for solvents tailored to each exfoliation, and most importantly, the use of any additional contaminants like surfactants is simply avoided. Wen et al. [220] synthesized a new 2D core–shell nanostructure of $CuFeSe_2$@gold. This amorphous core is produced, while the crystalline shell is generated. The sample's unique design exhibits high near-infrared absorption and demonstrates an outstanding H_2 evolution rate of 3.48 $mmolh^{-1}\ g^{-1}$.

6.3.3 Oxygen Evolution Reaction

When water is split, the oxidative half-reaction is frequently slow, resulting in significant kinetic overpotential loss [17]. The most often utilized high-efficiency OER catalysts, such as IrO_2 and RuO_2, are rare and costly, limiting their practical applicability. As a result, the development of low-cost, earth-abundant OER electrocatalysts with excellent efficiency has gained considerable interest in the recent past [221]. Earth-abundant TMOs/transition metal (oxy)hydroxides have extensively been researched as OER catalysts in alkaline environments during the last several decades, and a correlation between intrinsic activity and local structures and compositions was already harnessed [222]. Further, LDHs have shown themselves as viable options as OER catalysts [223]. Indeed, Fe, Co, or Ni-based LDHs have demonstrated potential OER activity in alkaline environments [224, 225]. For example, Zou et al. [226] demonstrated that a bimetallic Zn–Co LDH NS acts as an effective OER electrocatalyst with a lower overpotential and a larger time-of-flight than monometallic Co-based LDHs. TM-based LDHs have demonstrated the promise for OER; their performance needs to be further improved to improve their coupling to HER. Two major disadvantages of LDHs for OER are their low electronic conductivity and limited active sites. To compensate for the poor electrical conductivity and improve charge transfer, LDH NS are frequently produced directly on conductive substrates, most notably GR, MWCNTs, rGO, and 3D GR [227, 228]. By increasing the electrical conductivity and generating new routes for chemical transfer, the development of LDH NS on conductive C materials can substantially increase their OER activities. For example, Long et al. [228] showed over potential of strongly linked GR and Fe–Ni LDH hybrid could be decreased to as little as 200 mV and the time-of-flight could reach up to 1 s^{-1} at 0.3 V. The synergistic action of the conductive C and LDH may result in a change in the oxidation state of the TMs in the LDH, therefore increasing its activity as an active species. The greater oxidation state of Ni species in the $Ni(OH)_2$/MWCNT composite than in the physically mixed was already verified by XPS that may account for composites' lower onset potential for OER [229]. The discovery of a synergistic interaction between LDHs and C materials lays the groundwork for further optimizing its activity through the use of different doped C materials. Another method for increasing the electrical conductivity of LDHs is to grow them directly on 3D conductive substrates. For example, Lu and Zhao grew an amorphous mesoporous Ni–Fe composite on nickel foam using an electrodeposition technique, and the resulting composite electrode demonstrated significant activity for OER in alkaline conditions with a low over potential of 200 mV [230]. Surprisingly, the electrode was capable of delivering a high J_{sc} of 500 $mAcm^{-2}$ at a 240 mV overpotential, meeting the criteria for commercial water electrolyzers [230]. To further enhance the loading of active LDH NS, a hierarchical structure based on the nickel foam nanowire array was utilized to produce NiCoFe LDH nanoplates, and the resulting hierarchical electrode demonstrated efficient and sustained catalytic activity for OER with a Tafel slope of 53 mV/dec [231]. LDHs are a family of ionic-layered compounds composed of intercalated balanced anions and solvation molecules inside positively brucite-like layers. Because these anions are easily exchangeable, the interlayer distance of LDHs may be easily adjusted by selecting anions with a range of ionic diameters [232]. Indeed, it has been demonstrated that the OER activity of LDHs correlates with the interlayer spacing, i.e., the larger the interlayer spacing, the greater the activity. This may be as a result of more exposed active sites and enhanced exchange capacity of the reactive anion (OH) in interlamination [233]. Expanding LDH active sites by exfoliating them to several layers or even an ML is very effective.

Song and Hu [224] demonstrated that an SL-LDH exhibited significantly increased OER activity compared to its bulk counterpart and even outperformed IrO_2 in alkaline solution. With a Tafel slope of 40 mV/dec and time-of-flight of 0.05 s^{-1} (Fig. 6.37a–d), the exfoliated SL-NiFe LDH NS demonstrated more activity than CoCo and NiCo LDHs [224]. According to their study, by employing bromine or iodine as an oxidant to convert Co^{2+} (Fe^{2+}) in transition metal double hydroxides to Co^{3+} (Fe^{3+}), resulting in Br^{2-}-or I^{3-}-intercalated LDHs that were subsequently exfoliated into charged NS [234, 235]. In some cases, an additional oxidation step is necessary to convert Co^{2+} to Co^{3+} during the preparation process, as hydrated Co^{3+} ions are more active in OER than Co^{2+} ions. To circumvent the tedious synthesis procedure,

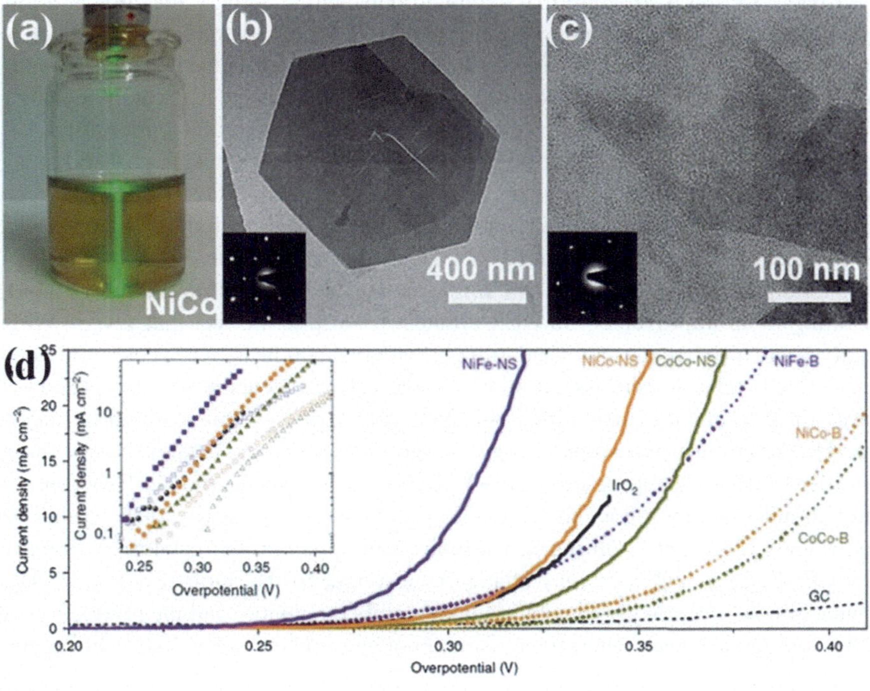

Fig. 6.37 **a** Exfoliated CoNi LDH-NS as colloidal solution (optical image), Tyndall effect was visible upon irradiation with a laser beam, **b** bulk NiCo LDH crystal (TEM image), inset is corresponding SAED pattern, **c** exfoliated SL-NiCo LDH NS (TEM image). *Inset* Corresponding SAED pattern shows crystalline [110] plane family, **d** polarization curves of exfoliated LDH NS and IrO_2 NPs, inset shows Tafel plots (Reproduced with permission from [224])

Jin et al. [236] reported the direct production of Ni–Co LDH by oxidizing the originally produced $[Co(NH_3)_6]^{2+}$ to $[Co(NH_3)_6]^{3+}$ using ammonia. The as-prepared LDH crystal was subsequently exfoliated into LDH NS exhibited excellent water oxidation activity as well. By reducing the thickness of LDHs, their electrical surface area may be significantly increased. When exfoliated LDH NS were used for OER, the increase in catalytic activity was frequently greater than the increase in electrical surface area when relative to as-prepared bulk LDHs [236]. The significantly increased catalytic activity of OER might have resulted from the distinctive chemical and electrical characteristics of exfoliated LDH NS or the more exposed edge sites and defects generated during exfoliation. Apart from LDHs, other OER catalysts such as transition metal double hydroxides and oxy(hydroxide)s were extensively studied [237, 238]. As with LDHs, the composition and shape of double hydroxides have a significant effect on their intrinsic activity. For example, it was established that NiFe, NiCr, and NiMn double hydroxides have better catalytic activity when compared to other TM-doped $Ni(OH)_2$ catalysts for OER. NiFe and NiMn double hydroxides have active centers of Fe and Mn, respectively, whereas NiCr double oxides have an active core of Ni [238]. The SL-$Ni_xCo_{1-x}(OH)_2$ was more active than its ML equivalents [239]. Wang et al. [240] synthesized a novel spiral-structured bimetallic hydroxide NS spiral structure that exhibited outstanding activity for OER at a low overpotential of 210 mV. In addition to oxides with spinel, structures are effective catalysts for OER [241]. Due to the significant correlation between the accessible active surface and flaws in spinel structured oxides and their OER catalytic activities, the porous nanoplate structure is a morphologically favorable option for water oxidation. It has a high surface area and improves charge transfer probability by decreasing GBs [242]. According to Xie et al. [243], the $NiCo_2O_4$ NS rich in oxygen vacancies exhibited better activity for OER, with an overpotential of 320 mV and a Tafel slope of 30 mV/dec. About 850 DFT calculations have predicted that oxygen vacancies included in ultrathin NS might reduce the adsorption energy of H_2O, resulting in an improvement in OER efficiency. Despite the fact that several OER catalysts are also being established, little is known about the active species produced during the oxidation process. Similarly, some active species have been presented based on theoretical calculations [244, 245], although it is still required to understand the relationship between intrinsic activity, structure, and catalyst composition in order to develop a more effective OER electrocatalysis candidate. Experiments have shown that amorphous materials are significantly more active than crystalline equivalents [224, 246]. During the arduous OER process, the catalyst's structure and surface chemical state may change in response to the applied potential. The connection between oxygen evolution under applied potential and the production of active species has been determined using an in situ grazing angle X-ray diffraction technique [247].

The production of metal oxyhydroxide in an active phase of OER confirms theoretical predictions. Notably, the visible surface also alters the electrocatalyst's inherent activity while determining adsorption energy for the intermediates on the surface. An entire knowledge owing to electrocatalyst's structure depending activity may be used in turn to reasonable designing and constructing ultrathin 2D NS toward enormously operative OER catalysts. Elemental doping by a typical technique in enhancing catalytic activity for OER catalysts is assured. However, it makes complexity while identifying active species in oxidation process. Experimental research has explored $Ni(OH)_2$ films aging into impure KOH solution that may assimilate significantly specific quantity of Fe, leading to revealing enhancement in catalytic activity while comparing to $Ni(OH)_2$ films matured in Fe-free-solution [248]. It is extremely notable for Fe incorporation that may occur at relatively low Fe presence (< 1 ppm Fe). Research on similar ground has demonstrated for contaminated electrolyte associated with nickel impurity, sufficiently act as an efficient catalyst on OER substrates in alkaline solution [249]. These findings increase the burden of establishing the active species in OER, and more stringent control tests should be conducted to rule out the potential of contamination affecting the catalytic activity of OER catalysts. Additionally, 2D chalcogenides (sulfide and selenide) are being studied as OER catalysts [250, 251]. For example, Xie et al. [252] found that ZrS_3 and Co_3S_4 NS exhibited excellent water oxidation activity. Selective exposure surface containing octahedral coordination cations, in particular, may be used to control the spin states of the Co^{3+}. As a result of the atomic structural and electrical configuration modulations, the Co_3S_4 NS exhibits an OER onset potential as low as 310 mV under neutral circumstances [253]. They proved that the $CoSe_2$ NS is a potential electrocatalyst for water oxidation due to its metallic properties allowing for rapid electron transport [254, 255]. The introduction of cobalt vacancies into the $CoSe_2$ NS significantly increases the number of active sites for OER, with OER overpotential as low as 0.32 V at 10 $mAcm^{-2}$ in pH 13 medium [255]. While these TMDCs have very strong activity for OER, their mechanistic activity requires more research due to their susceptibility to oxidation during the severe oxidation process. TM nitrides are another type of material that shows promise for OER because of their metallic characteristics, which increase electrical conductivity and reduce internal potential loss. A 3-nm-thick Ni_3N NS displayed good OER activity with a modest Tafel slope of 45 mV/dec in both O_2-saturated and basic conditions (1 M KOH) was demonstrated. By connecting the OER to a carbon cloth, it was possible to achieve a high catalytic J_{sc} of 52.3 $mAcm^{-2}$ at a low overpotential of 350 mV [256]. Shalom et al. [257] conducted a similar OER research; the Ni_3N layer was placed over nickel foam. The significant catalytic activity of Ni_3N was attributable to the development of a $Ni(OH)_2$ surface layer during the electrochemical process was verified by XPS. The discovery reinforced the need for comprehensive characterization of the catalysts' compositions and chemical states following OER, as it aids in elucidating the active species produced during the OER process. Due to their low cost and ease of scaling up, doped graphitic carbon nanomaterials have also become an appealing and promising catalyst for OER in recent years [258, 259]. Although certain issues continue, such as limited OER activity and poor stability at high applied voltage. As a result, more research is required to validate their potential applicability in zinc-O_2 batteries.

6.3.4 Reduction of CO_2

Aside from H_2 evolution via photocatalysis, photocatalysts can also be used to reduce CO_2 and make fuels containing hydrocarbons were shown to be an efficient approach to address energy problems and global warming simultaneously. Essentially, fossil fuel burning produces a large amount of CO_2 can increase the greenhouse effect owing to the unabated growth of CO_2. Notably, the conversion of CO_2 to useful fuels has become an alarmingly difficult task in recent years. While CO_2 was unquestionably an important C1 building block for the chemical industry, its thermodynamic stability and high kinetics precluded its widespread commercial usage. The unmistakable solution to this conundrum is to develop carbon-neutral fuels for the low-carbon market in order to secure a long-term future with little environmental impact [260]. Furthermore, CO_2 activation involves a number of proton-coupled transfers of electron mechanisms, and the pathways of reaction were complex and varied depending on the synthesis of different compounds. The products are CO, HCHO, CH_3OH, and CH_4OH, depending on the number of e^-s injected ($2e^-$, $4e^-$, $6e^-$, and $8e^-$). However, rivalry occurs between CO_2 photoreduction and reduction of H_2O, because reduction of H_2O to establish H_2 is more energy-efficient, limiting product selectivity. According to a recent study, ultrathin photocatalysts based on 2DMs have also shown exceptional CO_2 photoreduction. An intermediate design of lamellar hybrid, for example, was used to manufacture Bi_2WO_6 layers with a thickness of a single unit cell [261]. Sodium oleate was utilized to generate ions of oleate and therefore establish an electrostatic bond with Bi^{3+}. Following that, self-assembly of oleate ions in a head-to-head or tail-to-tail BL array has established lamellar Bi oleate complexes. While Na_2WO_4 was injected and hydrothermally treated, a single unit cell layer of Bi_2WO_6 was generated and self-exfoliated. Photocatalyst for CO_2 photoreduction using synthesized single unit-cell Bi_2WO_6 layer and 300 W Xe lamp. Recent studies indicate that an outstanding photocatalytic performance was achieved by the use of a distinctive geometrical configuration of Bi_2WO_6 single unit cell layers, as follows:

1. The use of super large SSAs of single unit cell Bi_2WO_6 layers resulted in a twofold increase in adsorption of CO_2 and capacitance at first that was a critical point for CO_2 activation and reduction.
2. As demonstrated by time-resolved fluorescence emission spectrum calculations, charge separation and carrier lifespan are aided by the single unit cell thicknesses.
3. The thickness of the single unit cell carries a nearly increased DOS at the CB edge in addition to an increased surface charge density, therefore increasing the 2D conductivity.

There is no doubt about the benefits of 2DMs photocatalyst for photoreduction of CO_2. CO_2 Adsorption is a critical requirement for photoreduction of CO_2, since it has a major effect on the e transfer mechanism. Here, it is needed to develop a suitable approach for expanding the CO_2 adsorption site and generating solid contact to ease es$^-$ transfer in order to activate effectively. Another possibility, according to research, is to establish surface fault locations [262]. Although controlled synthesis is required, ZnAl-LDH ultrathin NSs having thicknesses 2.7 and 4.1 nm and of bulk ZnAl-LDH with thicknesses of almost 210 nm may be produced and are designated $ZnAl^{-1}$, $ZnAl^{-2}$, and $ZnAl^{-3}$, respectively. As a result, Zn^+-V_o compounds are formed in ZnAl-LDH NSs, as demonstrated by XAFS, ESR, and calculation of PAS spectrum. In the photoreduction of CO_2 pathway, produced Zn^+-V_o complexes can act as entrapping sites, therefore promoting adsorption of CO_2. Further, DFT and EIS simulations demonstrated that complexes like Zn^+-V_o can act as e$^-$s entrapping locations, enhancing the effectiveness of charge separation and facilitating e$^-$ transfer to CO_2. Thus, defective ZnAl-LDH NSs exhibited excellent PCA for the conversion of CO_2 to CO through the 2e$^-$ technique, with a CE of 7.6 μmolg^{-1} h^{-1} for $ZnAl^{-1}$. More improvements in CO_2 conversion competence, selectivity, and stability were found [263]. Converted V_v to o-$BiVO_4$ in a single unit cell and then utilized it as a photocatalyst for reduction of CO_2. To establish layers of o-$BiVO_4$ a lamellar hybrid halfway design was used with a thickness of single unit cell using C-TAB. Atomic layers of o-$BiVO_4$ (V_v-rich and V_v-poor) were produced in the [001] direction to control reaction time and temperature. AFM and TEM findings on atomically thick o-$BiVO_4$ NSs were obtained. As demonstrated by XRF and PAS, V_v with distinct levels developed on both the V_v-deficient o-$BiVO_4$ and V_v-rich o-$BiVO_4$ NSs surfaces. V_v exhibited a major role in photocatalytic CO_2 reduction. To begin with, V_v introduced a fresh defect concentration into BG and increased the hs$^+$ level to approach Fermi levels. As a result, light harvesting and electrical conductivity of V_v-rich o-$BiVO_4$ were enhanced. In comparison to V_v-poor o-$BiVO_4$, the V_v-rich o-$BiVO_4$ exhibited greater adsorption of CO_2 capacitance and greater hydrophilicity of surface; this was undoubtedly beneficial to the reduction of CO_2 process. Lastly, V_v enhanced charge separation and significantly increased carrier lifespan, providing more e$^-$s for CO_2 photoreduction. To reap the benefits stated before, V_v-rich o-$BiVO_4$ demonstrated increased conversion efficacy, producing methanol at a rate of 398.3 μmolg^{-1} h^{-1}. The V_v-rich o-$BiVO_4$ may endure continuous photoreduction for up to 96 h without a discernible drop in photocatalytic effectiveness. The aforementioned results demonstrate that 2D photocatalysts are an efficient substitute for achieving high-proficiency CO_2 removal. In the case of pure MXene, it was necessary to examine surface closures (–OH and –O functional groups) in order to reveal the CO_2RR steps of reaction. In the case of Mo_3C_2 and Cr_3C_2, energy inputs of 1.05 and 1.31 eV were required for CO_2 to CH_4 conversion in MXenes without surface termination groups (Fig. 6.38a). By terminating the Mo_3C_2 surface with O or OH, the energies were decreased even more to 0.35 and 0.54 eV, respectively. Thus, in comparison to unfunctionalized MXene, OH or terminated MXene facilitated CO_2 conversion. The photocatalytic reduction of CO_2 at V_o was theoretically investigated using first-principles DFT simulations on Ti_2CO_2, $Ti_3C_2O_2$, and V_2CO_2 [263]. The total energy barrier for CO_2 hydrogenation into HCOOH at O_v in Ti_2CO_2 SL was rather low (0.53 eV), compared to other products of reduction that necessitated superior kinetics. For instance, O_v on O-terminated MXene was an active location in CO_2RR, resulting in excellent HCOOH selectivity; hence, MXene NSs apply to

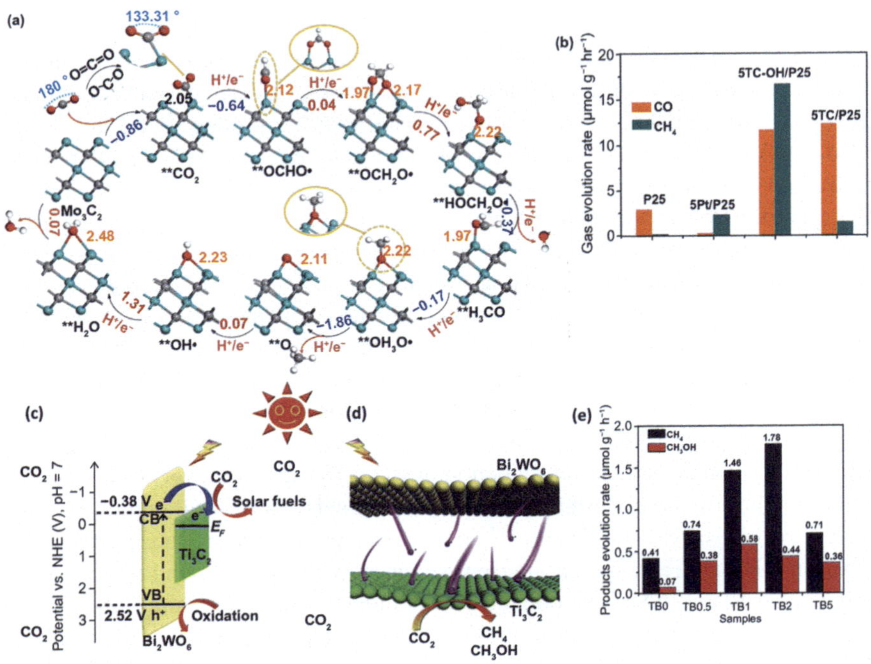

Fig. 6.38 **a** Calculated PBE/DFTD3 routes for CO_2 to $*CH_4$ and $**H_2O$ conversion catalyzed by Mo_3C_2. Gray, lilac, red, and white spheres represent C, Mo, O, and H atoms (Reproduced with permission from [268]), **b** CO and CH_4 evolution rates photocatalytically over P25, 5Pt/P25, 5TC/P25, and 5TC-OH/P25 (Reproduced with permission from [264]), **c** energy level diagram of Bi_2WO_6 and Ti_3C_2, **d** at the hybrids' interface, a photo-induced electron transfer mechanism occurs, **e** PCA of Ti_3C_2/Bi_2WO_6 (Reproduced with permission from [269])

a wide variety of applications. A simple mechanical mixing approach to combine surface alkalinized Ti_3C_2 MXene as co-catalysts with commercially available P25, resulting in a significant increase in photocatalytic CO_2RR [264]. Following surface alkalinization, 5 wt% $Ti_3C_2(OH)_2$-doped P25 (5T-COH/P25) demonstrates a significant increase in CH_4 release when compared to unmodified 5TC/P25 (Fig. 6.38b). As a result, CO_2 molecules were readily adsorbed onto the TC-OH (OH-termination) surface, resulting in the formation of activated CO_3^{2-}. Encouragement of charge separation, excellent electrical conductance, sufficient CO_2 adsorption, and activation locations on alkalinized MXene were thus essential variables in photocatalytic progress. These results demonstrated the critical role of the alkalinization of MXene's surface as a helpful co-catalyst of metal-free for synthetic photosynthesis. Likewise, Low et al. [265] investigated in situ generated TiO_2 NPs on Ti_3C_2 is conductive to produce TiO_2/Ti_3C_2 hybrids (TT-*x*, where *x* denotes the calcination temperature) by thermal annealing for CO_2RR CH_4 generation. After removing, -F functional groups at high temperatures, oxidation of Ti_3C_2 tuned with -O functional groups occur. The shape of TT550 and TT650 was obviously distinct from that of Ti_3C_2, but similar to that of rice crust (Fig. 6.38c). Ti_3C_2's high conductivity facilitates electron transport from TiO_2, while its unique rice crust analog shape and high active site density significantly increase PCA. Referring to the preceding study [264], it was difficult to compare the two $TiO_2/Ti_3C_2T_x$ photocatalytic systems directly as a result of a variety of synthesis procedures, the absence of the TiO_2 phase, the discrete variation in morphology, and the diversity of MXene surface modification. As a result, it makes sense to use AQY for CO_2RR as a standardized method for describing future experiment circumstances. Recent work by the same group of researchers has resulted in the formation of ultrathin 2D/2D Ti_3C_2/Bi_2WO_6 HJ hybrid nanocomposites (Fig. 6.38d) [266]. The 2D/2D HJ greatly improves charge carrier transit and partitioning, minimizing charge recombination, due to electrical coupling and strong physical effects. 2 wt% Ti_3C_2 improved Bi_2WO_6 NSs (TB_2) have the greatest CH_4 rate of release compared to other stoichiometry NSs (Fig. 6.38e). Intimate 2D/2D HJs with high interfacial contact surfaces lead to faster charge mobility than 0D/2D and 1D/2D HJs. HJs due to a reduction in the transfer of charge route [267]. Thus, such enticing research on parallel 2D/2D HJs opens up fresh opportunities in material science for the construction of stacked HJs for energy conversion applications such as electrocatalysis and photocatalysis.

6.4 Energy Storage Devices

6.4.1 Supercapacitors

Supercapacitors are a form of possible future energy storage device with a high power density (P_D), fast charge/discharge (numerous seconds), lengthy life cycle (example 10,000 cycles), low price, and reliable handling. As illustrated in Fig. 6.39, supercapacitors also have a far greater P_D (e.g., several thousands of Wkg^{-1}) than conventional capacitors and batteries, but with a lower energy density [270]. As a result, one of the major research challenges for supercapacitors is to establish an optimal balance of energy density and P_D. In other words, an ideal high-performance supercapacitor should have increased energy density without sacrificing P_D. The supercapacitor performance of electrode materials is influenced by two essential variables: surface area and electrical conductivity. With this in mind, extensive attempts have been made to develop progressive electrode materials via a variety of strategies, including engineering of surface and electrode materials functionalization, and the production of compounds employing highly electrically conductive materials are all examples of surface engineering [270, 271]. Activated carbon was one of the first commercially available supercapacitor electrode materials. Activated carbon utilizes the standard electric double-layer capacitor (EDLC) technology to store charges [272]. Depending on the preparation and activation techniques utilized, activated carbon can be generated as a porous-like structure with a very high surface area (3000 m^2/g) [273]. Despite their large surface area, the particular capacitance of activated carbon is

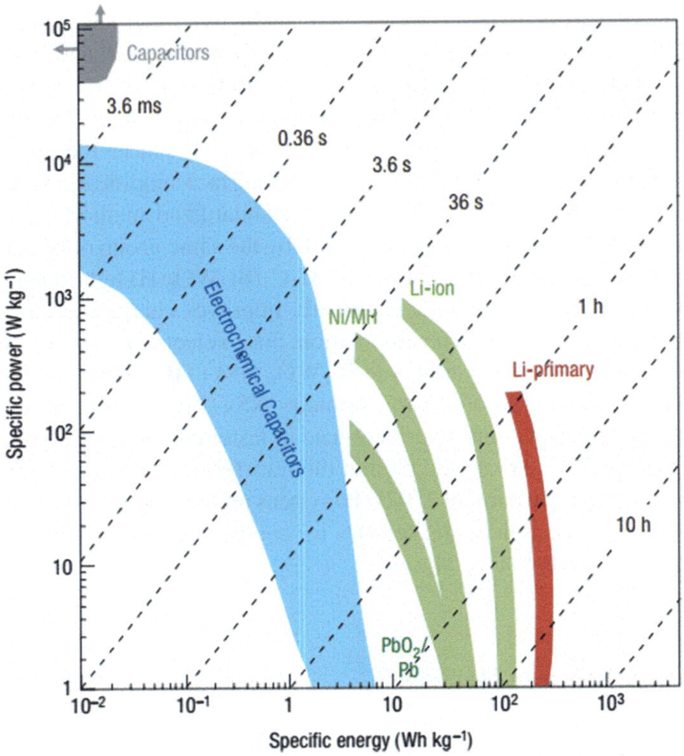

Fig. 6.39 An example of a Ragone plot of conventional energy storage device (Reproduced with permission from [270])

low (e.g., 10 F/cm^2) and much smaller than the theoretical value of capacitance of EDLCs (1525 mF/cm^2) [274, 275]. Numerous functional groups of surface formed during the process of activation of activated carbons further contribute to their low electrical conductivity and uneven electrochemical performance, when organic electrolytes are used, this is especially true [276, 277]. Other EDLC materials, like CNTs [278] and activated carbon fibers [279], are often costly and complex to fabricate, limiting their practical use. For example, CNTs are often produced through the CVD technique, which requires further purification to eliminate metallic catalysts, CNTs as electrodes of supercapacitors are becoming more expensive. Electrode materials with faradaic redox activity, in addition to EDLC materials, such as RuO_2 [280] and MnO_2 [281] are among the most prevalent supercapacitor materials are those that store charges on the electrode surface via a faradaic redox reaction and contribute to pseudocapacitance.

Although this type of electrode material may provide significantly greater capacitance than EDLCs, due to its low electrical conductivities, it has poor cycle stability. The high cost of RuO_2 is another barrier to everyday application. Generating nanostructured faradaic electrode materials with a low cost, active sites high density, an easy access surface area, and a short electrolyte ion diffusion path, similar to EDLC materials, is potential techniques to creating supercapacitors of high performance. When creating such a method, the price, scalability, and repeatability of these nanostructure materials are also taken into account. For supercapacitors, there has been a lot of research into ultrathin 2DMs [282]. As previously stated, the large surface area of electrode materials and electrical conductivity had a major effect on their supercapacitor performance. In theory, 2DMs with a single or FL have a large surface area that is more readily reachable from both sides than other nanostructured materials due to their ultra-thinness and large lateral dimensions. Certain 2DMs, such as GR and MXenes, also have very high mobility of electrons contributing to their high-rate performance, making them attractive supercapacitor materials. We will discuss current developments in developing 2DMs having lowest thickness for supercapacitors in this part.

6.4.1.1 Graphene-Based Supercapacitors

GR is a typical 2DM with a large area of surface (up to 2630 m^2/g) and high mobility of charge (~ 10,000 $cm^2\ V^{-1}\ s^{-1}$) (see Chap. 3), making it ideal for supercapacitors [283]. Due to its unique characteristics and special structure, its theoretical capacitance is up to 550 Fg^{-1} [284, 285] exceeding the theoretical capacitance of many other types of carbon materials. Furthermore, except for vdW contact between its sheets, GR appears to have a great proclivity for aggregation and the formation of a dense graphitic structure, resulting in the loss of some of the distinctive characteristics associated with GR's 2D shape, such as large surface area. The combination of GR also reduces the efficiency of mass transfer and the available surface area for the electrolyte, resulting in low rate performance. In other words, the rapid charge/discharge rate and high P_D that characterize GR-based supercapacitors are degrading owing to GR aggregation. As a result, many attempts have been devoted recently to resolve the GR restacking problem and to develop high-performance GR-based supercapacitor electrodes. These approaches include the following: (i) fabrication of GR electrodes with a variety of porous morphologies, including foams, fibers, spheres, networks, and hydro/aerogels [286, 287]; (ii) incorporation of a variety of "spacer" materials, including carbon nanomaterials, [288, 289] nanodiamonds, [290] and metallic nanocrystals, [291] to detached GR sheets within electrodes of thin-film; and (iii) GR preparation [292]. In contrast to stacked GR films, the surface area of the porous GR electrodes is greater (e.g., 850 m^2/g for the 3D GR linkages prepared via CVD [293]), provides additional electrolyte's pathways ion transport, and facilitates their contact to the surface of each GR sheet, resulting in improved supercapacitor performance. Using polystyrene as a template, a macroporous GR sheet was fabricated and then used as a supercapacitor electrode [294]. At a 1 A/g current density, the resulting porous electrode of GR had a particular capacitance of 202 Fg^{-1} this was higher than the stacked GR film electrodes (93 Fg^{-1}). The porous GR electrode retains a specific high capacitance of 97.7% of the value recorded at 0.5 Ag^{-1} even at a charge/discharge current density of 35 Ag^{-1}. Conversely, another intriguing method for fabricating high-performance supercapacitor electrodes is to design IP pores on GR sheets. These IP pores on GR sheets enable ion transfer via the layered GR film's top layer to its lowest layer, hence increasing the rate capability. As an illustration, Ruoff et al. [13] described IP porous sheets of GR were produced by chemically activating GO sheets at 800 °C with KOH, resulting in the highest surface area of 3100 m^2/g, well over the theoretical value of GR (2630 m^2/g) [295]. The regular supercapacitor constructed from these porous GR sheets has a particular capacitance of 150 Fg^{-1} having current density of 0.8 Ag^{-1} of TEA BF4/AN, an organic electrolyte. A method for fabricating IP porous sheets of GR for supercapacitors that is significantly more convenient was discovered [296]. H_2O_2 was used as an etchant in this study, and it reacted with GR sheets in a hydrothermal tank to produce GR hydrogels with IP porous sheets of GR. The specific capacitance of the symmetrical supercapacitor made from these porous GR sheets is 298 Fg^{-1} at 1 Ag^{-1} in the electrolyte of EMIMBF4/AN. Other types of nanoporous sheets of GR were also produced as possible building blocks for GR-based supercapacitors [292, 297]. It's realistic to anticipate that significantly improved performance of supercapacitors will be acquired by improving the relationship between the IP porous GR sheets' efficiency and their intrinsic qualities, such as size of pore and distribution, wettability, and functional groups of surface. Porous GR electrodes manufactured employing the aforementioned methodologies can successfully solve the restacking problem of GR sheets and make electrically conductive networks for quick charge transfer by utilizing GR's unique qualities, such as high surface area and electrical conductivity. Due to the porous nature of GR, these materials often have a low volumetric energy density and P_D. As a result, fabricating high-performance GR electrodes with a high density of packing remains a challenge for small energy storage and conversion devices.

High Volumetric Capacitance Graphene-Based Materials

In addition to particular gravimetric-volumetric capacitance is a critical performance metric for determining the viability of supercapacitor [298]. After all, materials with a high SSA are often comprised of shaggy sheets and have a low density of packing $<$ 0.5 g/cm^3, leading to capacitance having low volume and energy density. To address this challenge, Yang et al. [299] compressed chemically converted GR hydrogel films in the existence of a non-impulsive electrolyte to produce flexible liquid electrolyte-mediated chemically converted graphene (EM-CCG) films with a high density of packing $\sim$ 1.33 g/cm^3 (Fig. 6.40a, b). The sheets of GR in the films are tightly packed almost face to face, and the packing density can be easily modified by adjusting the volatile to non-volatile liquids ratio (Fig. 6.40c, d). More significantly, even at a high density of packing, EM-CCG films formed an unbroken ion and transport of electron network. As a result, it has been reported that EM-CCG film-based supercapacitors can attain a volumetric energy density of $\sim$ 60 Wh/L is comparable to that of lead-acid batteries. To address the issue of porous GR with low density of packing, Duan et al. [296] established a free-standing holey graphene framework (HGF) in a single

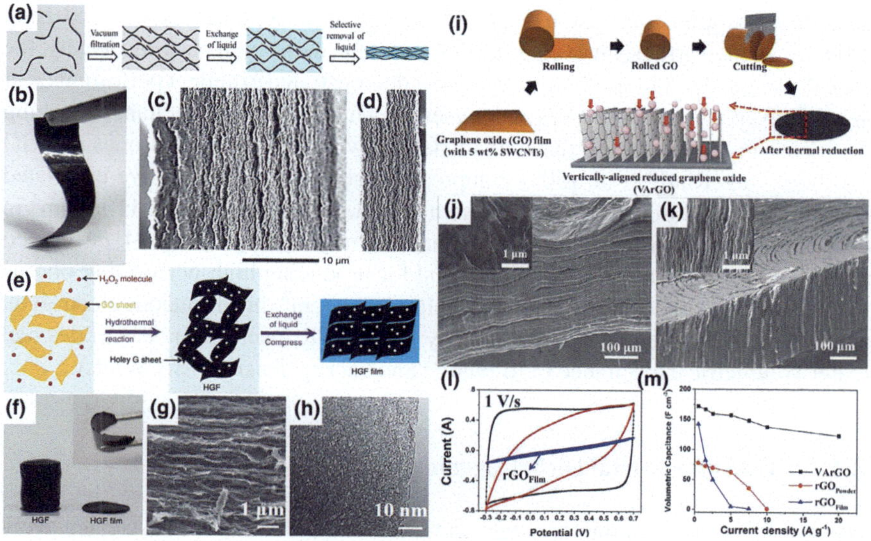

Fig. 6.40 **a** EM-CCG films fabrication from dispersion of chemically converted GR is schematically depicted, **b** photograph showing film flexibility, **c**, **d** cross-section SEM images of EM-CCG films (Reproduced with permission from [299]), **e** preparation procedure of HGF films and HGF schematic, **f** photograph of before and after mechanical compression of HGF, **g** cross-sectional SEM image of the HGF films, **h** holey GR sheets in HGF (TEM image) (Reproduced with permission from [296]), **i** the manufacture of vertically aligned rGO electrodes and the diffusion of electrolyte ions along vertically aligned rGO electrode sheets are depicted schematically, **j**, **k** cross-sectional SEM images of film electrodes of rGO and vertically aligned rGO electrodes, **l** cyclic voltammetry curves, **m** volumetric capacitance as a function of current density at electrodes (Reproduced with permission from [302])

step by simultaneously etching nanopores in GR with H_2O_2 and self-assembling GR into a 3D network structure and then using a hydraulic press to compress the HGF to synthesize a free-standing compact of HGF film (Fig. 6.40e, f). HGF films have a packing density ~ 60 times that of bulk HGF (Fig. 6.40g, h). The HGF films electrode's numerous pores of IP of a few nanometers form an effective electron and ion transportation networks, and high density of packing resulted in a high volume capacitance of 212 F/cm^3 in an electrolyte (organic) and an outstanding volumetric energy density of ~ 49 Wh/L when fully packaged. Likewise, Duan's group [300] synthesized holy GO that could be solution-processed holey GR. In comparison to nanopore-free GO, holey GR featured an abundance of IP nanopores that facilitated diffusion of ions and access to the surface of GR. As a result, in a 2M $EMIMBF_4$ electrolyte, an HGP-based supercapacitor may exhibit increased specific capacitance and an improved performance rate. Markedly, GR and its counterpart 2D-NS have an ultrathin plane structure that enables a well-designed planar supercapacitor this facilitates rapid flow of electrolyte ions along the 2D planes, therefore reducing ionic diffusion resistance [301]. Yoon et al. [302] were inspired by this and devised a simple method for fabricating very dense vertically aligned rGO using hand-rolling and cutting operations, as seen in Fig. 6.40i. 1.33 g/cm^3 of support was provided using vertically aligned arrays, an open-edge GR structure, and a high packing density, the vertically aligned rGO-based supercapacitors demonstrated excellent performance of capacitor even at a rate of 20 V/s and presented a high volume capacitance reaching 171 F/cm^3 is significantly greater than rGO powder and film electrodes of rGO (Fig. 6.40j–m). Ruoff et al. [303] compressed activation of microwave exfoliated GO-based material of electrode to obtain a volumetric capacitance of 110 F/cm^3 in two-electrode systems by utilizing an organic electrolyte. Yang et al. [304] developed an easy evaporation-induced drying method for producing high-performing porous microstructures with a high density of packing 1.58 g/cm^3. In an aqueous electrolyte, supercapacitor produced with high-density porous GR macroforms electrodes had greater volumetric capacitance (376 F/cm^3) and energy density (13.1 Wh/L).

6.4.1.2 Transition Metal Dichalcogenides

Because of their semiconductor nature and metallicity, 2D TMDCs have extensively been studied in the recent past for applications in hydrogen evolution reactions, opto-electronics, energy storage, and sensors [305, 306]. Due to the interlayer spacing of TMDCs, form ideal electrode materials for allowing electrolyte ions easy contact with their surface. Among this group of materials, Because of its greater intrinsic conductivity and theoretical capacity to graphite, MoS_2 is the most studied [307–309]. Soon and Loh produced MoS_2 formula films that are edge-oriented for micro-supercapacitor applications and observed that their performance was found to be comparable to that array electrodes of CNT [310]. Choudhary et al. [311] synthesized high-performance MoS_2 thin-film supercapacitor electrodes with an exceptional capacitance of ~ 330 Fcm^{-3} and capacitance retention of greater than 97% after 5000 charging/discharging cycles. Additionally, MoS_2 electrodes with a variety of configurations and moderate performance have also been investigated with promising findings reported [312, 313]. It is widely established that MoS_2 contains three polytypes: 1T (metallic), 2H (semiconducting), both are metastable, and 3R (stable under normal circumstances). Energy storage requires an electrode with high electrical conductivity, making 1T MoS_2 a more appealing electrode [314]. NSs of MoS_2 that have been chemically exfoliated and have a great concentration of the metallic 1T phase have shown exceptional proficiency and values of capacitance having a range from ~ 400 to ~ 700 Fcm^{-3} in a range of aqueous electrolytes from a recent study [282]. Moreover, as demonstrated in Fig. 6.41, this material is shown to be suitable for high-voltage (3.5 V) function in non-aqueous organic electrolytes and had 5000-cycle stability. This is due to 1T MoS_2's inherent hydrophilicity and strong conductivity, in count to the capacity of the exfoliated NS to intercalate different cations [282]. In comparison to MoS_2, WS_2 has a greater inherent electrical conductivity and hence may be a better choice for electrochemical electrodes [315]. However, because of the restacking of the NS and the relative brittleness of WS_2, its future uses are limited [316].

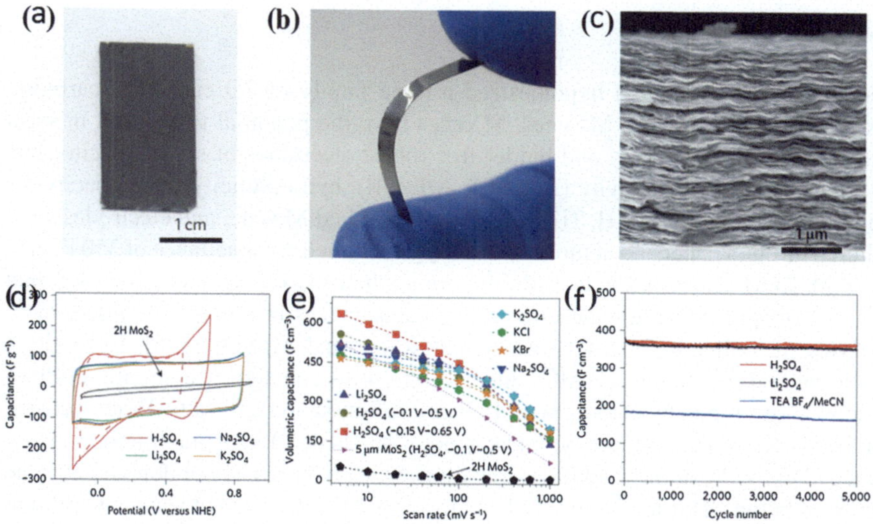

Fig. 6.41 **a** Chemically exfoliated 1T MoS_2 film vacuum filtered onto stiff glass schematic, **b** malleable substrate of polyimide, **c** electrode's side view from SEM presenting layered film prepared by restacking of exfoliated MoS_2 NS, **d** 1T phase's MoS_2 NS, **e** the 1T phase MoS_2 electrode's volumetric capacitance has changed over time, **f** retention of capacitance after 5000 cycles in AN with 0.5 M Li_2SO_4, H_2SO_4, and 1M $TEABF_4$ (Reproduced with permission from [282])

There is a dearth of literature on pure WS_2 capacitor electrodes. Highly conductive carbon materials were employed to prevent WS_2 from restacking to overcome the aforementioned issue and to improve the electrode's mechanical characteristics. In earlier work, designed a porous WS_2/SWCNT hybrid electrode with a sandwich-like structure by evenly inserting conductive SWCNT into lamellar WS_2 NS films [317]. The cyclic voltammetry findings showed one set of redox couples as a result of the electrochemical testing method converting W^{6+} to W^{4+} [318]. The WS_2/SWCNT compounds had the greatest capacitance i.e., 240.0 Fg^{-1} at a scan rate, i.e., 2 mVs^{-1}. SWCNT's distinctive lamellar structure substantially decreased WS_2 sheet restacking and increased electrolyte diffusion, while also improving the electrode's conductivity [317]. Newly, a very conductive 2D WS_2/rGO study demonstrated significantly increased conductivity and connectivity in WS_2 networks [319]. VS_2 is another member of the TMDC family, unlike WS_2; however, ML is metallic and possesses a spin-polarized ground state, making it ideal for use as IP supercapacitor electrodes [320]. The Xie group [321] established a novel ammonia-assisted approach for massive VS_2 flakes are exfoliated into ultrathin VS_2 NS with less than five S–V–S layers. The VS_2 thin films are extremely conductive were successfully constructed for use as electrodes in IP supercapacitors, reaching a particular capacitance of 4760 μFcm^{-2}, and even after 1000 cycles, there are no discernible decreases. Also, other TMDCs such as TiS_2 and CoS_2 have been investigated as supercapacitor electrodes [322].

6.4.1.3 MXenes

Gogotsi et al. [153, 323] hypothesized a large family of 2D early TM carbides, as well as C_3N_4, dubbed MXenes. MXenes have the potential to be used in great capacitance, mechanically and binder-free robust electrodes of supercapacitor due to their metallic conductivity (similar to ML-GR), hydrophilicity, and capacity for ion intercalation [207, 324]. $Ti_3C_2T_x$ is a well-studied MXene, and electrodes were filtered through paper and demonstrated a large volumetric capacitance of 350 Fcm^{-3} in 1 M KOH electrolytes. In addition, comprehensive research has revealed that $Ti_3C_2T_x$ displayed significant intercalated capacitances in a range of watery electrolytes including K_2SO_4, $Al_2(SO_4)_3$, $Al(NO_3)_3$, and $MgSO_4$ [325]. Recently, a safer method for etching Ti_3AlC_2 into $Ti_3C_2T_x$ films containing ammonium bifluoride was developed by etching with HCl acid solution and LiF salts. When verified in 1M H_2SO_4, The volumetric capacitance (900 Fcm^{-3}) and gravimetric capacitance (246 Fg^{-1}), of these additive-free films were both high, about double the value reported before. This technique is advantageous for the insertion of water and lithium ions and imparts a texture that is clay-like to the material that enables it to roll into free-standing films [326]. As previously stated, the Ti_3C_2 layers produced are often ended with F or/and OH groups, resulting in a greater diffusion barrier and reduced conductivity. However, when delaminated MXenes are employed as electrodes in supercapacitors, aggregation typically results in a significant reduction in specific area of surface and capacity. An approach for resolving the above issue is to incorporate conductive C-compounds. Gogotsi's further study demonstrated that three different sandwich-like films types, $Ti_3C_2T_x$/onion-like C, $Ti_3C_2T_x$/SWCNTs, and $Ti_3C_2T_x$/GR, all had an increased volumetric capacitance of 397 Fcm^{-3}, 390 Fcm^{-3}, and 435 Fcm^{-3}, respectively. The results are higher than those obtained with pure $Ti_3C_2T_x$ (360 Fcm^{-3}) in 1M $MgSO_4$ [327].

6.4.1.4 Other 2D Materials

Apart from the materials listed above, additional types of 2D NS such as Fe_2O_3, Nb_2O_5, V_2O_5, $Na_2Ti_3O_7$, and $VOPO_4$ have been utilized for high performance in electrodes [328, 329]. Xie et al. [330] synthesized ultrathin $VOPO_4$ NS by ultrasonically exfoliating bulk $VOPO_4 \cdot 2H_2O$ in 2-propanol. The synthesized $VOPO_4$ was capable of approaching with a high areal capacitance of 8360.5 $mFcm^{-2}$ and an electrochemical potential window of 1.23 V. Yu's group has demonstrated an urchin-like $Na_2Ti_3O_7$ anode and a carbon cathode form a sodium-ion capacitor in the form of quasi-solid-state with a sodium-ion-conducting gel polymer electrolyte. The proposed device-generated 111.2 $Whkg^{-1}$ of energy and 11,200 Wkg^{-1} of power [331]. The ultrasonicated $VOPO_4$ NS demonstrated a specific capacity of 136 $mAhg^{-1}$ for devices of sodium storage [332]. Due to their beneficial electrochemical properties, the designed SIB with nanostructured $Na_2Ti_3O_7$ anodes and $VOPO_4$ cathodes has a high working voltage of around 2.9 V, a huge reversible capacity of 114 $mAhg^{-1}$, good cycle stability, and a high energy density of 220 $Whkg^{-1}$ [332].

Vajtai et al. [333] established an effective method for large-scale manufacturing of V_2O_5 NS and then used a freeze-drying technique to make a 3D architecture. The resulting extremely flexible V_2O_5 structures demonstrated a high specific capacitance of 451 Fg^{-1} in Na_2SO_4 electrolyte and capacitance retention of more than 90% after 4000 cycles.

6.4.2 Rechargeable Batteries

2DMs offer intriguing materials of electrode for rechargeable batteries owed to their high surface: volume ratio and excellent electrochemical properties. The majority of rechargeable batteries store and release energy via the alkali metal ions' intercalation and deintercalation. A greater interlayer separation results in fewer volume changes and a smaller intercalation barrier. Extensive work has been done to include extended 2DMs into LIBs, SIBs, LSBs, and other types of batteries. The findings so far indicate that opening 2DMs is a critical approach for enhancing the efficiency of batteries that are rechargeable. We will mostly cover their use in LIBs and SIBs in this section.

6.4.2.1 Lithium–Ion Batteries

Rechargeable LIBs have seen significant growth as an essential form of energy storage device in recent years due to their extended life cycle, high definite energy, and lack effect of memory [334]. Because electrode materials are important to LIB electrochemical performance, several research teams have been working on developing novel materials of an electrode with high storage capacitances of lithium and excellent conductivity of electron and ion [335]. Expanded interlayer distance 2DMs feature open architectures that allow for efficient storage and movement of ions, electrons, and mass, resulting in better outcomes when employed as electrode materials for LIBs. Additionally, increasing the interlayer spacing benefits strain relaxation and lowers the Li intercalation barrier. Thus, increasing the interlayer distance is an operative method for 2DMs to enhance their LIB performance. For example, Liu et al. [336] produced extended, restacked MoS_2 along the *c*-axis by an exfoliation and restacking procedure. The restacked MoS_2 demonstrated an original capacity of 800 $mAhg^{-1}$ at a current density of 50 mAg^{-1}, and a capacity of 750 $mAhg^{-1}$ after 20 cycles. In contrast, the capacity of charge of massive MoS_2 electrodes fell dramatically after 50 cycles, from 800 to 226 $mAhg^{-1}$. The formation of nanocomposites by introducing poly(ethylene oxide) (PEO) inside interlayer space of MoS_2, demonstrated a reversible capacity of > 900 $mAhg^{-1}$ at a current density of 100 mAg^{-1} [337]. The MoS_2 composites' cycle stability and rate of performance, on the other hand, were not adequate and needed to be improved further. A straightforward and efficient approach for fabricating 3D nanospheres of MoS_2 with ultrathin NS walls is shown in Fig. 6.42a [338]. The resulting 3D-oriented nanospheres of MoS_2 enhance the (002) crystal plane's d-spacing that can offer sufficient room for rapid intercalation

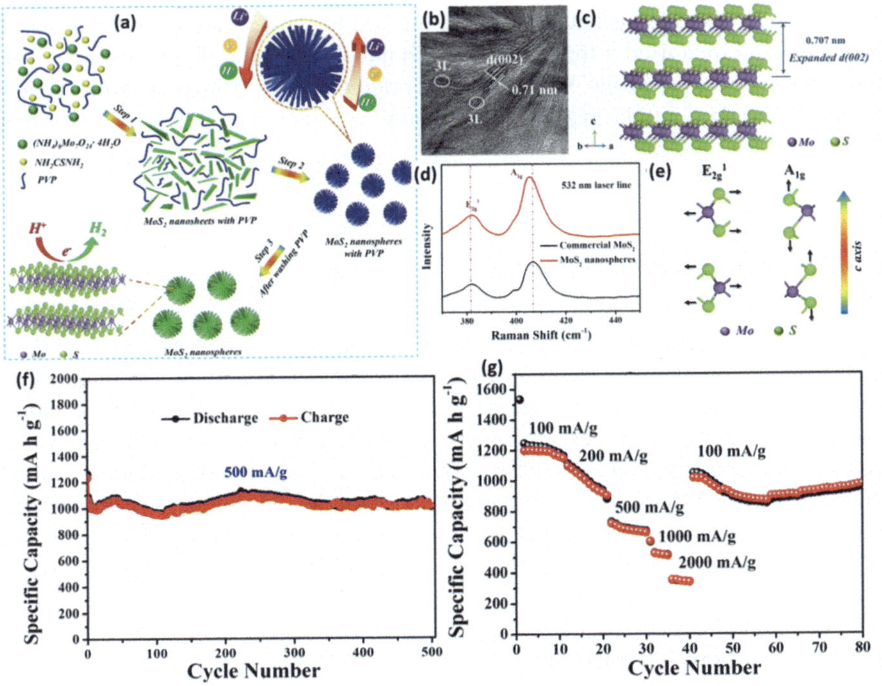

Fig. 6.42 **a** Diagrams illustrating the development of 3D-oriented nanospheres of MoS_2, **b** expanded MoS_2 (HRTEM images), **c** viewed from the [110] and [001] directions, a structural model of ultrathin MoS_2. The Mo (purple) and S atoms (green) are shown **d** expanded MoS_2 and commercial MoS_2 Raman spectra. **e** Atomic vibrations of the $E_{2g}{}^1$ and A_{1g} vibrational modes, **f** expanded MoS_2's cycling performance at a current density of 500 mAg^{-1}, **g** expanded MoS_2's rate capability (Reproduced with permission from [338])

of Li^+. As seen in Fig. 6.42b, c, the increased distance between the (002) plane and the (001) plane is approximately 0.71 nm ascribe to the presence of PVP throughout the creation procedure. The XRD pattern's shift of the (002) diffraction peak and the Raman spectrum's A_{1g} peak likewise demonstrated the interlayer distance expansion (Fig. 6.42d, e). As predicted, when these MoS_2 nanospheres were tested as LIB anode materials and demonstrated a high specific capacity of 1009.2 $mAhg^{-1}$ at density of current is 500 mAg^{-1} even after 500 cycles (Fig. 6.42f). MoS_2 nanospheres had outstanding rate performance due to their unique shape. The capacity was as high as 353.6 $mAhg^{-1}$ even at a density of current of 2000 mAg^{-1} (Fig. 6.42g) [338]. Despite the short orderings formed at the edges and deep surface defects, the regular structure of g-C_3N_4 features a large number of pyrrolic N "hole" defects that appear to enhance Li^+ adsorption and diffusion [339]. Zn_2GeO_4 NPs into g-C_3N_4 interlayer to sufficiently uncover the pyrrolic N "hole" defects for storage of lithium were determined [340]. Not only did the Zn_2GeO_4 NPs contribute great capacity but also acted as spacers, preventing restacking of the ultrathin layers of g-C_3N_4, resulting in a thicker interlayer and exposed voids with "hole" flaws for more Li^+ diffusion and storage. To begin with, the g-C_3N_4 in bulk form has been protonated in solution of HCL to facilitate layer exfoliation during ultrasonication. It served as a large-area scaffold for Zn_2GeO_4 NP hybridization, resulting in an intercalation structure. Taking advantage of the amazing effect of synergistic, hybrids of Zn_2GeO_4/g-C_3N_4 demonstrated an extremely reversible capacity of 1370 mAhg^{-1} at 200 mAg^{-1} after 140 cycles and an exceptional rate capability of 950 mAhg^{-1} at 2000 mAg^{-1}, revealing the intercalated structure's remarkable stability. In summary, increased interlayer distances in 2DMs can offer sufficient room for rapid Li^+ movement. Moreover, the increased interlayer spacing can relieve the tension and reduce the barrier to Li intercalation, leading to increased efficiency. A minor caveat is that intercalation may result in a drop in energy density is suggests that intercalating 2DMs with active components may become a trend.

6.4.2.2 Sodium–Ion Batteries

As a result of the high rate and insufficiency of lithium, it is extremely necessary to make new types of batteries with superior electrochemical performance at a cheaper cost. After lithium, sodium is the second-lightest and smallest alkali metal, with similar physical and chemical properties. Because sodium is abundant and inexpensive, as well as having a small redox potential (higher than lithium), for electric cars and grid-level energy storage, SIBs are a feasible alternative to LIBs [23]. Considering the fact that Li and Na are both alkali metals, it is frequently thought to be suitable to apply the visions obtained from the LIBs to the science of SIBs. Na^+, on the other hand, has a greater radius than Li^+ provide a direct effect on mass transit and storage throughout the electrochemical reaction. Arising from the fact Na^+ has a greater radius, many of the efficient LIB's anode materials, like graphite, are incompatible with SIBs. Although graphite has a regular layered structure having long-range, the interlayer distance is too small for the large Na^+ ion to pass through. Wang et al. [341] developed extended graphite (EG) as a better SIB anode material based on these findings. EG was a composite of graphite material makes using a two-step redox reaction that conserved graphite's regular layered structure having long-range, resulting in a large interlayer spacing. These characteristics favored the electrochemical intercalation of Na^+ ions. In addition to controlled oxidation, reduction processes may be used to alter the interlayer spacing of EG. It has been established that Na^+ could indeed reversibly be inserted into and removed from EG having interlayer distance of 0.43 nm after conducting a series of experiments. Electrochemical studies demonstrated that the EG has a great reversible capacity of 284 mAhg^{-1} at a density of current of 20 mAg^{-1}, 184 mAhg^{-1} at a current density of 100 mAg^{-1}, and retains 73.92% of capacity after 2000 cycles. As seen, EG had a capacity of 284 mAhg^{-1} at a current density of 20, and 184 mAhg^{-1} at 100 mAg^{-1}, 91 mAhg^{-1} at 200 mAg^{-1}, and 278 mAhg^{-1} at a current density of 20 mAg^{-1}, indicating EG's stability over a wide variety of charge–discharge conditions. The high interlayer spacing in TMDCs is advantageous for Na^+ storage. Furthermore, because of their poor electrical conductivity, their total performance, especially at high rates (kinetic

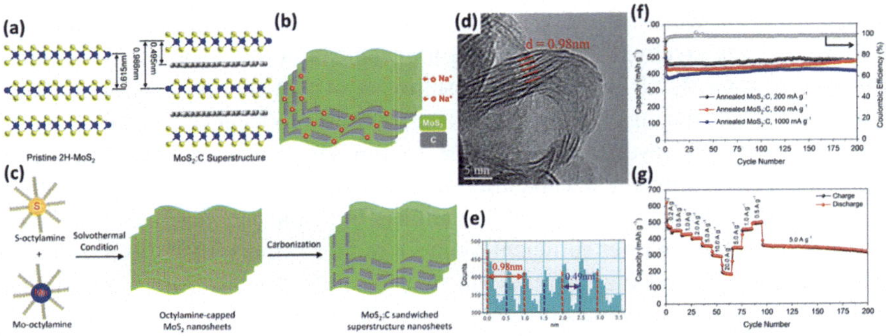

Fig. 6.43 **a** Expansion of interlayer space in pure 2H-MoS_2 and MoS_2/C models, **b** schematic representation of a superstructure of MoS_2/C with continuous C-MLs for insertion/extraction of Na^+, **c** growing the MoS_2/C superstructure NS shown schematically, **d** HRTEM images of superstructure of MoS_2/C, **e** calibration profile plot for measuring the spacing in (**d**), **f** cycling performance of electrode made up of MoS_2/C, **g** MoS_2/C electrode rate performance (Reproduced with permission from [342])

factors such as ion diffusivity and electron conductivity), is still far from satisfactory. As a result, opening TMDCs and adding conductivity-enhancing intercalates are potential approaches to resolving the problem. Lee et al. [342] described a unique hierarchical nanotube assembly method based on 2D superstructure NS composed of alternate MLs of MoS_2 and carbon. Models of the pristine 2H-MoS_2 and the MoS_2:C interlayer superstructure with improved interlayer spacing are shown in Fig. 6.43a. In comparison to previously reported MoS_2 and carbon compounds, the 2D superstructure exhibited a perfect contact of interface between MoS_2 and carbon with an increase between distances of layers, enabling quicker sodiation/desodiation kinetics (Fig. 6.43b). The interconnected superstructure of MoS_2:C NS was synthesized through a one-pot solvothermal production coupled with organic solvent carbonation in situ (Fig. 6.43c). As a result, the current superstructure of MoS_2:C exhibits a considerably increased (002) interlayer distance of up to 0.98 nm (Fig. 6.43d, e). Having current density of 200 mAg^{-1}, the as-made electrode produced a primary discharge capacity of 620 $mAhg^{-1}$ and a coulombic efficiency of 84%, and it could sustain a reversible discharge capacity of 477 $mAhg^{-1}$ after 200 cycles. Furthermore, at current densities of 500 and 1000 mAg^{-1}, the discharge capacities showed outstanding cycle stability (Fig. 6.43f). Figure 6.43g exhibits the MoS_2/C hierarchical nanotubes' remarkable rate behavior. At a very high current density of 5.0 Ag^{-1}, the MoS_2/C anode maintained a reversible capacity of 320 $mAhg^{-1}$ for 105 cycles without evident capacity fading, showing that the MoS_2:C inter-overlapped superstructure remained exceptionally robust during the sodiation/desodiation process. Along with graphite and TMDCs, BP is a host that has been intercalated. Cui et al. [343] fabricated a sandwiched FL BP sandwiched between GR sheets. The sheets of BP provided a small and efficient route of diffusion for sodium ions because to the increased interlayer distance given by GR intercalation. The hybrid materials demonstrated an exceptionally great capacity of 2440 $mAhg^{-1}$ at 50 mAg^{-1}, with

83% retention of capacity after 100 cycles. It was discovered that sheets of GR not only acted as a mechanical backbone and electrical conductor but also as an elastic buffer space to accommodate expansion (anisotropic) during cycling. Due to Na^+'s greater radius, the majority of anode materials with superior storage performance of lithium cannot be inserted by Na^+. The increased interlayer distance enabled by opening 2DMs enables electrochemical intercalation of Na^+, resulting in quicker sodiation/desodiation kinetics and improved salt storage enactment.

6.4.2.3 Other Batteries

Other than LIBs and SIBs, 2DMs with greater interlayer spacing might be used in LSB, MgRB, and LIB batteries. LSBs are among the most promising upcoming high energy density rechargeable batteries for all-electric and plug-in hybrid vehicles owing to their high capacity (1672 $mAhg^{-1}$) and energy density (2600 $Whkg^{-1}$) of sulfur compounds [344]. However, due to sulfur's low electrical conductivity and the diffusion of soluble lithium polysulfides intermediates that are formed during cycling, known as the polysulfide shuttle, practical applications are severely limited. Fortunately, 2DMs with increased interlayer spacing may be a suitable home for sulfur and decrease the shuttle effect. Sun et al. [345] produced the EG using an improved Hummers' technique served as an effective host and excellent conductor of the electronic current of sulfur cathode. A simple melt diffusion technique was used to fabricate a sulfur-expanded graphite nanocomposite. The larger interlayer space inside the EG can function as a container, trapping sulfur and polysulfides within the framework, increasing electrode electrical conductivity, and decreasing polysulfide dissolving in the liquid electrolyte. Thus, the electrode's life cycle and reversible capacity were increased, and the capacity is projected to reach 879.5 $mAhg^{-1}$ after completing 70 cycles at 280 mA g^{-1}. In addition to dendrite-free behavior of deposition for enhanced safety of device, availability of resource, great volumetric capacity (3833 $mAhcm^{-3}$), and small electrode potential (0.67 V vs. Li^+/Li), Mg metal anodes are appealing options for mobility of electrons and energy storage of grid [346]. MgRBs are effective and reliable battery technology, although they are restricted by the slow solid-state diffusion of divalent Mg^{2+}. Its divalent nature and tiny ionic size make it too polarizing for diffusion readily in most ion-intercalated materials that produce MgRBs useful for many applications. A widespread and effective method may be to open 2DMs to increase the diffusivity of Mg^{2+} in intercalation hosts having layer structure. Unlike Li^+ or Na^+, which requires only a little increase in distance between layers, Mg^{2+} requires an extensive boost to decrease the interaction between Mg^{2+} and 2DMs. Yao et al. [347] demonstrated the effectiveness of interlayer expansion in increasing the inherent Mg^{2+} diffusivity in 2DMs used as intercalation hosts (Fig. 6.44a). The necessary increase was accomplished experimentally by manufacturing adjustable interlayer distance PEO-MoS_2 nanocomposites. Chemical lithiation of commercially available MoS_2 to $LiMoS_2$ was followed by delamination into SL structures in aqueous PEO solutions. Three composites were obtained based on PEO content in solution: Res-MoS_2 with trapped water molecules in layers ($n$$MoS_2$/$n$PEO

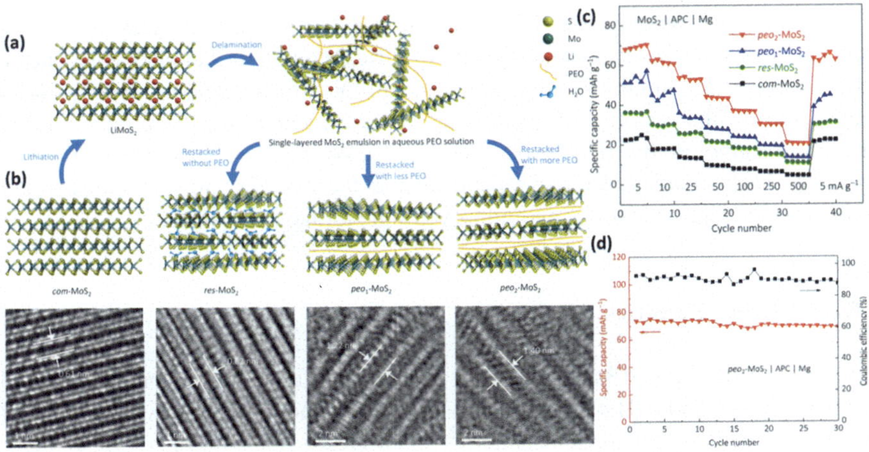

Fig. 6.44 **a** Synthesis of interlayer Li-expanded MoS_2 composites, **b** models of structure and corresponding TEM images (com-, res-, peol-, and peo-)MoS_2, **c** rate performance, **d** Peo2-MoS_2 cycling performance during deep discharge–charge at 5 mAg^{-1} (Reproduced with permission from [347])

= 1:1), peo1-MoS_2 with a modest quantity of PEO ($n$$MoS_2$/$n$PEO = 1:0.5), and peo2-$MoS_2$ with a maximum proportion of PEO ($n$$MoS_2$/$n$PEO = 1:1) (Fig. 6.44b).The composites' rate capability was compared in Fig. 6.44c. In Fig. 6.44d, the peo2-MoS_2's cycling performance under profound discharge–charge at 5 mAg^{-1} is shown. In this system, the capacity increased in the order peo2-MoS_2 > peo1-MoS_2 > res-MoS_2 > com-MoS_2 at all current densities tested, implying that the interlayer gap was kinetically limiting their electrode performance. The reversibility and stability of the Mg^{2+} storage in PEO-pillared MoS_2 were studied using profound intercalation/deintercalation at a low current density of 5 mAg^{-1}. 94% of the original capacity was retained, after 30 cycles, indicating that the composite exhibited high structural stability during the intercalation/deintercalation of Mg^{2+} process. Finally, interlayer extension is a very successful method for enhancing cation diffusion kinetics in 2DM hosts; it may also be utilized to alleviate strain produced by cation insertion. The open structure enables the efficient transit and storage of ions, electrons, and mass, enabling energy storage of high-performance. However, in some situations, 2DMs enlarged by intercalants lack electrochemical activity, lowering the energy density of batteries. Intercalating 2DMs with electrochemical active materials is one way to address this issue, creating a synergistic impact. These discoveries lay the way for future growth of a new class of sophisticated battery systems capable of delivering high P_D and energy density.

6.5 Wearable Energy Harvesting and Storage Devices

6.5.1 *Flexible Supercapacitors*

Due to their high P_D and cyclability, supercapacitors are potential energy storage technologies for rapid power supply. 2DMs have gained widespread acceptance as emerging materials for applications such as wearable supercapacitors. The flexibility of 2DMs on PET/PI/paper/CC/PDMS substrate is a critical technique for achieving 2D materials-based wearable supercapacitors (2DM-WSs) [348]. Usually, 2DMs may be incorporated directly or indirectly into flexible substrates using solution-based assembly methods and the resulting hybrid structures are being utilized as stretchy electrodes for 2DM-WSs (Fig. 6.45). Exfoliated MXene NS may be immediately merged with a flexible substrate of PET by coating of spray [349] or ink printing [350]. Conversely, an MXene film can be produced and subsequently transferred to the target substrate by vacuum filtering [351]. This section will describe achievements of 2DM-WSs' state of the art, such as cyclic stability and capacitance. In addition, the wearable supercapacitor's flexibility (bendability and stretchability) will be summarized and addressed.

For instance, Wang et al. [353] used ball milling to exfoliate 2D-conjugated MOF based on phthalocyanine ($Ni_2[CuPc(NH)_8]$) with a lateral dimension (160 nm) (Fig. 6.46b) and a mean thickness of 7 nm with 10 layers. The device (Fig. 6.46b) was fabricated using a $Ni_2[\text{-}CuPc(NH)_8]$/GR electrode and a PVA/LiCl electrolyte gel on a substrate of polyimide. It exhibits excellent cyclic stability (91.4% retention of capacitance after charging/discharging of 5000 cycles) and high capacitance in the area of 18.9 $mFcm^{-2}$ at a J_{sc} of 0.04 $mAcm^{-2}$. Xiao et al. [354] devised a one-step interdigital mask-assisted microsupercapacitor manufacturing method (Fig. 6.46c). The interdigital hybrid electrode based on layer-by-layer phosphorene deposition and GR may be immediately produced and moved toward a flexible substrate of PET as the flexible electrode using a tailored interdigital mask and vacuum filtering. In addition to the excellent areal capacitance of 9.8 $mFcm^{-2}$ at 5 mVs^{-1} that this simplified production process allows for, the manufactured flexible (Fig. 6.46a, c), displays good stability even when folded (Fig. 6.46c). In an ideal world, 2DM-WSs would have a high energy density and a high P_D. In fact, there is a cost associated with this.

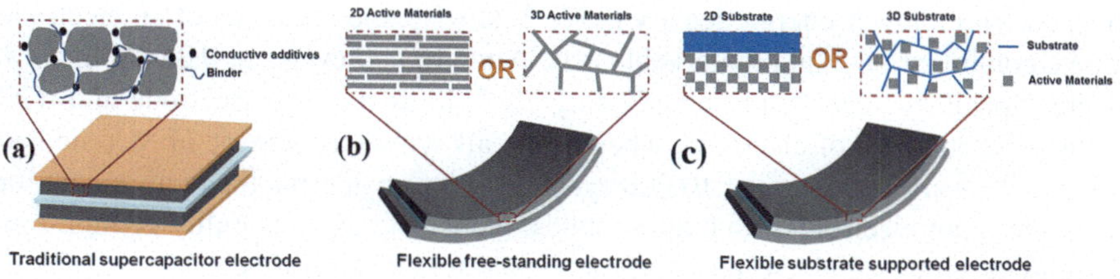

Fig. 6.45 Schematic is demonstrating differences of structures between **a** conventional and **b**, **c** supercapacitor's flexible electrodes (Reproduced with permission from [352])

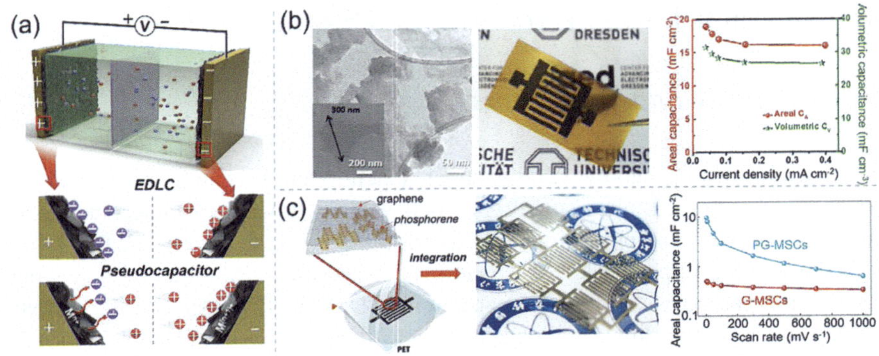

Fig. 6.46 Advances in 2DM-WSs achievement: **a** process of charging in basic supercapacitor and pseudocapacitor of EDLC schematic (Reproduced with permission from [348]). **b** Exfoliated MOF with ball milling (TEM image), invented flexible electrode (photo), and performance of device, respectively (Reproduced under the terms of CC BY 4.0 license [353], https://creativecommons.org/licenses/by/4.0/), **c** vacuum filtration assisted GR/phosphorene's electrode fabrication schematic, picture of interconnected supercapacitor pack in series, and performance of the device (Reproduced with permission from [354])

Increasing energy density is often accomplished through the use of pseudocapacitive materials. Pseudocapacitive materials, on the other hand, bring chemical reactions into the charging/discharging processes, resulting in a lowered charging/discharging rate and, as a result, a lower P_D. Orangi et al. [355] proved that maximum energy density (51.7 μWhcm^{-2}) can be obtained at a relatively low P_D (5.7 mWcm^{-2}) when $Ti_3C_2T_x$ is printed on a PET substrate. The Ni_2[-CuPc(NH)$_8$]/GR/PI system with a small energy density (1.1 μWhcm^{-2}) may reach the maximum P_D (168 mWcm^{-2}) [353]. To achieve synergy between high energy density and high P_D, persistent revolution in the discovery of material, structural design, and treating technologies is required. Depending on the energy storage technique, a potential approach may be creating, improving electronic conductivity, and opening channels for the rapid movement of ions and electrons in materials and structures [356]. Another significant aspect affecting the commercialization of 2D-WSs is the device's cycle stability. In general, cycle stability is dependent on 2DMs, substrate, electrolyte, mechanism of energy storage, and device design. Due to the absence of redox reaction and volume change in EDLC electrode materials typically have a longer cycling lifespan with a lower energy density [357]. Likewise, the materials of an electrode can be combined to obtain remarkable acts. After 10,000 cycles at 0.2 mAcm^{-2}, % of the capacitance can be preserved using $Ti_3C_2T_x$ as the $-$ve electrode and rGO as the + ve material of electrode, whereas the all symmetric device of MXene had faded to 75% after completing 10,000 cycles [349]. Considering the requirement for quick charging electronics and automobiles, increasing cycle stability of high-rate charging/discharging electronics and cars would remain a priority.

6.5.1.1 Flexible Stability of Wearable Supercapacitors

Surprisingly, 2DM-WSs are intended to be conformally connected to the human interface and capable of accommodating different deformations throughout daily activities. Thus, it is important to keep 2DM-WSs exceptional electrochemical performance under deformations caused by cycles such as bending and stretching. Resulting in high 2DMs mechanical toughness as 1 TPa for GR [358], ~ 3–75 GPa for MXene [359], the low stiffness of polymer, and nanomaterial-substrate interfacial microscopic cracks because of poor interfacial strength with 2DMs and the polymer substrate. Currently, 2DM-WSs are tested for flexibility by bending and stretching. The term "flexible stability" refers to device's capacity to retain its performance (capacitance retention) during bending or stretching cycles. Figure 6.47a illustrates a typical bending cycling test. The device's performance will be evaluated for various bending radius/angle combinations. These electrodes were used to test the device's performance at 0–180° bending angles utilizing TMOs (MoO_3, MoO_2, and RuO_2)/SWCNT complex electrodes (Fig. 6.47b) [360]. Aside from outperforming reported earlier supercapacitors based on TMO in terms of particular capacitance, the researchers found that devices produced with the TMOs NS/SWCNT composites are extremely stable throughout a range of angles of bending from 0 to 180° (Fig. 6.47c). To improve pattern adhesion to the polyimide substrate, PVA may be used with GR ink to obtain 94.6% retention of capacitance after 10,000 cycles of bending from 0 to 180°. Hyun et al. [361] conducted a bending test on the printed GR supercapacitor

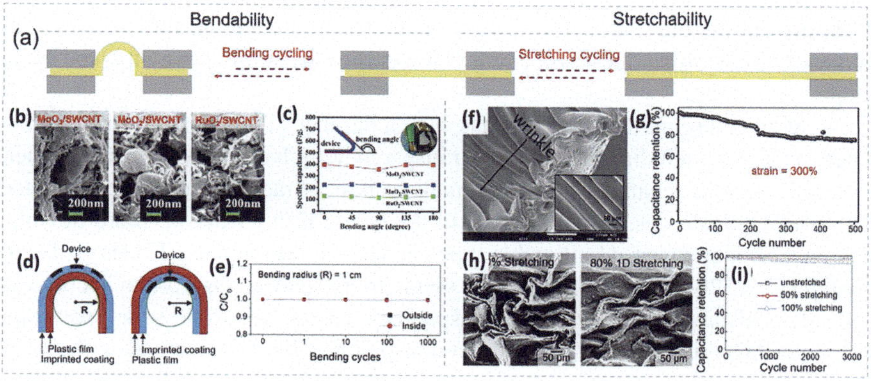

Fig. 6.47 2DM-WSs (flexible stability measurement): **a** schematic demonstrating cycling bending and stretching (Reproduced with permission from [348]), **b** different composites of TMOs/SWCNT (SEM images) **c** performance of matching flexible supercapacitor devices at various bending angles (Reproduced with permission from [360]), **d** outside and inside bending schematic, **e** relative particular capacitance of bending devices outside and inside the substrate during 1000 cycles at a bending $R = 1$ cm (Reproduced with permission from [361]), **f** PEDOT: PSS/rGO/Eco wavy like structure (SEM images), **g** for 500 cycles of stretching and releasing, flexible stability was tested (Reproduced with permission from [363]), **h** crumpled MXene nanocoating's top-down SEM images at 0 and 80% 1D states of stretching, **i** stretchable MXene electrodes have flexible stability (Reproduced with permission from [364])

in two distinct orientations (Fig. 6.47d). The devices' specific capacitance remained stable after 1000 cycles of bending at a 1 cm radius (Fig. 6.47e). Inappropriately, due to the varied sample thickness, bending radius and angle (0°–180°), and cycle number (100–10,000), there is no standard measurement. A rigorous comparison is difficult to get. This also emphasizes the critical necessity to develop a standard that can be generally accepted in this sector. Instead of localized deformation like bending, stretching causes global strain on 2DM-WSs and is primarily suitable for devices with polymer substrates. Skin and flesh simulation requires stretchability. To evaluate a device's performance, it is usually subjected to a cyclic stretching/releasing test. New designs and materials with increased stretchability despite maintaining high specific capacitance have been developed in significant amounts. Until now, 2DMs deposition onto substrates that are prestretched to fabricate a structure like buckled or wavy has been a promising and effective approach for achieving great stretchability [362]. Recently GO coated onto a prestretched substrate of Ecoflex sponge and then reduced it using hydriodic acid vapor [363]. A brief periodic wavy structure was fabricated by covering PEDOT: PSS and PVA/H_2SO_4 gel and liberating the prestretch (Fig. 6.47f). The constructed devices exhibit remarkable stretching stability, retaining 85.1% of capacitance at 300% strain. 75.5% of the capacitance is preserved after 500 cycles of 300% stretching (Fig. 6.47g). Chang and associates used prestretched acrylic tapes to produce crumbed MXene nanocoatings (Fig. 6.47h) [364]. Due to the accordion-like structure's ability to be reversibly folded/unfolded by IP deformations, it can withstand significant strain. The constructed gadget exhibits an extremely high level of electrochemical stability. After 3000 charge–discharge cycles, the specific capacitance (93%) may be intact under 100% strain is equivalent to the retention of 98% in an unstrained sample (Fig. 6.47i). Likewise, the test of stretching aches from a deficiency of a standardized test procedure that specifies the strain, geometry, and cycle number.

Being an emerging material, GR has such a number of advantages for wearable supercapacitors, including great optical transparency, high mobility of charge carrier, and exceptional mechanical flexibility, making it an attractive candidate to replace ITO or FTO electrodes. In comparison to the brittle ITO, GR is far more flexible, making it a better material for flexible supercapacitors. Additionally, ITO is visible to UV-light ($\lambda < 320$ nm) and infrared light owing to free carrier absorption; in comparison, ML-GR has a 97.7% optical transparency across the whole solar spectrum, with every layer adding a ~ 2.3% drop in transparency. For GR, increasing the layer count results in a loss in optical transparency but an increase in conductivity; hence, whenever it comes to GR layers, there will be a balance. GR's primary constraint is its absence of an inherent E_g. This restriction can be overcome by functionalization [29]. Thanks to their exceptional transparency, chemical stability, and high carrier mobility, GR and rGO were used as DSSCs, HJ Si supercapacitor, and OSCs (Fig. 6.48a–c) [365, 366]. Furthermore, GR was already utilized as the photoactive layer of OSCs for harvesting/accepting of light because its 2D structure and high ratio of aspect allow the creation of the appropriate donor/acceptor interface for charge separation and transfer [367, 368]. Because functionalized GR has lattice defects' high density and O_2-containing functional groups, it was utilized as a CE in DSSCs

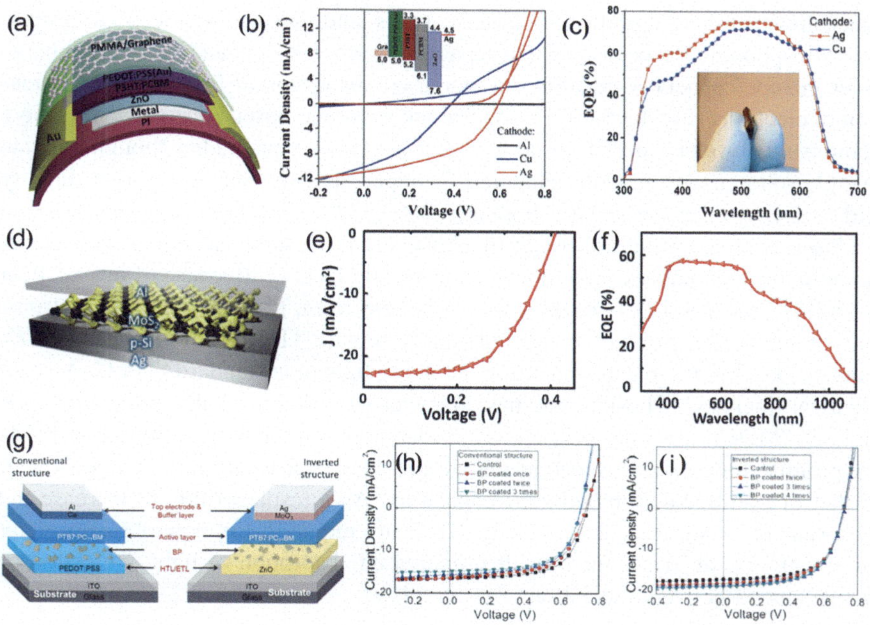

Fig. 6.48 **a** Schematic of wearable OSC structure, **b** electrical characteristics and **c** external quantum efficiency of supercapacitors (Reproduced with permission from [365]), **d** illustration of MoS_2/p-Si HJ supercapacitor structure, **e** *J–V* curve, **f** external quantum efficiency (Reproduced with permission from [110]), **g** schematic of organic supercapacitors structures with and without incorporation of BP, *J–V* characteristics of OSCs **h** without and **i** with BP (Reproduced with permission from [377])

[369]. The remarkable performance exhibited is attributed to the doping-prompted shift in the GR's chemical potential that enhances GR's carrier density and hence the cell's intrinsic potential [370]. The supercapacitors made of GR and its derivatives have been claimed to have a PCE of up to ~ 10.34%. As with GR, the E_g of some 2D TMDCs are layer-dependent (ML and bulk MoS_2 have E_g ~ 1.8 direct and indirect 1.3 eV, respectively), and this property enables them to be linked to a variety of donor/acceptor systems in OSCs. In comparison to GR, semiconducting 2D TMDCs such as MoS_2, WS_2, $MoSe_2$, and WSe_2 contain more appropriate E_g and are anticipated up to 5–10% of incident sunlight is expected to be absorbed, which is roughly one order in magnitude greater than Si or GaAs having same thickness (below 1 nm). In Schottky or p–n junctions, 2D TMDCs have been used as the photoactive layer (Fig. 6.48d–f) [107, 110, 112]. Arising from their high catalytic activity and hole-extraction capacity, 2D TMDCs have been utilized as the CE (catalytic) in DSSCs and as the layer of hole extraction in OSCs [371–373]. However, it should the noted that although supercapacitors based on 2D TMDCs are anticipated to have great solar ECEs, the reported actual PCEs (0.1–8.11%) are still modest in comparison to that of typical silicon supercapacitors (15–30%). Phosphorene and its variants have characteristics that are intermediate between GR and 2D TMDCs, including great optical

transparency, strong mobility of charge carriers, and outstanding electrical properties. Phosphorene has been shown to act as an absorber of light and transfer of charge layer in supercapacitors, resulting in increased absorption of light and recombination of electrons (Fig. 6.48g–i) [374]. The phosphorene-based supercapacitors were demonstrated to have a PCE of up to ~ 8.25%. Theoretical studies though indicate that phosphorene is a potential material for supercapacitors due to its great stability and efficiency of conversion (as high as ~ 20%) [375, 376]. However, experimental evidence demonstrating the usage of phosphorene in supercapacitors with correspondingly remarkable characteristics is currently lacking. Because of its excellent flexibility, transparency over the entire solar spectrum, high mobility of electrons, and chemical stability, GR could potentially replace ITO electrodes. Tunable E_g enables 2D TMDCs to captivate more incoming light than standard Si or GaAs with identical thickness. Phosphorene has characteristics that are halfway between GR and 2D TMDCs, allowing it to act as a bridge between the two. All of these 2DMs are promise for wearable supercapacitors, but their PCE is currently not equivalent to that of traditional Si-based supercapacitors. As a result, efforts to increase the performance of supercapacitors using 2DMs are required; for example, improving the integration of 2DMs may result in increased absorption of light, collection of charge, and effective dissociation of exciton [29].

6.5.2 Wearable Batteries

To store energy, batteries use Faradaic reactions on electrodes are generally resulting in a phase shift in electrode materials. Anode–cathode intercalation of Li^+ or Na^+ ions stores energy in Li^+ or Na^+ batteries, as illustrated in Fig. 6.49a. Higher electrical and diffusivity of ions and stability of structure during cycles of charging/discharging are required for improved energy storage efficiency [348, 378, 379]. Their extremely high conductivity, large specific areas of surface, and small diffusion barriers of surface allow 2D NS (GR, MoS_2, MXene, and MOF) the ideal candidates for future wearable batteries [380, 381]. However, hybrid structures comprising 2D NS were investigated as electrode materials with remarkable characteristics [382]. This section will focus on cutting-edge energy storage technologies. Following this, the 2DM-WBs flexibility will be investigated.

6.5.2.1 Energy Storage Performances

Formally, 2DMs may be built on a variety of flexible substrates, including carbon cloth, membrane of nylon, melamine foam, and metal foil, using solution-based techniques to produce battery wearable electrodes [383, 384]. Hierarchical nanostructures of hybrid nanomaterials are also meant to address the problems of weak force of bonding, effect of volume, and uncontrolled dendritic development. Among the various combinations, MXene-metal foil system had the greatest capacitance

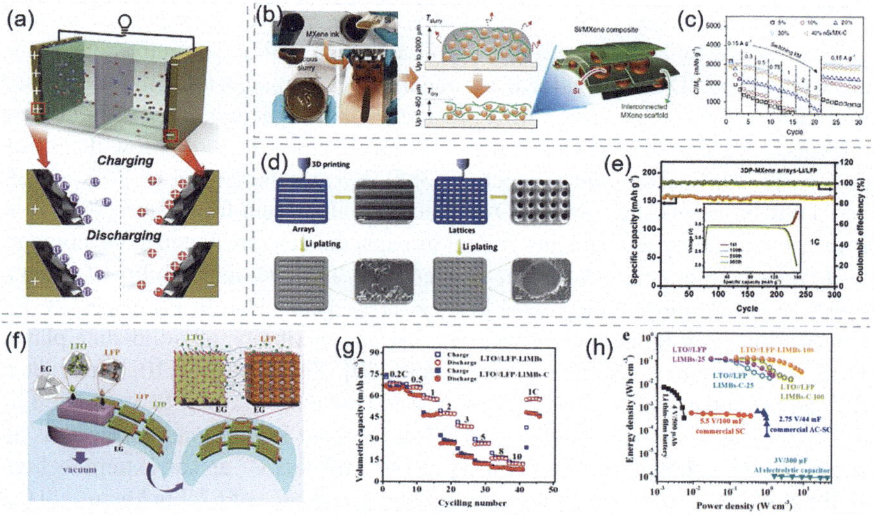

Fig. 6.49 2DM-WB energy storage performance benchmarks: **a** battery mechanism of energy storage depicted schematically (Reproduced with permission [348]) **b** Si/MXene complex electrode manufactured by method of slurry casting, **c** performance rate of *n*Si/MX-C electrodes (Reproduced under the terms of CC BY 4.0 license [385], https://creativecommons.org/licenses/by/4.0/), **d** 3D printing MXene arrays and lattices schematically, **e** stability of cycles of MXene arrays-Li/LFP (Reproduced with permission [386]), **f** schematic of mask-assisted manufacture of integrated LTO//LFP-Li–ion microbatteries, **g** rate capability of LTO//LFP-Li–ion microbatteries, **h** in contrast to other widely viable energy storage devices, a Ragone plot of LTO/LFP-LIMBs with and without EG cover at 25 and 100 °C is shown (Reproduced with permission from [384])

and most stable cycles of charge–discharge. In 2DM-WBs, metal foil substrates are frequently exploited which has their excellent electrical contact, low diffusion distances, and adequate tolerance for volume fluctuations. Zhang et al. [385], for example, showed the assembly of nSi/$Ti_3C_2T_x$ on Cu foil by slurry casting, achieving the greatest primary reversible capacitance (~ 3200 $mAhg^{-1}$ at 0.15 Ag^{-1}) (Fig. 6.49b). However, its capacitance stays constant after eight cycles (Fig. 6.49c). MXene acts as an efficient binder, forming a continuous metallic network that facilitates rapid charge transfer and offers enough mechanical support. Similarly, cycling performance is achieved by bonding $Ti_3C_2T_x$ to a substrate of stainless steel. For instance, Shen et al. [386] established that the $Ti_3C_2T_x$ with Li plating may be constructed via 3D printing on stainless steel (Fig. 6.49d). After 300 cycles at 1 °C, the MXene-Li/LFP battery retains 99.4% of its capacitance (Fig. 6.49e). MXene arrays with a high number of accessible interspaces not only promote lithium nucleation and regulate the lithium–ion flow and electric field, thereby preventing the formation of lithium dendrites, but also offer sufficient room for the creation of huge cobblestone-like lithium. Furthermore, other 2DMs with high conductivity, such as GR, can be utilized. Bai et al. [387] fabricated a Li-metal battery by spray coating GR directly onto lithium foil. The storage of energy performance demonstrates that a reversible capacitance of 130.1 $mAhg^{-1}$ at 1 °C is attained initially, but decreases to 5 mAcm^{-2} after 1000 cycles. Additionally, a large-scale and continuous electrode can be achieved by combining a roll-to-roll approach with a continuous electrode. Due to the strong 2D design and reduced dendritic development, these methods based on solutions offer the batteries with a great reversible capacitance and even cycling performance. Metal foil presents a number of difficulties, including low flexibility and a weak 2DM-substratum connection [388]. For instance, Fu et al. [388] developed a GR-based ink that can be 3D-printed onto aluminum foil. After 280 cycles, the integrated battery retained just 78% of its capacitance is attributed to the comparatively poor bonding that takes place at interfacial. To increase flexibility of battery, growth of 2DMs-on-polymer 2DM-WBs is at its peak. Due to the necessity for reduced leakage of liquid electrolyte, security, and flexibility, all-solid-state planar Li–ion microbattery constitutes a new trend [389]. Zheng et al. [110] constructed a Li–ion microbattery employing mask-assisted vacuum filtering to assemble GR on a nylon membrane (Fig. 6.49f) [384]. The resultant Li–ion microbattery has an extremely high volumetric energy density of 146.3 mWhcm^{-3} and an extremely high P_D of 8.8 Wcm^{-3}. Moreover, it has a reversible capacitance of 67.7 mAhcm^{-3} at 0.2 °C and no capacitance decay at 5 °C after 3300 cycles. The exceptional rate capacity is due to the bidirectional Li–ion diffusion process (Fig. 6.49g, h).

6.5.2.2 Flexible Stability of Wearable Batteries

For 2DM-WBs, maintaining a constant power supply during deformation is a big challenge, and, remarkably, just a few attempts have been made to over it. Though there is no uniform norm to follow, researchers are experimenting with several methods for demonstrating the devices' flexibility and stability. The majority of the tests center on bending. Several groups have attempted to measure the intensity of light of LEDs driven by 2DM-WBs at various angles of bending and after multiple cycles of bending [390, 391]. For instance, a 3D GR-NS/SiOC$_f$/LFP is a battery capable of sustaining steady illumination of linked LEDs at various bending angles (45°, 90°, 180°) [390]. Tang et al. [383] used the drop-filtration approach to construct $Ti_3C_2T_x$ on carbon cloth-I in order to produce a pseudo-solid-state battery. Even when the flat-bending-flat condition is repeated, a red LED bulb can be powered gradually, this could be because MXene and iodine modified carbon cloth have a strong bond. A flexible LIB complete cell (MoS_2@C/carbon cloth/$LiCoO_2$/Al) capable of powering an LED even after 300 bending cycles was developed [392]. Certain researchers attempted to measure capacitance retention. For example, electroplating $Ti_3C_2T_x$ MXene on carbon cloth results in bending-durable batteries [391]. After 300 cycles under 90° bending condition, the complete battery built with Na-$Ti_3C_2T_x$-carbon cloth electrode and $Na_3V_2(PO_4)_3$ retains 88% of its capacitance (current density ~ 500 mA/g) and seem identical to value in steady-state. This indicates that Na-$Ti_3C_2T_x$-carbon cloth has a high bending strength. Even at a bending angle of 270°, the capacitor retains a great capacitance of 89%. The procedure of electroplating strengthens the bond between $Ti_3C_2T_x$ and carbon cloth, causing $Ti_3C_2T_x$ to efficiently promote Na nucleation and laterally directed deposition while avoiding the formation of mossy/dendritic Na [393]. An interesting result was given by Zheng et al. [384], who developed a planar GR-based battery utilizing the vacuum filtering technique. The capacitance remains remarkably constant after 10,000 bending cycles at 180°, with just a small variation in capacitance. There is no deterioration of structure or delamination of the interface between the microelectrodes and the substrate in the integrated construction. In comparison to supercapacitors that are wearable, the of 2DM-WBs' flexible stability is challenged by phase shifts produced by dynamic charging and discharging processes, as well as the 2DMs' comparatively weak heterogeneous interface with the flexible substrate. For instance, employing $Ti_3C_2T_x$ as an electrode to prevent the formation of mossy/dendritic Na results in 89% retention of capacitance after 300 bending cycles [393]. After three 180° bending cycles, GR built on CNT paper retains 96.7% of its capacitance [394]. Clearly, research into the flexible stability of 2DMs-based wearable batteries is in its infancy and has tremendous potential. Collaboration across disciplines, its expansion will be aided by a community-wide discussion to determine the testing standard.

6.5.3 Photodetectors

Photodetectors are energy-conversion/harvesting devices that are capable of converting light impulses to electrical signals. As a result, such devices may be utilized in a number of practical applications imaging, communication, and different sensors for safety monitoring, and play a vital role in our everyday lives. Photodetectors operate on a similar principle as photovoltaic devices. When light is excited, bound e^-–h^+ pairs are produced by striking photons with energy greater than E_g of semiconductors, and the resulting photocurrent is proportional to the light intensity. In comparison to "Si" with an indirect E_g or GR without one, 2DMs with an adjustable direct E_g energy and excellent absorption efficiency were evolved as promising options for next-generation flexible photodetectors [395, 396]. ML-MoS_2, being a typical semiconductor with a straight E_g from the TMDCs family, may be a viable material for photodetectors because of its E_g high coefficient of absorption and efficient e^-–h^+ pair production upon photoexcitation. Lopez-Sanchez et al. [397] announced the development of an ultrasensitive ML-MoS_2 photodetector (Fig. 6.50a) with a photoresponsivity of 880 AW^{-1} (Fig. 6.50b) is about 6-orders of magnitude more relative to prior result for ML-GR. This enhancement has been brought about by the direct E_g and high electrostatic control degree may result in a low dark current, allowing the ML-MoS_2-based photodetectors to be switched off. Moreover, the devices were limited intrinsically by sluggish photoresponse dynamics, resulting in a response time of up to 4 s. The development of flexible photodetectors based on ultrathin Bi_2S_3 NS (Fig. 6.50c), with a reaction time of 10 ps (Fig. 6.50d) [398]. Bi_2S_3's high absorption coefficient and bulk direct E_g of 1.3 eV make attractive photoactive material for broadband photodetectors. The rapid response time may be ascribed to photoactive layer's contact quality and inherent characteristics. The Bi_2S_3 NS used in the devices was produced by reacting triphenyl bismuth with

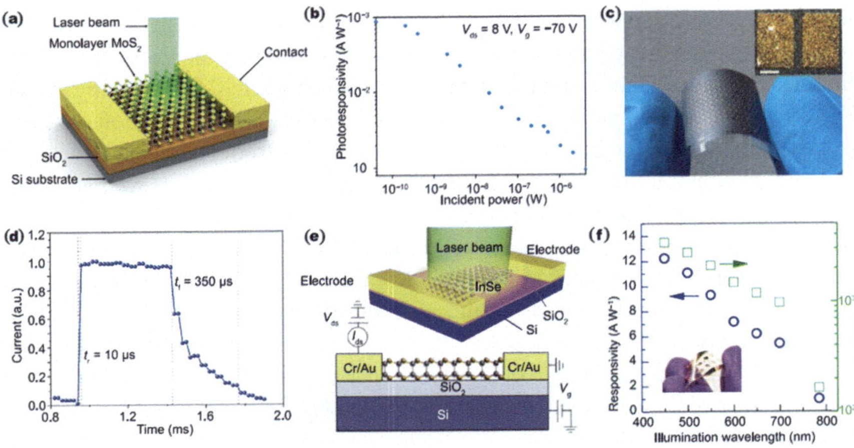

Fig. 6.50 **a** SL-MoS_2 PD (3D schematic), **b** device photoresponsivity showing high sensitivity (Reproduced with permission from [397]), **c** photograph of 2D Bi_2S_3 nanostructured flexible device with typical device micrograph (inset), **d** magnified representations of a single response cycle to display response time (Reproduced with permission from [398]), **e** schematic of InSe-based PD, **f** spectral responsivity and external quantum efficiency of device (Reproduced with permission from [400])

dibenzyl disulfide for the first time, and a thickness of 2.2 nm was obtained. Likewise, the Bi_2S_3 NS-based bendable photodetectors were fabricated using a simple drop-deposition technique on PET substrates. The device in its as-prepared state demonstrated outstanding photoresponse characteristics, with a response of 4.4 AW^{-1}. Photodetection range is another critical metric for evaluating photodetectors' performance [399]. Due to its shorter E_g and lower exciton decreased mass, InSe has a higher level of quantum confinement than the other members of the IIIA-VIA family. This enables controllability of the E_g of the materials used to fabricate the spectrally adjustable devices. Thus, photodetectors based on FLs of InSe were able to perform broadband photodetection from the visible range to NIR range (450–785 nm) having photoresponsivities of ~ 12.3 AW^{-1} [400]. Flexible devices have been fabricated using PET substrates and exfoliated InSe samples (Fig. 6.50e). The photoresponsivity and photoswitching stability of the devices have been evaluated prior to and after bending, and the observed values were comparable to those of the devices on SiO_2/Si substrates (Fig. 6.50f). However, the FL-InSe photodetectors demonstrated consistent photoswitching performance over a lengthy period. $Pb_{1-x}Sn_xSe$, a kind of topological crystalline insulator with a small E_g (0.43 eV), was used to construct photodetectors with a bored spanning UV's spectral response to infrared (375, 473, 632, 800, and 980 nm) [401]. The vdW epitaxy was used to produce $Pb_{1-x}Sn_xSe$ nanoplates having thickness of ~ 15–45 nm on mica sheets. After fabricating flexible photodetectors, even after being bent 100 times, the devices showed quick, stable, and reversible photoresponse. It was concluded that 2D $Pb_{1-x}Sn_xSe$ photodetectors on mica sheets had enormous promise for flexible and wearable optoelectronic applications. Further investigation of various 2DMs, we feel, is critical for improving the performance of flexible optoelectronic devices.

6.5.4 Other Types of Wearable Energy Harvesters

Graphene and its derivatives were employed to produce electricity from multiple kinds of energy, like water and moisture droplets [402–404]. Energy harvesters that generate electricity from ambient moisture use GO. An energy harvester developed on GO film generates electricity via moisture diffusion. When the humidity fluctuates, the gradient of concentration of groups containing O_2 in GO fabricates an ionic gradient results in a regular electric o/p (Fig. 6.51a) [402]. This energy harvester has a ~ 35 mV o/p voltage, a P_D ~ 4.2 mWm^{-2}, and an energy conversion efficiency of up to ~ 62%. It has been proven that electricity may be generated from human breathing (Fig. 6.51b–e). To be used as wearable power source, this energy harvester's electrical o/p must be increased, to directly power wearable gadgets, and should be utilized in conjunction with energy storage devices. Long-term cycle stability is necessary for usage as a sensor of self-powered respiration. Certain GO-based concentration cells were utilized to generate energy from moisture [405, 406]. A GO-based concentration cell using PEDOT, GO ink, and rGO through an insulating substrate (polymer sheet/paper) was designed (Fig. 6.51d–f) [406]. The amount of power fabricated by humidity is proportional to the device's size. When combined with a room temperature ionic liquid, a single unit cell of 0.5 cm in length may generate 30 AhL^{-1} of energy and a voltage of up to 0.7 V. These cells can operate an electrochromic device with P_D up to 0.4 Wcm^{-3} and an energy density of 4 WhL^{-1}. A wearable GO-based disposable energy source has been inscribed on a plastic glove to show its wearability. This study establishes new platforms for wearing electronics in the future. The relative motion of GR and ionic/non-ionic H_2O that was being demonstrated to generate electricity may be explained by a variety of processes. Dhiman et al. [404] observed that ~ 0.6 M HCl flows over GR with speed of ~ 0.01 ms^{-1}, an o/p power of ~ 85 nW is fabricated, and hypothesized that power generation is due to the surface ion hopping process. Yin et al. demonstrated the ability to produce electricity by flowing ionic H_2O over ML-GR. They hypothesized that Na^+ ions adsorption at the front border and Na^+ ions desorption at the back cause free electrons to flow through GR underneath the droplet. Falling ionic water onto a GR surface with a tilt of 70° results in a "*V*" of ~ 30 mV, an "*I*" of ~ 1.7 mA, and a power o/p of ~ 19.2 nW (Fig. 6.51f–h) [407]. A PTFE layer was added between the ionic water droplet and GR, resulting in an increase in performance with an o/p "*V*", "*I*", and power of ~ 1.1 V, ~ 8 mA, and ~ 1.9 mW, respectively [408]. Ho Lee et al. [407] discovered that when non-ionic H_2O moves across GR at a velocity of 0.1 ms^{-1}, a voltage of ~ 48.2 mV is produced. They proposed that distilled H_2O is a polar liquid capable of forming "ordered" structures along with hydrophobic interfaces that can interrelate with charge carriers in GR, resulting in

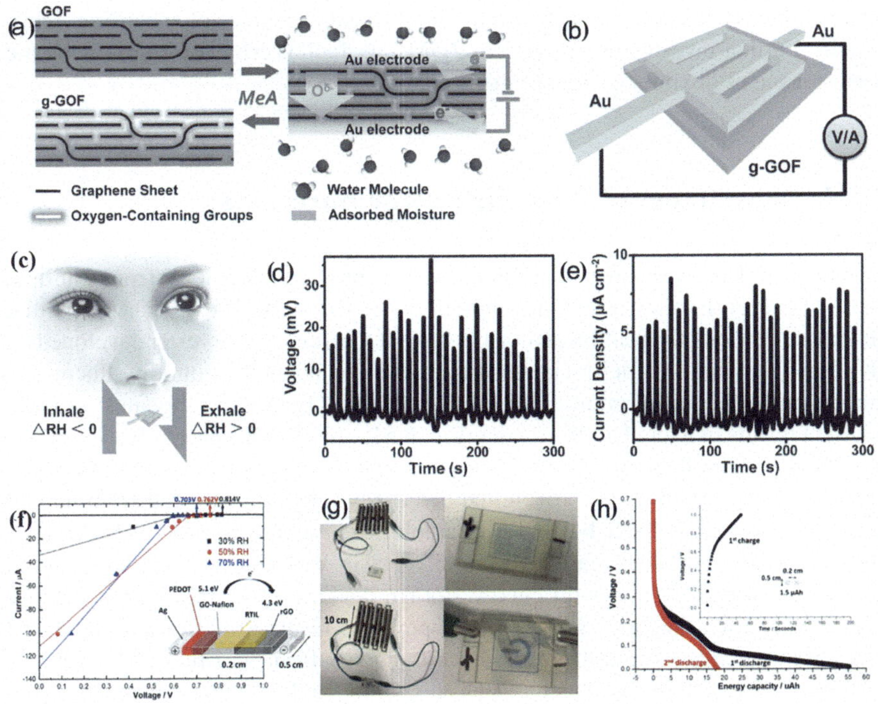

Fig. 6.51 **a** In a humid atmosphere, moisture-electric annealing of GO sheet with applied voltage is shown schematically, **b** schematic of energy harvester structure, **c** schematic of human breathing is used to power an energy harvester, **d** o/p *V* and **e** with a ΔRH of 21%, current density is fabricated by respiratory moisture-tide (Reproduced with permission from [402]), **f** electrical parameters of the PEDOT/GO-Nafion/rGO concentration cell. *Insets* GO concentration cell structure, **g** discharge of a series of GO concentration cells printed on a PEN substrate, **h** 1st and 2nd discharge at 1 mA and inset shows 1st charge at 100 μA with a size of 2 × 10 cm at an RH of 70% (Reproduced under the terms of CC BY 4.0 license [406], https://creativecommons.org/licenses/by/4.0/)

charge redistribution. This redistribution of charge in GR generates a difference of potential, and the distilled water flowing across GR generates a constant voltage. The o/p voltage of the NGs based on non-ionic water/GR is up to ~ 0.1 V [407, 409]. Even though process is unexplained, this type of energy harvester may well have wearable uses due to GR's exceptional flexibility. For instance, raincoats made of GR may be fabricated to capture energy from precipitation. However, the supplied power must be increased further, and appropriate device designs are required. To summarize, GO was used to fabricate electricity by capturing energy from human breathing or powering self-powered respiration sensors via ambient moisture diffusion. Water and GR may produce energy through their relative motion, indicating the possibility for wearable power sources.

References

1. Wang ZL, Wu W (2012) Angew Chem Int Ed 51:11700
2. Wang ZL, Chen J, Lin L (2015) Energy Environ Sci 8:2250
3. Jost K, Dion G, Gogotsi Y (2014) J Mater Chem A 2:10776
4. Bowen CR, Kim HA, Weaver PM, Dunn S (2014) Energy Environ Sci 7:25
5. Beeby SP, Tudor MJ, White NM (2006) Meas Sci Technol 17:R175
6. Law M, Greene LE, Johnson JC, Saykally R, Yang P (2005) Nat Mater 4:455
7. Zhang T, Lin W (2014) Chem Soc Rev 43:5982
8. Wang ZL, Song J (2006) Science 312:242
9. Wang ZL (2017) Mater Today 20:74
10. Radisavljevic B, Radenovic A, Brivio J, Giacometti V, Kis A (2011) Nat Nanotechnol 6:147
11. Wang QH, Kalantar-Zadeh K, Kis A, Coleman JN, Strano MS (2012) Nat Nanotechnol 7:699
12. Xia F, Wang H, Xiao D, Dubey M, Ramasubramaniam A (2014) Nat Photonics 8:899
13. Liu K, Wu J (2016) J Mater Res 31:832
14. Zhang G, Zhang Y-W (2017) Chin Phys B 26:034401 (2017)
15. Judek J, Gertych AP, Świniarski M, Łapińska A, Dużyńska A, Zdrojek M (2015) Sci Rep 5:12422
16. Bonaccorso F, Colombo L, Yu G, Stoller M, Tozzini V, Ferrari AC, Ruoff RS, Pellegrini V (2015) Science 347:1246501
17. Tan C et al (2017) Chem Rev 117:6225
18. Lee J-H, Lee KY, Kumar B, Tien NT, Lee N-E, Kim S-W (2013) Energy Environ Sci 6:169
19. Kim S et al (2014) Adv Mater 26:3778
20. Zhang Z, Li X, Yin J, Xu Y, Fei W, Xue M, Wang Q, Zhou J, Guo W (2018) Nat Nanotechnol 13:1109
21. Ye M, Zhang Z, Zhao Y, Qu L (2018) Joule 2:245
22. Wang X et al (2015) Nat Nanotechnol 10:517
23. Xue Y, Zhang Q, Wang W, Cao H, Yang Q, Fu L (2017) Adv Energy Mater 7:1602684
24. Fan F-R, Tian Z-Q, Lin Wang Z (2012) Nano Energy 1:328
25. Fan FR, Tang W, Wang ZL (2016) Adv Mater 28:4283
26. Paradiso JA, Starner T (2005) IEEE Pervasive Comput 4:18
27. Granstrom J, Feenstra J, Sodano HA, Farinholt K (2007) Smart Mater Struct 16:1810
28. Cui C, Xue F, Hu W-J, Li L-J (2018) npj 2D Mater Appl 2:18
29. Yi F, Ren H, Shan J, Sun X, Wei D, Liu Z (2018) Chem Soc Rev 47:3152
30. Chhowalla M, Shin HS, Eda G, Li L-J, Loh KP, Zhang H (2013) Nat Chem 5:263
31. Masmanidis SC, Karabalin RB, De Vlaminck I, Borghs G, Freeman MR, Roukes ML (2007) Science 317:780
32. Kingon AI, Srinivasan S (2005) Nat Mater 4:233
33. Wu W et al (2014) Nature 514:470
34. Ganatra R, Zhang Q (2014) ACS Nano 8:4074
35. Duerloo K-AN, Ong MT, Reed EJ (2012) J Phys Chem Lett 3:2871
36. Zhu H et al (2015) Nat Nanotechnol 10:151
37. Kim SK et al (2016) Nano Energy 22:483
38. Lee J-H et al (2017) Adv Mater 29:1606667
39. Michel KH, Verberck B (2009) Phys Rev B 80:224301
40. Li Y, Rao Y, Mak KF, You Y, Wang S, Dean CR, Heinz TF (2013) Nano Lett 13:3329
41. Zelisko M, Hanlumyuang Y, Yang S, Liu Y, Lei C, Li J, Ajayan PM, Sharma P (2014) Nat Commun 5:4284
42. Fei R, Li W, Li J, Yang L (2015) Appl Phys Lett 107:173104
43. Li W, Li J (2015) Nano Res 8:3796
44. Blonsky MN, Zhuang HL, Singh AK, Hennig RG (2015) ACS Nano 9:9885
45. Zeng Z et al (2012) Angew Chem Int Ed 51:9052
46. da Cunha Rodrigues G, Zelenovskiy P, Romanyuk K, Luchkin S, Kopelevich Y, Kholkin A (2015) Nat Commun 6:7572

47. Cui C et al (2018) Nano Lett 18:1253
48. Lu A-Y et al (2017) Nat Nanotechnol 12:744
49. Zhou Y et al (2017) Nano Lett 17:5508
50. Jin H-J, Yoon WY, Jo W (2018) ACS Appl Mater Interfaces 10:1334
51. Zhou Y, Liu W, Huang X, Zhang A, Zhang Y, Wang ZL (2016) Nano Res 9:800
52. Dai M et al (2020) Nano Lett 20:201
53. Cheng J et al (2015) Adv Mater 27:4069
54. Dai M et al (2019) Nano Lett 19:5410
55. Hsieh K, Kochat V, Zhang X, Gong Y, Tiwary CS, Ajayan PM, Ghosh A (2017) Nano Lett 17:5452
56. Najmaei S et al (2013) Nat Mater 12:754
57. van der Zande AM et al (2013) Nat Mater 12:554
58. Habibur RM, Yaqoob U, Muhammad S, Uddin ASMI, Kim HC (2018) Mater Chem Phys 215:46
59. Xue F et al (2018) ACS Nano 12:4976
60. Yuan S, Io WF, Mao J, Chen Y, Luo X, Hao J (2020) ACS Appl Nano Mater 3:11979
61. Yu S, Rice Q, Tabibi B, Li Q, Seo FJ (2018) Nanoscale 10:12472
62. Zheng W et al (2019) Adv Sci 6:1802204
63. Mohanta MK, Rawat A, Jena N, Dimple, Ahammed R, De Sarkar A (2020) ACS Appl Mater Interfaces 12:3114
64. Watanabe K, Taniguchi T, Kanda H (2004) Nat Mater 3:404
65. Kuang H et al (2021) Nano Energy 80:105561
66. Michel KH, Verberck B (2011) Phys Rev B 83:115328
67. Lee G-J, Lee M-K, Park J-J, Hyeon DY, Jeong CK, Park K-I (2019) ACS Appl Mater Interfaces 11:37920
68. Suo G, Yu Y, Zhang Z, Wang S, Zhao P, Li J, Wang X (2016) ACS Appl Mater Interfaces 8:34335
69. Šutka A et al (2018) Energy Environ Sci 11:1437
70. Sriphan S, Charoonsuk T, Maluangnont T, Vittayakorn N (2019) ACS Appl Energy Mater 2:3840
71. Cho H-B, Shoji M, Fujiwara T, Nakayama T, Suematsu H, Suzuki T, Niihara K (2010) J Ceram Soc Jpn 118:66
72. Cho H-B, Nakayama T, Tokoi Y, Endo S, Tanaka S, Suzuki T, Jiang W, Suematsu H, Niihara K (2010) Compos Sci Technol 70:1681
73. Kim K-B, Jang W, Cho JY, Woo SB, Jeon DH, Ahn JH, Hong SD, Koo HY, Sung TH (2018) Nano Energy 54:91
74. Cho S, Choi JW (2011) Jpn J Appl Phys 50:07HG01
75. Kar E, Bose N, Dutta B, Banerjee S, Mukherjee N, Mukherjee S (2019) Energy Convers Manage 184:600
76. Maity K, Mahanty B, Sinha TK, Garain S, Biswas A, Ghosh SK, Manna S, Ray SK, Mandal D (2017) Energ Technol 5:234
77. Roy K, Ghosh SK, Sultana A, Garain S, Xie M, Bowen CR, Henkel K, Schmeiβer D, Mandal D (2019) ACS Appl Nano Mater 2:2013
78. Lien D-H et al (2015) Nano Lett 15:1356
79. Das S, Pandey D, Thomas J, Roy T (2019) Adv Mater 31:1802722
80. Fang H, Chuang S, Chang TC, Takei K, Takahashi T, Javey A (2012) Nano Lett 12:3788
81. Huang Y et al (2014) ACS Nano 8:10743
82. Xia F, Wang H, Jia Y (2014) Nat Commun 5:4458
83. Castellanos-Gomez A (2015) J Phys Chem Lett 6:4280
84. Akinwande D et al (2017) Extreme Mech Lett 13:42
85. Tomasi A, Paviet-Salomon B, Lachenal D, Nicolas SMD, Descoeudres A, Geissbühler J, Wolf SD, Ballif C (2014) IEEE J Photovoltaics 4:1046
86. Taguchi M, Yano A, Tohoda S, Matsuyama K, Nakamura Y, Nishiwaki T, Fujita K, Maruyama E (2014) IEEE J Photovoltaics 4:96

87. Roy AB, Das S, Kundu A, Banerjee C, Mukherjee N (2017) Phys Chem Chem Phys 19:12838
88. Louwen A, van Sark W, Schropp R, Faaij A (2016) Sol Energy Mater Sol Cells 147:295
89. Castellanos-Gomez A (2016) Nat Photonics 10:202
90. Kowalczewski P, Andreani LC (2015) Sol Energy Mater Sol Cells 143:260
91. Das S, Kundu A, Saha H, Datta SK (2015) Plasmonics 10:1895
92. Das S, Kundu A, Saha H, Datta SK (2015) J Opt 18:015903
93. Petersen KE (1982) Proc IEEE 70:420
94. Zhang T, Hou Z, Johnson RW, Castillo LD, Moussessian A, Greenwell R, Blalock BJ (2009) IEEE Trans Electron Packag Manuf 32:291
95. Akinwande D, Petrone N, Hone J (2014) Nat Commun 5:5678
96. Marinova N, Valero S, Delgado JL (2017) J Colloid Interface Sci 488:373
97. Li C, Sleppy J, Dhasmana N, Soliman M, Tetard L, Thomas J (2016) J Mater Chem A 4:11648
98. Green MA, Ho-Baillie A (2017) ACS Energy Lett 2:822
99. Qiu L, Ono LK, Qi Y (2018) Mater Today Energy 7:169
100. Abdulrazzaq OA, Saini V, Bourdo S, Dervishi E, Biris AS (2013) Part Sci Technol 31:427
101. Li G, Zhu R, Yang Y (2012) Nat Photonics 6:153
102. Tamai Y, Ohkita H, Benten H, Ito S (2015) J Phys Chem Lett 6:3417
103. Vezie MS et al (2016) Nat Mater 15:746
104. An Q, Zhang F, Zhang J, Tang W, Deng Z, Hu B (2016) Energy Environ Sci 9:281
105. An Q, Zhang F, Yin X, Sun Q, Zhang M, Zhang J, Tang W, Deng Z (2016) Nano Energy 30:276
106. Qin X, Zhao Z, Wang Y, Wu J, Jiang Q, You J (2017) J Semiconductors 38:011002
107. Bernardi M, Palummo M, Grossman JC (2013) Nano Lett 13:3664
108. Britnell L et al (2013) Science 340:1311
109. Mak KF, Lee C, Hone J, Shan J, Heinz TF (2010) Phys Rev Lett 105:136805
110. Tsai M-L, Su S-H, Chang J-K, Tsai D-S, Chen C-H, Wu C-I, Li L-J, Chen L-J, He J-H (2014) ACS Nano 8:8317
111. Rehman AU et al (2016) ACS Appl Mater Interfaces 8:29383
112. Shanmugam M, Durcan CA, Yu B (2012) Nanoscale 4:7399
113. Wi S, Kim H, Chen M, Nam H, Guo LJ, Meyhofer E, Liang X (2014) ACS Nano 8:5270
114. Shanmugam M, Bansal T, Durcan CA, Yu B (2012) Appl Phys Lett 100:153901
115. Li X, Zhang W, Wu Y, Min C, Fang J (2013) ACS Appl Mater Interfaces 5:8823
116. Le QV, Nguyen TP, Jang HW, Kim SY (2014) Phys Chem Chem Phys 16:13123
117. Yang X et al (2014) J Mater Chem A 2:7727
118. Kakavelakis G et al (2017) ACS Nano 11:3517
119. Hu X, Chen L, Tan L, Zhang Y, Hu L, Xie B, Chen Y (2015) Sci Rep 5:12161
120. Yang X, Liu W, Xiong M, Zhang Y, Liang T, Yang J, Xu M, Ye J, Chen H (2014) J Mater Chem A 2:14798
121. Anaraki EH et al (2016) Energy Environ Sci 9:3128
122. Giordano F et al (2016) Nat Commun 7:10379
123. Yang WS et al (2017) Science 356:1376
124. Tan H et al (2017) Science 355:722
125. Yoon H, Kang SM, Lee J-K, Choi M (2016) Energy Environ Sci 9:2262
126. Correa Baena JP et al (2015) Energy Environ Sci 8:2928
127. Ke W et al (2015) J Am Chem Soc 137:6730
128. Bati ASR, Batmunkh M, Shapter JG (2020) Adv Energy Mater 10:1902253
129. Wang P, Wu Y, Cai B, Ma Q, Zheng X, Zhang W-H (2019) Adv Func Mater 29:1807661
130. Roose B, Wang Q, Abate A (2019) Adv Energy Mater 9:1803140
131. Leijtens T, Eperon GE, Pathak S, Abate A, Lee MM, Snaith HJ (2013) Nat Commun 4:2885
132. Bati ASR, Yu L, Tawfik SA, Spencer MJS, Shaw PE, Batmunkh M, Shapter JG (2019) iScience 14:100
133. Wang Y, Yue Y, Yang X, Han L (2018) Adv Energy Mater 8:1800249
134. Yeo J-S et al (2015) Nano Energy 12:96
135. Urbani M, de la Torre G, Nazeeruddin MK, Torres T (2019) Chem Soc Rev 48:2738

136. Li T, Pan Y, Wang Z, Xia Y, Chen Y, Huang W (2017) J Mater Chem A 5:12602
137. Wang X, He J, Zhou B, Zhang Y, Wu J, Hu R, Liu L, Song J, Qu J (2018) Angew Chem Int Ed 57:8668
138. Cai Y, Zhang G, Zhang Y-W (2014) Sci Rep 4:6677
139. Mariano M, Mashtalir O, Antonio FQ, Ryu W-H, Deng B, Xia F, Gogotsi Y, Taylor AD (2016) Nanoscale 8:16371
140. Tang G et al (2019) Adv Mater 31:1807689
141. Huang P, Wang Z, Liu Y, Zhang K, Yuan L, Zhou Y, Song B, Li Y (2017) ACS Appl Mater Interfaces 9:25323
142. Zhao X et al (2019) Adv Func Mater 29:1805168
143. Yin G, Zhao H, Feng J, Sun J, Yan J, Liu Z, Lin S, Liu S (2018) J Mater Chem A 6:9132
144. Rudenko AN, Katsnelson MI (2014) Phys Rev B 89:201408
145. Fu N et al (2018) J Mater Chem A 6:8886
146. Batmunkh M, Bat-Erdene M, Shapter JG (2016) Adv Mater 28:8586
147. Bat-Erdene M et al (2017) Small Methods 1:1700260
148. Singh R, Giri A, Pal M, Thiyagarajan K, Kwak J, Lee J-J, Jeong U, Cho K (2019) J Mater Chem A 7:7151
149. Huang P, Yuan L, Zhang K, Chen Q, Zhou Y, Song B, Li Y (2018) ACS Appl Mater Interfaces 10:14796
150. Huang P, Chen Q, Zhang K, Yuan L, Zhou Y, Song B, Li Y (2019) J Mater Chem A 7:6213
151. Song HS, Li SL, Gao L, Xu Y, Ueno K, Tang J, Cheng YB, Tsukagoshi K (2013) Nanoscale 5:9666
152. Zhao E, Gao L, Yang S, Wang L, Cao J, Ma T (2018) Nano Res 11:5913
153. Naguib M, Kurtoglu M, Presser V, Lu J, Niu J, Heon M, Hultman L, Gogotsi Y, Barsoum MW (2011) Adv Mater 23:4248
154. Ghidiu M, Lukatskaya MR, Zhao M-Q, Gogotsi Y, Barsoum MW (2014) Nature 516:78
155. Yang L et al (2019) J Mater Chem A 7:5635
156. Calió L, Kazim S, Grätzel M, Ahmad S (2016) Angew Chem Int Ed 55:14522
157. Wang S, Jiang Y, Juarez-Perez EJ, Ono LK, Qi Y (2016) Nature Energy 2:16195
158. Kato Y, Ono LK, Lee MV, Wang S, Raga SR, Qi Y (2015) Adv Mater Interfaces 2:1500195
159. Gupta A, Arunachalam V, Vasudevan S (2016) J Phys Chem Lett 7:4884
160. Capasso A, Matteocci F, Najafi L, Prato M, Buha J, Cinà L, Pellegrini V, Carlo AD, Bonaccorso F (2016) Adv Energy Mater 6:1600920
161. Najafi L et al (2018) ACS Nano 12:10736
162. Muduli SK, Varrla E, Kulkarni SA, Han G, Thirumal K, Lev O, Mhaisalkar S, Mathews N (2017) J Power Sources 371:156
163. Wang Y, Zhang H, Zhang T, Shi W, Kan M, Chen J, Zhao Y (2019) Solar RRL 3:1900197
164. Gong X, Li M, Shi X-B, Ma H, Wang Z-K, Liao L-S (2015) Adv Func Mater 25:6671
165. Liang P-W, Liao C-Y, Chueh C-C, Zuo F, Williams ST, Xin X-K, Lin J, Jen AKY (2014) Adv Mater 26:3748
166. Guo Z, Gao L, Xu Z, Teo S, Zhang C, Kamata Y, Hayase S, Ma T (2018) Small 14:1802738
167. Long R, Fang W, Akimov AV (2016) J Phys Chem Lett 7:653
168. Luo D et al (2018) Science 360:1442
169. Winder C, Sariciftci NS (2004) J Mater Chem 14:1077
170. Jeon NJ, Noh JH, Yang WS, Kim YC, Ryu S, Seo J, Seok SI (2015) Nature 517:476
171. Wang Y, Wang S, Chen X, Li Z, Wang J, Li T, Deng X (2018) J Mater Chem A 6:4860
172. Kim YG, Kwon KC, Le QV, Hong K, Jang HW, Kim SY (2016) J Power Sources 319:1
173. Dai R, Wang Y, Wang J, Deng X (2017) Chemsuschem 10:2869
174. Dasgupta U, Chatterjee S, Pal AJ (2017) Sol Energy Mater Sol Cells 172:353
175. Kakavelakis G et al (2018) Adv Energy Mater 8:1702287
176. Lian X et al (2019) J Mater Chem A 7:18980
177. Yang W et al (2019) Solar RRL 3:1900132
178. Ganguly P, Harb M, Cao Z, Cavallo L, Breen A, Dervin S, Dionysiou DD, Pillai SC (2019) ACS Energy Lett 4:1687

179. Chen J, Wu X-J, Yin L, Li B, Hong X, Fan Z, Chen B, Xue C, Zhang H (2015) Angew Chem Int Ed 54:1210
180. Li Q, Zhou B, McBride JR, Lian T (2017) ACS Energy Lett 2:174
181. Li Q, Wu K, Chen J, Chen Z, McBride JR, Lian T (2016) ACS Nano 10:3843
182. Nasilowski M, Mahler B, Lhuillier E, Ithurria S, Dubertret B (2016) Chem Rev 116:10934
183. Li Q, Xu Z, McBride JR, Lian T (2017) ACS Nano 11:2545
184. Simon T et al (2014) Nat Mater 13:1013
185. Han Z, Qiu F, Eisenberg R, Holland PL, Krauss TD (2012) Science 338:1321
186. Zhukovskyi M, Tongying P, Yashan H, Wang Y, Kuno M (2015) ACS Catal 5:6615
187. Xie YP, Zheng Y, Yang Y, Jiang R, Wang G, Zhang Y, Zhang E, Zhao L, Duan C-Y (2018) J Colloid Interface Sci 514:634
188. Dou Y, Zhang L, Xu X, Sun Z, Liao T, Dou SX (2017) Chem Soc Rev 46:7338
189. Cheng Z, Shifa TA, Wang F, Gao Y, He P, Zhang K, Jiang C, Liu Q, He J (2018) Adv Mater 30:1707433
190. Wulan B-R, Yi S-S, Li S-J, Duan Y-X, Yan J-M, Jiang Q (2018) Appl Catal B 231:43
191. Kong L, Ji Y, Dang Z, Yan J, Li P, Li Y, Liu S (2018) Adv Func Mater 28:1800668
192. Meng N, Ren J, Liu Y, Huang Y, Petit T, Zhang B (2018) Energy Environ Sci 11:566
193. Zeng D, Xu W, Ong W-J, Xu J, Ren H, Chen Y, Zheng H, Peng D-L (2018) Appl Catal B 221:47
194. He Z, Kim C, Lin L, Jeon TH, Lin S, Wang X, Choi W (2017) Nano Energy 42:58
195. Wu K-H, Huang Y-J, Teng H (2018) Part Part Syst Charact 35:1700221
196. Cai X, Zhang J, Fujitsuka M, Majima T (2017) Appl Catal B 202:191
197. Chen X, Chen H, Guan J, Zhen J, Sun Z, Du P, Lu Y, Yang S (2017) Nanoscale 9:5615
198. Lin B, Li H, An H, Hao W, Wei J, Dai Y, Ma C, Yang G (2018) Appl Catal B 220:542
199. Gomes Silva C, Luz I, Llabrés i Xamena FX, Corma A, García H (2010) Chem Eur J 16:11133
200. Horiuchi Y, Toyao T, Saito M, Mochizuki K, Iwata M, Higashimura H, Anpo M, Matsuoka M (2012) J Phys Chem C 116:20848
201. He J, Yan Z, Wang J, Xie J, Jiang L, Shi Y, Yuan F, Yu F, Sun Y (2013) Chem Commun 49:6761
202. Shen L, Luo M, Huang L, Feng P, Wu L (2015) Inorg Chem 54:1191
203. Tilgner D, Kempe R (2017) Chem Eur J 23:3184
204. Gao H, Zhen W, Ma J, Lu G (2017) Appl Catal B 206:353
205. Chen X et al (2018) Nanoscale 10:1111
206. Halim J et al (2014) Chem Mater 26:2374
207. Anasori B, Xie Y, Beidaghi M, Lu J, Hosler BC, Hultman L, Kent PRC, Gogotsi Y, Barsoum MW (2015) ACS Nano 9:9507
208. Yuan W, Cheng L, An Y, Lv S, Wu H, Fan X, Zhang Y, Guo X, Tang J (2018) Adv Sci 5:1700870
209. Sun Y, Jin D, Sun Y, Meng X, Gao Y, Dall'Agnese Y, Chen G, Wang X-F (2018) J Mater Chem A 6:9124
210. Su T, Peng R, Hood ZD, Naguib M, Ivanov IN, Keum JK, Qin Z, Guo Z, Wu Z (2018) Chemsuschem 11:688
211. Ran J, Gao G, Li F-T, Ma T-Y, Du A, Qiao S-Z (2017) Nat Commun 8:13907
212. Peng C et al (2018) Nano Energy 53:97
213. Nishino H et al (2017) J Am Chem Soc 139:13761
214. Ida S, Takashiba A, Koga S, Hagiwara H, Ishihara T (2014) J Am Chem Soc 136:1872
215. Okamoto Y, Ida S, Hyodo J, Hagiwara H, Ishihara T (2011) J Am Chem Soc 133:18034
216. Guan Z, Xu Z, Li Q, Wang P, Li G, Yang J (2018) Appl Catal B 227:512
217. Kong X, Xu Y, Cui Z, Li Z, Liang Y, Gao Z, Zhu S, Yang X (2018) Appl Catal B 230:11
218. Pilarski M, Marschall R, Gross S, Wark M (2018) Appl Catal B 227:349
219. Wang Y et al (2017) Sci Adv 3:e1701500
220. Wen H, Li H, He S, Chen F, Ding E, Liu S, Wang B, Peng Y (2018) Nanoscale 10:2380
221. Burke MS, Enman LJ, Batchellor AS, Zou S, Boettcher SW (2015) Chem Mater 27:7549
222. Trotochaud L, Ranney JK, Williams KN, Boettcher SW (2012) J Am Chem Soc 134:17253

223. Zou X, Goswami A, Asefa T (2013) J Am Chem Soc 135:17242
224. Song F, Hu X (2014) Nat Commun 5:4477
225. Song F, Hu X (2014) J Am Chem Soc 136:16481
226. Tang D et al (2014) Dalton Trans 43:15119
227. Chen S, Duan J, Jaroniec M, Qiao SZ (2013) Angew Chem Int Ed 52:13567
228. Long X, Li J, Xiao S, Yan K, Wang Z, Chen H, Yang S (2014) Angew Chem Int Ed 53:7584
229. Zhou X, Xia Z, Zhang Z, Ma Y, Qu Y (2014) J Mater Chem A 2:11799
230. Lu X, Zhao C (2015) Nat Commun 6:6616
231. Yang Q, Li T, Lu Z, Sun X, Liu J (2014) Nanoscale 6:11789
232. Wang Q, O'Hare D (2012) Chem Rev 112:4124
233. Xu Y, Hao Y, Zhang G, Lu Z, Han S, Li Y, Sun X (2015) RSC Adv 5:55131
234. Liang J, Ma R, Iyi N, Ebina Y, Takada K, Sasaki T (2010) Chem Mater 22:371
235. Ma R, Liang J, Takada K, Sasaki T (2011) J Am Chem Soc 133:613
236. Liang H, Meng F, Cabán-Acevedo M, Li L, Forticaux A, Xiu L, Wang Z, Jin S (2015) Nano Lett 15:1421
237. Diaz-Morales O, Ledezma-Yanez I, Koper MTM, Calle-Vallejo F (2015) ACS Catal 5:5380
238. Burke MS, Zou S, Enman LJ, Kellon JE, Gabor CA, Pledger E, Boettcher SW (2015) J Phys Chem Lett 6:3737
239. Wang L, Lin C, Huang D, Zhang F, Wang M, Jin J (2014) ACS Appl Mater Interfaces 6:10172
240. Ni B, Wang X (2015) Chem Sci 6:3572
241. Cheng F, Shen J, Peng B, Pan Y, Tao Z, Chen J (2011) Nat Chem 3:79
242. Zhou X, Shen X, Xia Z, Zhang Z, Li J, Ma Y, Qu Y (2015) ACS Appl Mater Interfaces 7:20322
243. Bao J et al (2015) Angew Chem Int Ed 54:7399
244. García-Mota M, Bajdich M, Viswanathan V, Vojvodic A, Bell AT, Nørskov JK (2012) J Phys Chem C 116:21077
245. Bajdich M, García-Mota M, Vojvodic A, Nørskov JK, Bell AT (2013) J Am Chem Soc 135:13521
246. Leng X, Zeng Q, Wu K-H, Gentle IR, Wang D-W (2015) RSC Adv 5:27823
247. Tung C-W, Hsu Y-Y, Shen Y-P, Zheng Y, Chan T-S, Sheu H-S, Cheng Y-C, Chen HM (2015) Nat Commun 6:8106
248. Klaus S, Cai Y, Louie MW, Trotochaud L, Bell AT (2015) J Phys Chem C 119:7243
249. Roger I, Symes MD (2015) J Am Chem Soc 137:13980
250. Alexander A-M, Hargreaves JSJ (2010) Chem Soc Rev 39:4388
251. Xie J, Xie Y (2016) Chem Eur J 22:3588
252. Xie J, Wang R, Bao J, Zhang X, Zhang H, Li S, Xie Y (2014) Inorg Chem Front 1:751
253. Liu Y, Xiao C, Lyu M, Lin Y, Cai W, Huang P, Tong W, Zou Y, Xie Y (2015) Angew Chem Int Ed 54:11231
254. Liang L et al (2015) Angew Chem Int Ed 54:12004
255. Liu Y et al (2014) J Am Chem Soc 136:15670
256. Xu K et al (2015) J Am Chem Soc 137:4119
257. Shalom M, Ressnig D, Yang X, Clavel G, Fellinger TP, Antonietti M (2015) J Mater Chem A 3:8171
258. Zhang J, Zhao Z, Xia Z, Dai L (2015) Nat Nanotechnol 10:444
259. Lin Z, Waller GH, Liu Y, Liu M, Wong C-P (2013) Carbon 53:130
260. Vasileff A, Xu C, Jiao Y, Zheng Y, Qiao S-Z (2018) Chem 4:1809
261. Liang L, Lei F, Gao S, Sun Y, Jiao X, Wu J, Qamar S, Xie Y (2015) Angew Chem Int Ed 54:13971
262. Zhao Y et al (2015) Adv Mater 27:7824
263. Gao S et al (2017) J Am Chem Soc 139:3438
264. Ye M, Wang X, Liu E, Ye J, Wang D (2018) Chemsuschem 11:1606
265. Liu C, Xu Q, Zhang Q, Zhu Y, Ji M, Tong Z, Hou W, Zhang Y, Xu J (2019) J Mater Sci 54:2458
266. Kong XY, Tan WL, Ng B-J, Chai S-P, Mohamed AR (2017) Nano Res 10:1720

267. Li N, Chen X, Ong W-J, MacFarlane DR, Zhao X, Cheetham AK, Sun C (2017) ACS Nano 11:10825
268. Bai Y, Ye L, Chen T, Wang L, Shi X, Zhang X, Chen D (2016) ACS Appl Mater Interfaces 8:27661
269. Cao S, Shen B, Tong T, Fu J, Yu J (2018) Adv Func Mater 28:1800136
270. Simon P, Gogotsi Y (2008) Nat Mater 7:845
271. Yan J, Wang Q, Wei T, Fan Z (2014) Adv Energy Mater 4:1300816
272. Qu D, Shi H (1998) J Power Sources 74:99
273. Raymundo-Piñero E, Kierzek K, Machnikowski J, Béguin F (2006) Carbon 44:2498
274. Barbieri O, Hahn M, Herzog A, Kötz R (2005) Carbon 43:1303
275. Zhang LL, Zhao XS (2009) Chem Soc Rev 38:2520
276. Kierzek K, Frackowiak E, Lota G, Gryglewicz G, Machnikowski J (2004) Electrochim Acta 49:515
277. Azaïs P, Duclaux L, Florian P, Massiot D, Lillo-Rodenas M-A, Linares-Solano A, Peres J-P, Jehoulet C, Béguin F (2007) J Power Sources 171:1046
278. Futaba DN et al (2006) Nat Mater 5:987
279. Xu B, Wu F, Chen R, Cao G, Chen S, Zhou Z, Yang Y (2008) Electrochem Commun 10:795
280. Hu C-C, Chang K-H, Lin M-C, Wu Y-T (2006) Nano Lett 6:2690
281. Zhang H, Cao G, Wang Z, Yang Y, Shi Z, Gu Z (2008) Nano Lett 8:2664
282. Acerce M, Voiry D, Chhowalla M (2015) Nat Nanotechnol 10:313
283. Zhao G, Wen T, Chen C, Wang X (2012) RSC Adv 2:9286
284. Xia J, Chen F, Li J, Tao N (2009) Nat Nanotechnol 4:505
285. Wang Y, Wu Y, Huang Y, Zhang F, Yang X, Ma Y, Chen Y (2011) J Phys Chem C 115:23192
286. Xu Y, Wu Q, Sun Y, Bai H, Shi G (2010) ACS Nano 4:7358
287. Mao S, Wen Z, Kim H, Lu G, Hurley P, Chen J (2012) ACS Nano 6:7505
288. Yu D, Dai L (2010) J Phys Chem Lett 1:467
289. Lei Z, Christov N, Zhao XS (2011) Energy Environ Sci 4:1866
290. Sun Y, Wu Q, Xu Y, Bai H, Li C, Shi G (2011) J Mater Chem 21:7154
291. Liu C, Wang K, Luo S, Tang Y, Chen L (2011) Small 7:1203
292. Fang Y, Lv Y, Che R, Wu H, Zhang X, Gu D, Zheng G, Zhao D (2013) J Am Chem Soc 135:1524
293. Chen Z, Ren W, Gao L, Liu B, Pei S, Cheng H-M (2011) Nat Mater 10:424
294. Choi BG, Yang M, Hong WH, Choi JW, Huh YS (2012) ACS Nano 6:4020
295. Zhu Y et al (2011) Science 332:1537
296. Xu Y, Lin Z, Zhong X, Huang X, Weiss NO, Huang Y, Duan X (2014) Nat Commun 5:4554
297. Zeng Z et al (2012) Adv Mater 24:4138
298. Gogotsi Y, Simon P (2011) Science 334:917
299. Yang X, Cheng C, Wang Y, Qiu L, Li D (2013) Science 341:534
300. Xu Y et al (2015) Nano Lett 15:4605
301. Yoo JJ et al (2011) Nano Lett 11:1423
302. Yoon Y et al (2014) ACS Nano 8:4580
303. Murali S, Quarles N, Zhang LL, Potts JR, Tan Z, Lu Y, Zhu Y, Ruoff RS (2013) Nano Energy 2:764
304. Tao Y et al (2013) Sci Rep 3:2975
305. Jha RK, Guha PK (2016) Nanotechnology 27:475503
306. Mak KF, Shan J (2016) Nat Photonics 10:216
307. Cao L et al (2013) Small 9:2905
308. Ramadoss A, Kim T, Kim G-S, Kim SJ (2014) New J Chem 38:2379
309. Clerici F, Fontana M, Bianco S, Serrapede M, Perrucci F, Ferrero S, Tresso E, Lamberti A (2016) ACS Appl Mater Interfaces 8:10459
310. Soon JM, Loh KP (2007) Electrochem Solid-State Lett 10:A250
311. Choudhary N, Patel M, Ho Y-H, Dahotre NB, Lee W, Hwang JY, Choi W (2015) J Mater Chem A 3:24049
312. Eda G, Yamaguchi H, Voiry D, Fujita T, Chen M, Chhowalla M (2011) Nano Lett 11:5111

313. Papageorgopoulos CA, Jaegermann W (1995) Surf Sci 338:83
314. Joensen P, Crozier ED, Alberding N, Frindt RF (1987) J Phys C Solid State Phys 20:4043
315. Chen D, Ji G, Ding B, Ma Y, Qu B, Chen W, Lee JY (2013) Nanoscale 5:7890
316. Shiva K, Ramakrishna Matte HSS, Rajendra HB, Bhattacharyya AJ, Rao CNR (2013) Nano Energy 2:787
317. Liu Y, Wang W, Huang H, Gu L, Wang Y, Peng X (2014) Chem Commun 50:4485
318. Hu B, Qin X, Asiri AM, Alamry KA, Al-Youbi AO, Sun X (2013) Electrochem Commun 28:75
319. Tu C-C, Lin L-Y, Xiao B-C, Chen Y-S (2016) J Power Sources 320:78
320. Masikhwa TM, Barzegar F, Dangbegnon JK, Bello A, Madito MJ, Momodu D, Manyala N (2016) RSC Adv 6:38990
321. Feng J, Sun X, Wu C, Peng L, Lin C, Hu S, Yang J, Xie Y (2011) J Am Chem Soc 133:17832
322. Wang Q, Jiao L, Du H, Yang J, Huan Q, Peng W, Si Y, Wang Y, Yuan H (2011) CrystEngComm 13:6960
323. Naguib M, Mashtalir O, Carle J, Presser V, Lu J, Hultman L, Gogotsi Y, Barsoum MW (2012) ACS Nano 6:1322
324. Gogotsi Y (2015) Nat Mater 14:1079
325. Lukatskaya MR et al (2013) Science 341:1502
326. Gan Z, Gui Q, Shan Y, Pan P, Zhang N, Zhang L (2016) J Appl Phys 120:104503
327. Liu Y, Peng X (2017) Appl Mater Today 8:104
328. Zheng X, Han Z, Chai F, Qu F, Xia H, Wu X (2016) Dalton Trans 45:12862
329. Li H, Zhu Y, Dong S, Shen L, Chen Z, Zhang X, Yu G (2016) Chem Mater 28:5753
330. Wu C, Lu X, Peng L, Xu K, Peng X, Huang J, Yu G, Xie Y (2013) Nat Commun 4:2431
331. Li H, Peng L, Zhu Y, Zhang X, Yu G (2016) Nano Lett 16:5938
332. Zhu Y, Peng L, Chen D, Yu G (2016) Nano Lett 16:742
333. Zhu J et al (2013) Nano Lett 13:5408
334. Mahmood N, Tang T, Hou Y (2016) Adv Energy Mater 6:1600374
335. Tang Y, Zhang Y, Li W, Ma B, Chen X (2015) Chem Soc Rev 44:5926
336. Du G, Guo Z, Wang S, Zeng R, Chen Z, Liu H (2010) Chem Commun 46:1106
337. Xiao J, Choi D, Cosimbescu L, Koech P, Liu J, Lemmon JP (2010) Chem Mater 22:4522
338. Zhang S, Chowdari BVR, Wen Z, Jin J, Yang J (2015) ACS Nano 9:12464
339. Wu M, Wang Q, Sun Q, Jena P (2013) J Phys Chem C 117:6055
340. Li X, Feng Y, Li M, Li W, Wei H, Song D (2015) Adv Func Mater 25:6858
341. Wen Y, He K, Zhu Y, Han F, Xu Y, Matsuda I, Ishii Y, Cumings J, Wang C (2014) Nat Commun 5:4033
342. Shi Z-T et al (2016) Nano Energy 22:27
343. Sun J, Lee H-W, Pasta M, Yuan H, Zheng G, Sun Y, Li Y, Cui Y (2015) Nat Nanotechnol 10:980
344. Manthiram A, Fu Y, Chung S-H, Zu C, Su Y-S (2014) Chem Rev 114:11751
345. Wang Y-X et al (2012) J Mater Chem 22:4744
346. Aurbach D, Lu Z, Schechter A, Gofer Y, Gizbar H, Turgeman R, Cohen Y, Moshkovich M, Levi E (2000) Nature 407:724
347. Liang Y, Yoo HD, Li Y, Shuai J, Calderon HA, Robles Hernandez FC, Grabow LC, Yao Y (2015) Nano Lett 15:2194
348. Zhou D, Zhao L, Li B (2021) J Energy Chem 62:27
349. Couly C, Alhabeb M, Van Aken KL, Kurra N, Gomes L, Navarro-Suárez AM, Anasori B, Alshareef HN, Gogotsi Y (2018) Adv Electron Mater 4:1700339
350. Zhang C et al (2019) Nat Commun 10:1795
351. Xie W, Wang Y, Zhou J, Zhang M, Yu J, Zhu C, Xu J (2020) Appl Surface Sci 534:147584
352. Shao Y, El-Kady MF, Wang LJ, Zhang Q, Li Y, Wang H, Mousavi MF, Kaner RB (2015) Chem Soc Rev 44:3639
353. Wang M et al (2020) Adv Func Mater 30:2002664
354. Xiao H, Wu Z-S, Chen L, Zhou F, Zheng S, Ren W, Cheng H-M, Bao X (2017) ACS Nano 11:7284

355. Orangi J, Hamade F, Davis VA, Beidaghi M (2020) ACS Nano 14:640
356. Lukatskaya MR et al (2017) Nat Energy 2:17105
357. Miller EE, Hua Y, Tezel FH (2018) J Energy Storage 20:30
358. Lee C, Wei X, Kysar JW, Hone J (2008) Science 321:385
359. Come J, Xie Y, Naguib M, Jesse S, Kalinin SV, Gogotsi Y, Kent PRC, Balke N (2016) Adv Energy Mater 6:1502290
360. Dutta S, Pal S, De S (2019) New J Chem 43:12385
361. Hyun WJ, Secor EB, Kim C-H, Hersam MC, Francis LF, Frisbie CD (2017) Adv Energy Mater 7:1700285
362. An T, Cheng W (2018) J Mater Chem A 6:15478
363. Moon IK, Ki B, Oh J (2020) Chem Eng J 392:123794
364. Chang T-H, Zhang T, Yang H, Li K, Tian Y, Lee JY, Chen P-Y (2018) ACS Nano 12:8048
365. Liu Z, Li J, Yan F (2013) Adv Mater 25:4296
366. Wu Y, Zhang X, Jie J, Xie C, Zhang X, Sun B, Wang Y, Gao P (2013) J Phys Chem C 117:11968
367. Yan X, Cui X, Li B, Li L-S (2010) Nano Lett 10:1869
368. Liu Z, Liu Q, Huang Y, Ma Y, Yin S, Zhang X, Sun W, Chen Y (2008) Adv Mater 20:3924
369. Roy-Mayhew JD, Bozym DJ, Punckt C, Aksay IA (2010) ACS Nano 4:6203
370. Lin Y et al (2013) Energy Environ Sci 6:108
371. Liu W, He S, Wang Y, Dou Y, Pan D, Feng Y, Qian G, Xu J, Miao S (2014) Electrochim Acta 144:119
372. Song D, Li M, Jiang Y, Chen Z, Bai F, Li Y, Jiang B (2014) J Photochem Photobiol A 279:47
373. Gu X, Cui W, Li H, Wu Z, Zeng Z, Lee S-T, Zhang H, Sun B (2013) Adv Energy Mater 3:1262
374. Yang Y et al (2016) Adv Mater 28:8937
375. Dai J, Zeng XC (2014) J Phys Chem Lett 5:1289
376. Hu W, Lin L, Yang C, Dai J, Yang J (2016) Nano Lett 16:1675
377. Lin S et al (2016) Adv Func Mater 26:864
378. Zhou Z, Si W, Lu P, Guo W, Wang L, Zhang T, Hou F, Liang J (2020) J Energy Chem 47:29
379. Guo W et al (2019) Carbon 152:888
380. Son IH et al (2015) Nat Commun 6:7393
381. Wang L, Zhang Q, Zhu J, Duan X, Xu Z, Liu Y, Yang H, Lu B (2019) Energy Storage Mater 16:37
382. Lu P et al (2019) Small 15:1805064
383. Tang X, Zhou D, Li P, Guo X, Wang C, Kang F, Li B, Wang G (2019) ACS Cent Sci 5:365
384. Zheng S, Wu Z-S, Zhou F, Wang X, Ma J, Liu C, He Y-B, Bao X (2018) Nano Energy 51:613
385. Zhang C, Park S-H, Seral-Ascaso A, Barwich S, McEvoy N, Boland CS, Coleman JN, Gogotsi Y, Nicolosi V (2019) Nat Commun 10:849
386. Shen K, Li B, Yang S (2020) Energy Storage Mater 24:670
387. Bai M et al (2018) Adv Mater 30:1801213
388. Zhao F et al (2019) Ionics 25:3099
389. Janek J, Zeier WG (2016) Nat Energy 1:16141
390. Sang Z, Yan X, Wen L, Su D, Zhao Z, Liu Y, Ji H, Liang J, Dou SX (2020) Energy Storage Mater 25:876
391. Fang Y et al (2020) ACS Nano 14:8744
392. Deng Z, Jiang H, Hu Y, Liu Y, Zhang L, Liu H, Li C (2017) Adv Mater 29:1603020
393. Zhao Y et al (2018) Small 14:1703717
394. Aliahmad N, Liu Y, Xie J, Agarwal M (2018) ACS Appl Mater Interfaces 10:16490
395. Chen G, Wang W, Wang C, Ding T, Yang Q (2015) Adv Sci 2:1500109
396. Liu J, Cao H, Jiang B, Xue Y, Fu L (2016) Sci China Mater 59:459
397. Lopez-Sanchez O, Lembke D, Kayci M, Radenovic A, Kis A (2013) Nat Nanotechnol 8:497
398. Chen G, Yu Y, Zheng K, Ding T, Wang W, Jiang Y, Yang Q (2015) Small 11:2848
399. Tian W et al (2014) Adv Mater 26:3088

400. Tamalampudi SR, Lu Y-Y, Sankar RKUR, Liao C-D, Cheng KMBC-H, Chou FC, Chen Y-T (2014) Nano Lett 14:2800
401. Wang Q et al (2015) Nano Lett 15:1183
402. Zhao F, Cheng H, Zhang Z, Jiang L, Qu L (2015) Adv Mater 27:4351
403. Zhao F, Liang Y, Cheng H, Jiang L, Qu L (2016) Energy Environ Sci 9:912
404. Dhiman P, Yavari F, Mi X, Gullapalli H, Shi Y, Ajayan PM, Koratkar N (2011) Nano Lett 11:3123
405. Wei D (2019) Adv Mater Technol 4:1900226
406. Wei D (2015) Sci Rep 5:15173
407. Ho Lee S, Jung Y, Kim S, Han C-S (2013) Appl Phys Lett 102:063116
408. Kwak SS, Lin S, Lee JH, Ryu H, Kim TY, Zhong H, Chen H, Kim S-W (2016) ACS Nano 10:7297
409. Zhong H, Xia J, Wang F, Chen H, Wu H, Lin S (2017) Adv Func Mater 27:1604226

Chapter 7
Concluding Remarks and Outlook

The field of 2DMs has advanced dramatically in the last decade, from basic research to the development of next-generation technologies. The importance of dimensionality in defining the intrinsic properties and possible uses of nanomaterials has obviously been emphasized. In this book, we discuss current advances in composition and crystal structure, synthetic techniques, and exciting prospective applications in energy conversion and storage. After GR, 2DMs have emerged, permitting researchers to obtain materials with desired characteristics for a particular application. These 2DMs may be synthesized using a variety of techniques, each with its own benefits and drawbacks. Different synthetic techniques may be employed to develop 2DMs with variable structural characteristics like size, thickness, crystallinity, crystal phase, defect, doping, strain, and surface property. Their unique features were studied for a number of applications, and several 2DMs results showed exceptional performance, with the potential to replace present commercial sources/technologies. Their application performance is a consequence of their carefully developed structural characteristics. The newly developed ways such as constructions of HSs and HJs are deeply discussed with possible opportunities. As consequence, several energy conversions and storage applications are deliberated and address the current progress in various device engineerings and how various synthesis techniques are affecting the device performance.

7.1 Challenges

1. Extensive research into 2DMs presents new challenges. One issue is that 2DMs currently do not meet the standards necessary for industrial or commercialization in terms of yield, quality, quantity, and rate of synthesis. Determining how to mass-produce 2DMs in large quantities is one of the significant challenges in this emerging field.

M. Ikram et al., *2D-Materials for Energy Harvesting and Storage Applications*, Nanostructure Science and Technology,
https://doi.org/10.1007/978-3-030-96021-6_7

2. The physical, chemical, and electrical characteristics of 2DMs are strongly reliant on their structural characteristics, which define their application performance. Most existing well-developed techniques are unable to control the synthesis of 2DMs with desired geometries. It is still difficult to produce 2DMs with the required structural properties.
3. However, because of their low physical and/or chemical stability, most existing 2DMs have limited application possibilities. Their poor stability is caused by:

- The 2DMs distributed in liquid cannot be preserved. The high susceptibility to irreversible aggregation results in considerable loss of 2D structural progress.
- Most 2DMs easily oxidize in air, causing structural degradation.
- In applications like electrocatalysis and LIBs, structural change, breakdown, or collapse can occur during a reaction. The stability of 2D-nanomaterials is essential, not only during storage and processing but also in applications. Thus, one of the most significant problems in this research is to find simple but dependable techniques to stabilize these 2DMs.

7.2 Suggestions

Regardless of the fact that significant progress was already achieved in this promising field, current research on 2DMs, particularly those beyond GR, is far from mature. To begin, because 2DMs are defined by their dimensionality instead of their composition:

1. We think that any type of 2DMs may be produced if their development can be confined to two dimensions and down to SL or FLs under the appropriate experimental circumstances. As a consequence, the easiest approach is to employ established technologies or to create new ones for the preparation of novel 2DMs with varied single- or multicompositions that are predicted to display novel characteristics and functions. Micromechanical cleavage, sonication-assisted liquid exfoliation, and ion intercalation-assisted liquid exfoliation are all well-developed exfoliation techniques that might be utilized to exfoliate novel layered compounds into SL and/or FLs-NSs
2. Second, it is important noting that each substance has unique drawbacks. The simplest method of overcoming a material's disadvantage is to hybridize it with other materials to produce composites or hetero-nanostructures. GR has been frequently utilized as a highly conductive matrix to hybridize with other low-conductive materials, such as TMDs and metal oxides, in order to increase their electrical conductivity and therefore optimize their performance in certain applications (e.g., electrocatalysis, LIBs, and other energy storage devices).
3. More interestingly, the synergistic impact between multiple components may provide new attractive characteristics or functionalities. Cobalt metal NSs and cobalt oxide NSs had modest electrocatalytic activities; however, their composite had high CO_2 reduction activity. Keeping this in mind, one of the most interesting future research areas is to develop hybrid nanomaterials

or hetero-nanostructures from 2DMs, thereby improving their characteristics and/or functionalities.

4. Lastly, material preparation research is often investigated with regard to the investigation of their prospective applications. Although several types of ultra-thin 2D-nanomaterials were synthesized in recent years, the majority of their potential uses remain unexplored. For instance, the excellent catalytic activity of MoS_2 toward electrochemical HER was discovered just approximately a decade ago, despite the fact that MoS_2 research dates back decades. Given the possible superior performance of these newly synthesized 2DMs in certain undiscovered applications, another interesting future path is to determine the most appropriate use for each 2DM

Index

M. Ikram et al., *2D-Materials for Energy Harvesting and Storage Applications*, Nanostructure Science and Technology,
https://doi.org/10.1007/978-3-030-96021-6

The manufacturer's authorised representative in the EU is Springer Nature Customer Service Centre GmbH, Europaplatz 3, 69115 Heidelberg, Germany. If you have any concerns regarding our products, please contact ProductSafety@springernature.com

Printed and bound by CPI Group (UK) Ltd, Croydon, CR0 4YY
02/10/2026
02227130-0005